# Soils of Tropical Forest Ecosystems

## Characteristics, Ecology and Management

Springer
*Berlin*
*Heidelberg*
*New York*
*Barcelona*
*Budapest*
*Hong Kong*
*London*
*Milan*
*Paris*
*Santa Clara*
*Singapore*
*Tokyo*

Andreas Schulte · Daddy Ruhiyat (Eds.)

# Soils of Tropical Forest Ecosystems

## Characteristics, Ecology and Management

With 73 Figures and 97 Tables

Springer

*Editors*

Professor Dr. Andreas Schulte
University of Paderborn
Forest Ecology and Soil Science
An der Wilhelmshöhe 44
D-37671 Höxter
Germany

Dr. Daddy Ruhiyat
Mulawarman University
Faculty of Forestry
Institute of Soil Science
P.O. Box 1013
75123 Samarinda
Indonesia

Cataloging-in-Publication data applied for

Die Deutsche Bibliothek - CIP-Einheitsaufnahme

**Soils of tropical forest ecosystems :** characteristics, ecology and management ; with 97 tables /
Andreas Schulte ; Daddy Ruhiyat (ed.). - Berlin ; Heidelberg ; New York ; Barcelona ; Budapest ;
Hong Kong ; London ; Mailand ; Paris ; Santa Clara ; Singapore ; Tokyo :
Springer, 1998
ISBN 3-540-63607-2

ISBN 3-540-63607-2 Springer-Verlag Berlin Heidelberg New York


Printed in Germany

Coverdesign: *design & production* GmbH, Heidelberg
Typesetting: Camera ready by editors
SPIN: 10646654 30/3136 – 5 4 3 2 1 – Printed on acid-free paper.

# Foreword

**Djamaludin Surjohadikusumo**
**Minister of Forestry, Republic of Indonesia**

*This book is entirely dedicated to soils of tropical rain forests. The tropical rain forest all over the world has been receiving serious attention by all parties, and viewed from all aspects of life. The environmentalists, who claim that tropical rain forests should be kept intact to maintain the climate of the world, the ecologists who prefer the forest not to be exploited due to our limited understanding of the wealth of biological diversity contained within the forests, the economists who will pay anybody to keep away from the tropical forest, the non government organizations who advocate the empowerment of the local people. There a lot of issues raised by diverse groups, and that contributes to the increasing attention paid to tropical forests.*

*Tropical rain forests, as any forests are, undergo the dynamics of growth, namely from seedling to maturity and finally to products of decay. No exploitation leaves the resources unused and there goes another waste of economic opportunities. As with any other natural resource, the tropical rain forest should be wisely exploited and utilized for the maximum benefit for human kind, while simultaneously taking into account the ecological consequences and the well being of the people who live within and surrounding the forests. The jargon of today is sustainable forest management, which requires a relatively complete understanding of the dynamics of its elements, the impact of association and disassociation, their response to chaotic external stimuli.*

*Indonesia has decided to utilize the tropical forests wisely, since it will determine the course of development of the country. Tropical forest lands had to yield to various external pressures, namely land for settlements, for establishing factories and industrial estates, or for conversion to agricultural lands for food and industrial crops. At the same time management has determined to increase the area for forest plantation to reduce the pressure of demand for lumber from the natural tropical rain forests. Thus forest management has to work within four limiting constraints, namely the pressure for more land to achieve further development stages, the ever increasing demand for timber and other non-timber products, the improvement of the economy of the local people and the pressure from environmentalists to keep away from the forests.*

*Indonesia has embarked on the expansion of timber estates, to supplement the supply of logs from natural forests. The objectives of establishing timber plantation are to rehabilitate degraded forest lands, to product from marginal forest lands woods and other forest products as inputs for forest industries, to provide employment for the people and to improve the environment of the area. The challenges faced for the success of timber plantations depend mainly from the past treatment on soil and regeneration. The timber plantations really need completely improved methods of exploitation of the primary forest exploitation, to ensure that the next management regime will result in good stands.*

*The scientists believe that one crucial factor which will determine the success of forest plantations is soil and soil management. It is important to point out that mistakes in the management of a forest soil do not become apparent as quickly as they do in agricultural soils. Such mistakes, however, can not be corrected promptly. In this special regard, I would like to take this opportunity of acknowledging the scientists who have painfully investigated and analyzed the soil as an important element in the forest health. This book is a proof that a little yet essential work has been done, and a lot more has to be done, to improve our understanding of forest soils, to enable us to manage forest ecosystems sustainable. The scientists indicated that forest soils are an interesting subject to investigate.*

# Preface

With a few exceptions, tropical rain forests are found in hot, perhumid areas between 23°N and 23°S. During the last few decades these forest ecosystems have been exploited at unprecedented rates, and concern about the ecological and economical impacts is growing. In many regions of the humid tropics man is today confronted with a situation where the natural primary forests have either already disappeared or will do so in the very near future. What remains are residual or secondary forests, or even so-called unproductive wasteland, which is considerably different from the primary forests with respect to species composition, structure, dynamics and soil characteristics. In order to maintain the supply of raw materials to the forest industry and to safeguard the important social and environmental functions of forest ecosystems, it is of fundamental importance to change from exploitation and conversion of the forest to sustainable management practices. But, can there be anything left to say on the subject of sustainable forestry?

The growing interest in the concept of sustainability was given added stimulus at the United Nations Conference on Environment and Development (UNCED), held in Rio de Janeiro in 1992. Agenda 21, a major action plan developed at UNCED, focused attention on the need to make development more economically and environmentally sustainable, and socially acceptable. Chapter 10 of Agenda 21 is concerned with the planning and management of land resources. For these reasons sustainable management of soil resources is now receiving considerable attention from forest experts and policy makers. Nevertheless, the role of soil as a compartment of forest ecosystems, has not yet been taken into account enough in practical forest management, not only in the humid tropics. Often the view in forest research and management still center around the crop, e.g., a tree and its pest, a tree and its growth and yield, etc. The failure to recognize the importance of the soil resource in natural forest and plantation management has been, and still often is, a major impediment to successful development of sustainable forestry.

Since there are millions of hectares of so-called unproductive wasteland in the humid tropics, and as we can not renounce highly productive forest ecosystems, we are forced to accelerate the recovery process of the forest ecosystems by technical measures. Such technical measures include planting of trees (reforestation), but also the application of lime and fertilizer as well as soil manipulation. As most of the natural forests will be gone in the near future, consequently, **physical, chemical and biological soil amelioration together with reforestation will be the main task for forestry in the humid tropics during the coming decades.**

In this regard a knowledge of the characteristics and ecology of forest soils is essential. Nowhere is this more important than in the humid tropics where soil productivity is often (not always) limited by poor soil fertility and where the need for food production is large – reason enough for a long-term forest cooperation project in Indonesia to organize a conference on the role of soils in sustainable forestry.

The present book is the result of the "International Congress on Soils of Tropical Forest Ecosystems/3rd Conference on Forest Soils (ISSS-AISS-IBG)" held in Balikpapan, Indonesia. Almost 300 participants representing a wide range of disciplines and interests, came from some 35 countries in primarily tropical areas. The conference was organized by the editors of the book in close cooperation with the International Society of Soil Science (ISSS-AISS-IBG)/Subcommission Forest Soils as one activity of the Indonesian-German Forestry Project (IGFP-GTZ) at the Mulawarman University in Samarinda/Indonesia, and would not have been possible without the assistance and cooperation of

**H.E. The Minister of Forestry, Republic of Indonesia, Djamaludin Surjohadkusumo**
**The Secretary General of the International Society of Soil Science, Prof. Dr. Dr. h.c. mult. W. E. H. Blum**

and the authorities of East Kalimantan, Balikpapan and the Mulawarman University in Samarinda, as well as the team of the Indonesian-German Forestry Project (IGFP/GTZ).

Financial support from the

**German-Indonesian Governmental Cooperation**
**Deutsche Gesellschaft für Technische Zusammenarbeit (GTZ) GmbH**

as well as the

**Center for International Forest Research (CIFOR), Bogor/Indonesia**

is gratefully acknowledged.

The book is divided into four chapters. Chapter 1 is introductory dealing with characteristics and classification of forest soils in the humid tropics. Chapter 2 presents papers on chemical and hydrological changes in tropical forest ecosystems after utilization and/or conversion into other land uses. Chapter 3 – a special session during the conference sponsored by CIFOR (Center of International Forest Research) – covers aspects of soil fertility and fertilization in fast growing plantations and agroforestry systems. Chapter 4 focuses on the Dipterocarp forest region of Southeast Asia, mainly Indonesia as host country of the conference. Ecosystem studies, such as the relation between the soil and tree species diversity, as well as a discussion of soil under alternatives to slash-and-burn agriculture are presented.

The papers in this volume were presented during the conference as Oral or Poster Presentation. The editors are indebted to all authors for timely submitting their texts, and for prompt response to questions and comments by reviewers. We are also indebted to many colleagues, mainly the chairman of the seven working sessions of the conference, who selected the papers for publication and critically read the manuscripts and made suggestions for major improvement.

Finally we would like to extend our thanks to our Indonesian hosts and forestry colleagues and to all participants, with whom our encounters were encouraging, a pleasure, and a privilege.

The editors

Prof. Dr. Andreas Schulte
University of Paderborn
An der Wilhelmshöhe 44
D-37671 Höxter
Germany

Dr. Daddy Ruhiyat
Mulawarman University
Faculty of Forestry
Samarinda 75123
Indonesia

# Content

**CHAPTER 3**

**Soil Fertility and Fertilization in Forest Plantations and Agroforestry Systems**



**CHAPTER 4**

**Ecosystem Studies in Dipterocarp Forests of Southeast Asia**

# Contributors

*Asio,* Victor B. — Visayas State College of Agriculture (VISCA), Department of Agronomy and Soil Science, Baybay, Leyte, 6521-A Philippines

*Blum,* Winfried, E. H. — University of Agriculture, Institute of Soil Science, Gregor-Mendel-Str. 33, A-1180 Vienna, Austria
E-mail: ISSS@EDV1.BOKU.AC.AT

*Brandstetter,* A. — University of Agriculture, Institute of Soil Science, Gregor-Mendel-Str. 33, A-1180 Vienna, Austria

*Brouwer,* Leo O. — Utrecht University, Department of Physical Geography, P.O. Box 80115, 3508 TC Utrecht, The Netherlands
E-mail: I.BROUWER@FRW.RUU.NL

*Bruijnzeel,* Sampurno L. A. — Vrije Universiteit, Faculty of Earth Science, De Boelelaan 1085, 1081 HV Amsterdam, The Netherlands
E-mail: BRUL@GEO.VU.NL

*Deckers,* Jozef — Catholic University of Leuven, Institute for Land and Water Management, Vital Decosterstraat 102, B-3000 Leuven, Belgium
E-mail: SEPPE.DECKERS@AGR.KULEUVEN.AC.BE

*Drechsel,* Pay — University of Bayreuth, Institute of Soil Science and Soil Geography, D-95440 Bayreuth, Germany

*Fassbender,* Hans W. — Technical Forest Faculty Hildesheim/Holzminden, Büsgenweg 4, D-37077 Göttingen, Germany

*Hambleton,* Alison — Queensland Forest Research Institute, DPI-Forestry, Atherton, Australia

*Hairiah,* Kurniatun — Brawijaya University, Malang, Indonesia

*Hermansah* — Andalas University, Padang, Indonesia

*Hotta,* Mitsuru — Kagoshima University, Faculty of Science, Kagoshima 890, Japan

*Hulugalle,* Nilantha R. — Australian Cotton Research Institute, Myall Vale Mail Run, Narrabri NSW 2390, Australia

*Iskandar* — Mulawarman University, Faculty of Forestry, Institute of Soil Science, Samarinda 75123, Kaltim, Indonesia

*Jahn,* Reinhold — University of Hohenheim, Institute of Soil Science (310), D-70593 Stuttgart, Germany

*Johansson,* Eva — Swedish University of Agricultural Science, Department of Entomology, P.O. Box 7044, S-75007 Uppsala, Sweden
E-mail: EVA.JOHANSSON@ENTOM.SLU.SE

*Kadeba,* Olatunji — Federal University of Technology, Department of Forestry and Wood Technology, PMB 704, Akure, Nigeria

*Kadir,* Wan Rasidah — Forest Research Institute Malaysia (FRIM), Kepong, 52109 Kuala Lumpur, Malaysia
E-mail: RASHIDAH@FRIM.GOV.MY

*Kang,* Biauw T. — International Institute of Tropical Agriculture (IITA), PMB 5320, Ibadan, Nigeria

| | |
|---|---|
| *Kauffman,* Sjef | International Soil Reference and Information Centre (ISRIC),<br>P.O. Box 353, 6700 AJ Wageningen, The Netherlands<br>E-mail: KAUFFMAN@ISRIC.NL |
| *Keenan,* Rodney | Queensland Forest Research Institute, DPI-Forestry,<br>P.O. Box 210, Atherton, Australia<br>E-mail: KEENANR@DPI.QLD.GOV.AU |
| *Kluge,* Martina | Swedish University of Agricultural Science, Department of Forest Ecology,<br>S-90183 Umeå, Sweden |
| *Kubota,* Daisuke | Shimane University, Faculty of Life and Environmental Science,<br>Matsue 690, Japan |
| *Kuraz,* Václav | Czech Technical University, Department of Irrigation and Drainage,<br>Prague, Czech Republik |
| *Malmer,* Anders | Swedish University of Agricultural Science, Department of Forest Ecology,<br>S-90183 Umeå, Sweden<br>E-mail: ANDERS.MALMER@SEK.SLU.SE |
| *Mantel,* Stephan | International Soil Reference and Information Centre (ISRIC),<br>P.O. Box 353, 6700 AJ Wageningen, The Netherlands<br>E-mail: SOIL@ISRIC.NL |
| *Margraf,* Josef | Tropical Ecology Program VISCA-GTZ, Philippine-German Environmental<br>Research and Development Program,<br>Baybay, Leyte, 6521-A Philippines |
| *Masunaga,* Tsugiyuki | Shimane University, Faculty of Life and Environmental Science,<br>Matsue 690, Japan |
| *Mudiarso,* Daniel | University of Bogor, Faculty of Agriculture,<br>Bogor, Indonesia |
| *Murach,* Dieter | University of Göttingen, Faculty of Forestry, Institute of Silviculture,<br>Büsgenweg 1, D-37077 Göttingen, Germany<br>E-mail: DMURACH@GWDG.DE |
| *Nykvist,* Nils | Swedish University of Agricultural Sciences, Department of Forest Ecology,<br>Postal address (priv.), Skogsstigen 1, S-14462 Rönninge, Sweden |
| *O'Connor,* Patrick J. | University of Adelaide, Department of Soil Science,<br>SA 5005 Adelaide, Australia |
| *Otsamo,* Antti | University of Helsinki, Department of Forest Ecology/Tropical Silviculture,<br>P. O. Box 28, 00014 Helsinki, Finland |
| *Rachman,* Achmad | Center for Soil and Climate Research,<br>Bogor, Indonesia |
| *Rasyidin,* Azwar | Andalas University,<br>Padang, Indonesia |
| *Riezebos,* Hans Th. | Utrecht Univesity, Department of Physical Geography,<br>P. O. Box 80115, 3508 TC Utrecht, The Netherlands |
| *Robson,* Ken | Queensland Forest Research Institute, DPI-Forestry,<br>Atherton, Australia |
| *Ruhiyat,* Daddy | Mulawarman University, Faculty of Forestry, Institute of Soil Science,<br>Samarinda 75123, Kaltim, Indonesia |
| *Salako,* Felix K. | International Institute of Tropical Agriculture (IITA),<br>PMB 5320 Ibadan, Nigeria |
| *Schulte,* Andreas | University of Paderborn, FB 7, Forest Ecology and Soil Science,<br>An der Wilhelmshöhe 44, D-37671 Höxter, Germany<br>E-mail: GSCHU1@HUXOR.UNI-PADERBORN.DE |

| | |
|---|---|
| *Shinmura,* Yoshiaki | Shimane University, Faculty of Life and Environmental Science,<br>Matsue 690, Japan |
| *Simonrangkir,* Dicky | Mulawarman University, Faculty of Forestry, Institute of Soil Science,<br>Samarinda 75123, Kaltim, Indonesia |
| *Simpson,* John | Queensland Department of Primary Industries, Forestry Research Institute,<br>M/S 483 Fraser Road, Gympie 4570 Queensland, Australia |
| *Smith,* Andrew F. | University of Adelaide, Department of Botany,<br>SA 5005 Adelaide, Australia |
| *Smith,* Hilary F. | University of Adelaide, Department of Geography,<br>SA 5005 Adelaide, Australia |
| *Smith,* Sally, E. | University of Adelaide, Department of Soil Science,<br>SA 5005 Adelaide, Australia |
| *Sombroek,* Wim G. | International Soil Reference and Information Centre (ISRIC),<br>P.O. Box 353, 6700 AJ Wageningen, The Netherlands<br>E-mail: SOIL@ISRIC.NL |
| *Spaargaren,* Otto C. | International Soil Reference and Information Centre (ISRIC),<br>P.O. Box 353, 6700 AJ Wageningen, The Netherlands<br>E-mail: SPAARGAREN@ISRIC.NL |
| *Stahr,* Karl | University of Hohenheim, Institute of Soil Science (310),<br>D-70593 Stuttgart, Germany |
| *Tiessen,* Holm | University of Saskatchewan, Department of Soil Science,<br>Saskatoon S7N 5A8, Canada<br>E-mail: TIESSEN@SASK.USASK.CA |
| *Tomich,* Thomas P. | ICRAF – Southeast Asia,<br>P.O. Box 16001, Bogor, Indonesia<br>E-mail: ICRAF-INDONESIA@CGNET.COM |
| *Unterfrauner,* Hansjörg | University of Agriculture, Institute of Soil Science,<br>Gregor-Mendel-Str. 33, A-1180 Vienna, Austria |
| *van Cleemput,* O. | University of Gent, Faculty of Agricultural and Applied Biological Sciences,<br>Coupure Links 653, 9000 Gent, Belgium |
| *van Noordwijk,* Meine | ICRAF – Southeast Asia,<br>P.O. Box 16001, Bogor, Indonesia<br>E-mail: ICRAF-INDONESIA@CGNET.COM |
| *Wakatsuki,* Toshiyuki | Shimane University, Faculty of Life and Environmental Science,<br>Matsue 690, Japan |
| *Wasrin,* Upik Rosalina | SEAMEO-BIOTROP,<br>Bogor, Indonesia |
| *Webb,* Michael | CSIRO, Devision of Soils,<br>PMB, Aitkenvale, Australia |
| *Wenzel,* Walter | University of Agriculture, Institute of Soil Science,<br>Gregor-Mendel-Str. 33, A-1180 Vienna, Austria<br>E-mail: WAZI@EDV1.BOKU.AC.AT |
| *Zaharah,* A. R. | Universiti Pertanian Malaysia, Department of Soil Science,<br>43400 Serdang, Selangor, Malaysia |
| *Zech,* Wolfgang | University of Bayreuth, Institute of Soil Science and Soil Geography,<br>D-95440 Bayreuth, Germany |

# Soils of rainforests
# Characterization and major constraints of dominant forest soils in the humid tropics

**Sjef Kauffman**[1], **Wim Sombroek**[2] and **Stephan Mantel**[1]

[1] International Soil Reference and Information Centre (ISRIC). P.O.Box 353, 6700 AJ Wageningen, The Netherlands. E-mail: kauffman@isric.nl
[2] Formerly Food and Agriculture Organization of the United Nations. Via delle Terme di Caracalla, 00100 Rome, Italy.

## Abstract

Rational use of forest resources should be based upon accurate knowledge of land and soil properties. Selected characteristics of 149 soil profiles of the humid tropics are analyzed. General environmental characteristics, physical, chemical and mineralogical soil data have been derived from ISIS, the ISRIC Soil Information System. The humid zone includes 6 major soil groupings: Acrisols, Arenosols, Cambisols, Ferralsols, Luvisols and Podzols. For these soils the arithmetic mean and standard deviation of selected key analytical characteristics, relevant for agronomic and ecological research, are presented for a standardized topsoil and subsoil. In addition, their major agronomic constraints were assessed. About 64 % of the humid tropics is covered by Acrisols, Arenosols, Ferralsols and Podzols of low fertility, all presenting various degrees of limitations when forests are cleared for (low input) arable farming. A low plant nutrient content, a low nutrient retention capacity and a high toxic exchangeable aluminium content are major constraints. The overall data presented in this paper show that Ferralsols and Acrisols, covering 57 % of the humid tropics, have rather similar key characteristics. The taxonomic separation is principally based on a relatively small increase in clay content, which will not determine the major vegetation type or crop productivity level. When considering conservation of the natural vegetative cover and its biodiversity, several special properties of the soils have to be taken into account, such as depth of the litter layer, rootable depth and the mineral composition of the deeper subsoil and the substratum, as well as the geomorphologic-pedologic history of the terrain units and the degree of short-distance variation in soil conditions. It is concluded that land assessment studies and soil-vegetation/bio diversity research should not rely on a soil taxonomic approach alone, but should also be based on measured key pedological characteristics. The study illustrates the usefulness of the ISIS database to correlate soil characteristics and to determine agronomic-ecological constraints of major soil groupings.

## 1 Introduction

When the International Soil Reference and Information Centre (ISRIC) was established in 1966, its main task was to collect soil profiles, soil samples and associated information representative of the legend units of the FAO-Unesco Soil Map of the World (FAO, 1974). Suitable sites were selected with the national institutions concerned. The main selection criterion was the representativeness of a major soil type within any particular country. Furthermore, specific soil and land use features were taken into account such as sites with original vegetation versus cleared land. At present, ISRIC's soil reference collection holds over 900 soils from 64 countries of which the data of about 600 soils have been stored in the ISIS database (ISRIC, 1993, 1995). Reference soils have a comprehensive set of soil and environmental data, which are stored in a relational database system, the ISRIC Soil Information System (ISIS) (*Van Waveren* and *Bos*, 1988). Many reference soils and their data have been published in a series of 'Country Reports', which are now available for 13 countries (ISRIC, 1994, 1995). Additional information about these reference soils is provided in a series of 'Soil Briefs' also published by ISRIC, which discuss the formation, characterisation and agricultural suitability of one or more related soils in their agro-ecological setting.

Tropical forests are being cleared at an unprecedented rate. Global deforestation rates are estimated at 15.3 million ha yr$^{-1}$ for the period between 1981 and 1990. Deforestation of lowland moist rainforest, including moist deciduous forest, is estimated for the same period at 10.7 million ha yr$^{-1}$ (FAO, 1995). Assessment and evaluation of land is the first step in rational use of forest resources on a sustained basis. For example, some land may best be left under forest to maintain an ecological balance, because the combined limitations of climate, relief and soil make it unsuitable for sustainable agriculture, while other areas may be capable of sustained, intensive cropping. Other lands, in an intermediate situation, also should be kept under managed forest, or at least under forest fallow, to maintain an ecological balance (*Lal*, 1987). Many soils of tropical rainforests are formed in materials at an advanced stage of weathering. After clearing, the infertility of the soils often imposes considerable constraints upon arable farming.

The object of the current study is to characterize major forest soils, covering 72% of the humid tropics, and to assess their major limitations for agriculture. From the 600 reference profiles in ISIS, 149 are located in the humid tropics. This number is considered sufficient to make a valid statistical analysis of major soil groupings in these zones, the results of which are presented and discussed in this paper. For the seasonally dry (sub)tropic ecological zone a similar study was made and presented at the National Soil Reference Collection and Databases Workshop (*Kauffman* et al., 1996).

## 2 Materials and methods

### 2.1 ISIS Dataset, field and analytical methods

During the last decade three soil databases have been developed at ISRIC, viz. ISIS (ISRIC Soil Information System), SOTER (Soils and Terrain Digital Database) and WISE (World Inventory of Soil Emission Potentials). A review of these databases, and the possibilities of transfer between the databases is given by *Batjes* et al. (1994). Data collected for ISRIC's soil reference collection are stored in the relational database management system ISIS (*Van Waveren* and *Bos*, 1988), which later formed the basis for developing the FAO-ISRIC Soil Database (FAO, 1989). Version 4.0 of ISIS, which is written in dBase IV, permits handling and analysis of:

– Site data, which include about 60 mainly descriptive (coded) attributes on location, geology, landform, soil surface properties, hydrology, land use and vegetation.

– Quantitative synoptic climatic data from meteorological stations that are considered representative for the conditions prevailing at the profile site.

– Profile data, including:

i) soil profile descriptions based on FAO guidelines (FAO, 1977 and 1990).

ii) soil classification according to the legend of the FAO-Unesco Soil Map of the World (FAO, 1974 and 1988), USDA Soil Taxonomy (Soil Survey Staff, 1992) and the national system.

iii) physical, chemical and mineralogical attributes of each soil horizon.

In the field, soil horizons were described and sampled in pits of about 2 meters depth. Deeper layers were sampled using a soil auger to a maximum depth of 6 meters in appropriate cases. The effective, or 'rootable', soil depth was observed in the soil pits. It is > 2.0 m for 90% of the upland sites studied, while for the other soils it ranges between 0.5 and 2.0 m. Rootable depht is limited by physical root barriers such as hard rock, hard pans, high permament water table, or by chemical hindrance such as low content of one or more macro or micro nutrients, or toxic levels of exchangeable aluminium.

Chemical, physical and mineralogical data of ISRIC's world soil collection were determined in ISRIC's laboratory, using standardized analytical and quality assurance procedures (*Van Reeuwijk*, 1993). The following key soil characteristics, frequently used in agricultural land use assessment and soil vegetation/biodiversity research, were selected from ISIS for statistical analysis:

*Chemical characteristics*

– Soil reaction (pH-$H_2O$ in 1:2.5 soil-water solution, and pH-KCl likewise in 1 M KCl solution)

– Organic carbon (kg $kg^{-2}$ $10^2$, Walkley Black procedure)

– Organic nitrogen (kg $kg^{-2}$ $10^2$, micro Kjehldahl procedure)

– Sum of exchangeable bases ($cmol_c$ $kg^{-1}$, percolation with 1 M ammonium acetate buffered at pH 7)

– Exchangeable aluminium ($cmol_c$ $kg^{-1}$, extraction with 1 M KCl)

– Cation Exchange Capacity (CEC in $cmol_c$ $kg^{-1}$, percolation with 1 M ammonium acetate buffered at pH 7)

– Base saturation as percentage of CEC.

*Physical characteristics*

– Particle size (kg $kg^{-2}$ $10^2$). Sand fractions (2000–1000, 1000–500, 500–250, 250–100, 100–50 μm) were determined by sieving. The silt fractions (50–20 and 20–2 μm) and clay fraction (< 2 μm) were determined by pipetting from a sedimentation cylinder.

– pF or soil moisture retention data were obtained from undisturbed core samples equilibrated with water at various tension or pF values. The following parameters were derived: a) Bulk density (kg $dm^{-3}$); b) Potential Rootable Pore Volume, calculated as pF 0–pF 2.0 ($m^3$ $m^{-3}$ $10^2$); and c) Potential Plant Available Moisture, calculated as pF 2,0–pF 4.2 ($m^3$ $m^{-3}$ $10^2$).

In this paper most analytical results are presented for a standardized thickness of topsoil and subsoil to make comparison between the soil profiles possible. If no specific depth range is given in the text, topsoil refers to the depth range from 0 to 20 cm, and for subsoil from 70 to 100 cm. These limits are arbitrarily chosen to represent the topsoil with maximum concentration of roots and a subsoil to a depth where root growth of most annual crops ends. Calculated values of soil characteristics in the depth range of 20 to 70 cm in nearly all cases take intermediate values. Statistical analyses were made with Statistix (1994).

### 2.2 Ecological zones

The use of major climatic and soil-ecological zones for correlation of agricultural research is being promoted by several international agencies, such as the Food and

Agriculture Organisation of the United Nations (FAO, 1978–1981), the Consultative Group of International Agricultural Research (CGIAR, 1992), and the International Board for Soil Research and Management (IBSRAM) (*Greenland* et al., 1994). ISRIC applies the concept of major ecological zones in its projects.

In this paper we only consider the humid tropics with the following criteria to select reference soil profiles from the database: altitude less than 1000 m, a Length of Growing Period (LGP) between 270 and 365 days (LGP according to FAO, 1978–1981) and an average monthly temperature over 18 °C throughout the year. This coincides broadly with Köppen's climate classes Af, Am and part of the Aw class with an annual precipitation of more than 1500 mm. The distribution of the humid tropics and subtropics, with 4 LGP classes, is shown in figure 1.

### 2.3 Distribution of ISRIC reference profiles

Based on the criteria for the humid tropical zone mentioned before, 149 reference soils were selected from the ISIS database. They originate from 20 countries and are distributed as follows: Brazil (7), China (3), Colombia (14), Costa Rica (10), Côte d'Ivoire (7), Cuba (5), Ecuador (4), Gabon (6), India (1), Indonesia (29), Jamaica (4), Malaysia (18), Nicaragua (4), Nigeria (10), Peru (11), Philippines (6), Samoa (3), Sri Lanka (1), Thailand (3) and Zaire (2). The locations of the profiles are well distributed throughout the humid tropics, but soils from Brazil and Zaire are under-represented.

## 3 Soils of the humid tropics

### 3.1 Dominant soils

Six major soil groupings (FAO-Unesco, 1974) are dominant in the set of 149 soil profiles of the humid tropics: Acrisols, Ferralsols, Arenosols and Podzols, Cambisols, Luvisols (table 1). Arenosols and Podsols are grouped in this study because both are characterized by a sandy texture and usually occur in close spatial association in the humid tropics.

Compared to the revised legend of the Soil Map of the World (FAO, 1988) there are some shifts in classification for the considered dominant soils, for example the former Acrisols (FAO-Unesco, 1974) now include the Alisols (FAO, 1988) and the former Luvisols (FAO-Unesco, 1974) include the Lixisols (FAO, 1988). In addition some Ferralsols may have to be classified as Acrisols and Cambisols (see section 3.3.1). The representation of the six dominant soils of the humid tropics in ISRIC's database correlates well with the dominant extent of major soils groupings in the climate zone 'Humid Tropics and Subtropics' as summarized in FAO's World Soil Resources (FAO, 1993, p. 33). The six major soil groupings cover approximately 72 % of the total area of this climatic zone (table 1). It is observed that Gleysols (5 %), Nitisols (5 %) and Fluvisols (4 %) also cover important acreage (FAO, 1993), but these three soil groupings are not considered in this paper, because there are too few reference profiles to permit meaningful statistical analysis.

The most frequent soil units of the six dominant soils considered are:

| | |
|---|---|
| Ferralsols | Xanthic (40 %), followed by Acric and Orthic soil units. |
| Acrisols | Ferric (50 %) with an equal distribution over Gleyic, Humic, Orthic and Plinthic soil units. |
| Cambisols | Ferralic (50 %), followed by Dystric and Eutric soil units. |
| Arenosols and Podzols | no dominant soil units |

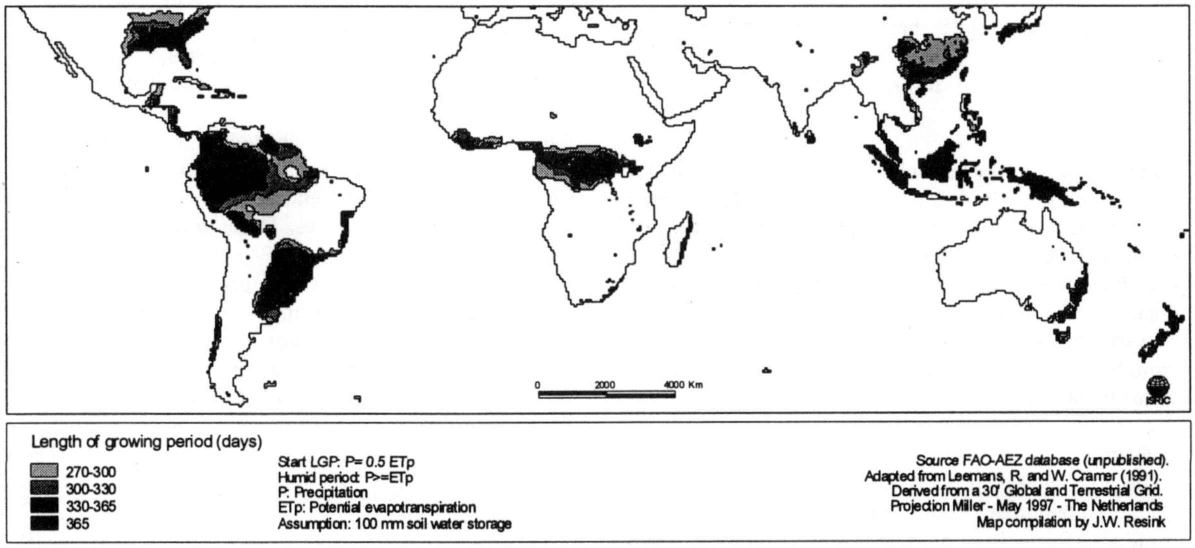


**Figure 1:** Distribution of the humid and perhumid zones in the tropics and subtropics.

**Table 1:** Dominant soil groupings in the humid tropics.

| DOMINANT SOIL GROUP (FAO, 1974) | ISIS database Number | (%) | Area in the (sub)tropics[1] million ha | (%) |
|---|---|---|---|---|
| Acrisols | 33 | 22 | 589 | 31 |
| Ferralsols | 30 | 20 | 507 | 26 |
| Arenosols and Podzols | 11 | 7 | 138 | 7 |
| Cambisols | 33 | 22 | 95 | 5 |
| Luvisols | 9 | 6 | 53 | 3 |
| Other | 33 | 22 | 544 | 28 |

[1] Including sub-tropics (FAO, 1993)

Similar area estimates, as found for the FAO-Unesco (1974) major soil groupings, are expected when the soils are classified according to the World Reference Base for Soil Resources (*Spaargaren*, 1994), although, some shifts between major soil groups are anticipated. For example, a part of the Ferralic Cambisols will be included with the Ferralsols and part of the Ferralsols, showing a distinct clay increase be accommodated with the Acrisols. For further information on this subject see *Spaargaren* and *Deckers* (this volume). Soil classification according to international systems such as FAO (1988) and Soil Taxonomy (Soil Survey Staff, 1992) is based on a limited number of characteristics used in the definition of diagnostic horizons and properties, which allows uniform classification. However, it is difficult to obtain quantitative information on key soil characteristics (frequently used in agricultural land use assessment) and their variation within each taxonomic soil group. The data available in ISIS can be used to provide quantitative information on the environmental and soil characteristics of the six dominant soils. In the remaining part of this paper these will be referred to as 'dominant soils' of the humid tropics.

### 3.2 Environmental characteristics

General environmental characteristics of the sites are described in this section, including precipitation, vegetation, parent material, landform and hydrology.

*Precipitation*

Precipitation at the 149 selected sites, ranges from 1500 to 4500 mm $yr^{-1}$. Monthly precipitation (P) exceeds evapotranspiration (ETp, Penman) in most months. Leaching Rainfall (LR), which is defined as the monthly excess of P over ETp, ranges from almost 0 to 3000 mm $yr^{-1}$ with an mean of 1400 mm $yr^{-1}$. LR is a parameter for the amount of water that potentially percolates through the soil to the zone below the roots and drains vertically to the groundwater or laterally to a river system.

*Vegetation*

The original vegetation at the considered sites is mainly evergreen rainforest. About 25% of the sites is under semi-deciduous forest. Rooting is in most cases > 2 m. The original forest vegetation has been cleared in about 70% of the sites studied, and secondary forest is frequently observed near these sites. Arable land and grazing land are dominant land-use types for cleared land. The perennial crops most frequently were rubber, stimulants, fruit and oil trees. Common annuals were cereals and root, oil and protein crops.

*Parent material*

The composition of parent material is an important determinant for the properties and agricultural potential of (tropical) soils. About 50 different types of parent material are considered in ISIS. For soil fertility and soil genesis correlation purposes, these have been grouped according to composition in five categories: acid, intermediate, basic, calcareous and unconsolidated parent materials. Acid refers to rocks with > 10% of free quartz (formerly defined as rock containing > 66% silica). Basic refers to quartz-free rock containing feldspars which are generally more calcic than sodic (formerly defined as rocks containing between 45 and 55% silica). Intermediate refers to rocks in between the acid and basic types (*Whitten* and *Brooks*, 1974; *Young*, 1976). The unconsolidated sediments/materials include predominantly alluvium with some colluvium and eolian sediments. All rock and sediment types, both consolidated and non-consolidated, which contain free calcium carbonate are grouped under the heading calcareous parent material.

The frequency of these five parent material groups for the dominant soils in the database is: acid (38%), intermediate (12%), basic (17%), calcareous (7%) and unconsolidated parent material (26%). Although the dominant soils are formed on all rock types, some trends can be observed:

| | |
|---|---|
| acid | all soil groups |
| intermediate | Luvisols dominant |
| basic | all soil groups, Ferralsols dominant |
| calcareous | all soil groups |
| unconsolidated | all soil groups, Acrisols, Cambisols and Arenosols/Podzols dominant |

*Landform, topography and position of site*

Over 30 landforms are described for the sites of the reference soils, which for correlation purposes are grouped in four major relief categories. These groups and their frequency are:

- lowland alluvial and coastal plains 22%
- upland dissected and non-dissected plains 38%
- hills, including for a minor part plateaus 35%
- mountains, including volcanoes 5%

Frequency of the five topography classes, according to FAO (1977 and 1990), at the site are:

- flat 26%
- undulating 38%
- rolling 17%
- hilly 10%
- steep 10%

On sloping landscapes, the sites are situated principally on middle or upper slope positions. Less than 25% is located on a lower slope or crest position.

The dominant soils are represented in all landform, topography or position classes, except for the Arenosols and Podzols, which are found in flat to undulating alluvial or coastal plains.

*Hydrology*

Most of the dominant soil sites studied are well drained (70%) or moderately well drained (20%), a minority is imperfectly (5%) or (somewhat) excessively drained (5%). In 95% of the sites no groundwater table was observed, which corresponds with the upland position of the dominant soils.

## 3.3 Soil characteristics

Selected key analytical characteristics, relevant for land use analysis and ecological research, are presented in this section. When appropriate, they are correlated with field information. The mean ($\bar{x}$) values and standard deviation ($s$) for the considered characteristics are given by dominant soil in table 2 and 3 and histograms were plotted.

For most physical and chemical characteristics considered in this paper, the mean was considered a suitable parameter to show trends and differences between major soil groupings. From the standard deviations ($s$) given in tables 2 and 3, it follows that ranges are large for nearly all characteristics of the dominant soils. Therefore, comparisons are only valid for the major soil grouping as a whole. The mean values of characteristics may be different for individual soil units of a major soil grouping.

### *3.3.1 Physical characteristics*

*Particle size distribution*

Data on the sand, silt and clay fractions are presented in table 2. Reflecting their sandy parent materials, Podzols and Arenosols have a very low silt and clay content. Ferralsols, Acrisols, Luvisols and Cambisols have a higher clay content. For these soils a correlation of clay content with parent material composition is more striking. The mean clay content of soils formed in acid parent materials is 23%, compared to 58% for soils formed in basic materials. Ferralsols, dominantly derived from basic parent materials, have a slightly higher clay content than the other major soil groupings.

A distinct increase in clay content between topsoil and subsoil is not a manifest feature in most dominant soils of the humid zone. The most pronounced increase is found in Acrisols, with a mean clay increase of 11% (from 29 to 40%) in the first metre of the soil. Although for Acrisols elsewhere in the world the increase in clay content takes place over a shorter depth range, the gradual increase of clay content in the Acrisols of the humid tropics will have only limited influence on root development and soil hydraulic properties. This is shown by the field morphology, where horizon boundaries are usually gradual or diffuse, as a result of biohomogenization.

The silt fraction is relatively low in soils of the humid tropics, being lowest in Ferralsols and highest in Cambisols. This is reflected in the silt/clay ratio. The mean ratio is 0.3 in subsoils of Ferralsols and double for the other major soil groupings. In the revised legend of the Soil Map of the World (FAO, 1988) a silt/clay ratio of < 0.2 is introduced as a diagnostic criterion in the classification of Ferralsols. When rigidly applied, some of the former Ferralsols (FAO-Unesco, 1974) will no longer be included as Fer-

**Table 2:** Physical characteristics of dominant soils in the humid tropics.

| | | FERR | | ACRI | | LUVI | | CAMB | | AREN | | PODZ | |
|---|---|---|---|---|---|---|---|---|---|---|---|---|---|
| | | ($\bar{x}$) | ($s$) | ($\bar{x}$) | ($s$) | ($\bar{x}$) | ($s$) | ($\bar{x}$) | ($s$) | ($\bar{x}$) | ($s$) | ($\bar{x}$) | ($s$) |
| Sand (%) | (*top*) | 35 | 27 | 52 | 23 | 40 | 32 | 25 | 23 | 92 | 5 | 86 | 12 |
| | (*sub*) | 29 | 25 | 36 | 21 | 30 | 21 | 26 | 25 | 94 | 4 | 85 | 8 |
| Silt (%) | (*top*) | 15 | 11 | 19 | 12 | 20 | 11 | 27 | 13 | 5 | 4 | 11 | 9 |
| | (*sub*) | 15 | 12 | 19 | 13 | 21 | 12 | 25 | 15 | 4 | 4 | 6 | 4 |
| Clay (%) | (*top*) | 49 | 25 | 29 | 18 | 35 | 19 | 46 | 22 | 3 | 3 | 3 | 4 |
| | (*sub*) | 52 | 27 | 40 | 21 | 40 | 20 | 40 | 28 | 7 | 7 | 4 | 5 |
| Silt/Clay | (*top*) | 0.4 | 0.3 | 0.9 | 0.7 | 0.6 | 0.2 | 0.9 | 0.8 | 5.1 | 6.2 | 18 | 32 |
| | (*sub*) | 0.3 | 0.3 | 0.6 | 0.7 | 0.5 | 0.4 | 0.8 | 0.6 | 1.7 | 2.3 | 3.0 | 3.6 |
| Bulk density (g. $cm^{-3}$) | (*top*) | 1.1 | 0.2 | 1.3 | 0.2 | 1.2 | 0.2 | 1.1 | 0.2 | 1.2 | 0.1 | 1.3 | 0.0 |
| | (*sub*) | 1.2 | 0.2 | 1.4 | 0.2 | 1.3 | 0.1 | 1.2 | 0.3 | 1.4 | 0.2 | 1.5 | |
| Rootable pore volume (%) | (*top*) | 16 | 7 | 13 | 4 | 16 | 5 | 14 | 7 | 23 | 9 | 14 | 3 |
| | (*sub*) | 15 | 8 | 9 | 3 | 8 | 3 | 13 | 7 | 26 | 13 | 11 | |
| Plant avail. moisture (%) | (*top*) | 12 | 5 | 14 | 6 | 14 | 5 | 17 | 5 | 18 | 6 | 22 | 1 |
| | (*sub*) | 10 | 3 | 10 | 6 | 11 | 4 | 8 | 1 | 12 | 6 | | |

($\bar{x}$) = Mean (*top*) = topsoil 0–20 cm
($s$) = Standard deviation (*sub*) = subsoil 70–100 cm
FERR = Ferralsols (22), ACRI = Acricols (22), LUVI = Luvisols (9), CAMB = Cambisols (30), AREN = Arenosols (5), PODZ = Podzols (6). The number between the brackets indicates the number of reference soils in ISIS database available for statistical analysis.

**Table 3:** Chemical characteristics of dominant soils in the humid tropics.

| | | FERR | | ACRI | | LUVI | | CAMB | | AREN | | PODZ | |
|---|---|---|---|---|---|---|---|---|---|---|---|---|---|
| | | $(\bar{x})$ | $(s)$ | $(\bar{x})$ | $(s)$ | $(\bar{x})$ | $(s)$ | $(\bar{x})$ | $(s)$ | $(\bar{x})$ | $(s)$ | $(\bar{x})$ | $(s)$ |
| pH-$H_2O$ (1:2.5) | *(top)* | 4.8 | 0.6 | 4.8 | 0.5 | 6.4 | 0.5 | 5.3 | 1.0 | 5.3 | 0.9 | 4.5 | 0.3 |
| | *(sub)* | 5.0 | 0.5 | 4.8 | 0.2 | 5.9 | 0.8 | 5.5 | 0.9 | 5.8 | 1.6 | 4.8 | 0.5 |
| pH-KCl (1:2.5) | *(top)* | 4.1 | 0.4 | 4.1 | 0.3 | 5.5 | 0.7 | 4.6 | 0.8 | 4.1 | 1.0 | 3.7 | 0.5 |
| | *(sub)* | 4.5 | 0.7 | 4.0 | 0.2 | 4.6 | 0.8 | 4.5 | 0.7 | 5.1 | 1.6 | 4.4 | 0.5 |
| Organic Carbon (%) | *(top)* | 2.3 | 1.5 | 2.0 | 1.0 | 2.2 | 1.0 | 2.3 | 1.4 | 0.8 | 1.8 | 5.0 | 8.3 |
| | *(sub)* | 0.4 | 0.2 | 0.4 | 0.2 | 0.3 | 0.1 | 0.4 | 0.4 | 0.1 | 0.0 | 0.7 | 0.7 |
| C/N ratio | *(top)* | 16 | 10 | 14 | 5 | 17 | 15 | 11 | 4 | 16 | 3 | 23 | 9 |
| | *(sub)* | 9 | 5 | 8 | 3 | 7 | 1 | 8 | 4 | 12 | | 11 | 6 |
| Sum of bases | *(top)* | 1.8 | 2.3 | 2.2 | 2.5 | 21.2 | 16.1 | 11.5 | 18.0 | 2.0 | 2.0 | 1.0 | 1. |
| ($cmol_c\ kg^{-1}$) | *(sub)* | 0.7 | 1.7 | 0.6 | 0.8 | 16.8 | 21.3 | 9.0 | 15.0 | 2.0 | 3.1 | 0.1 | 0.1 |
| Exch. Al ($cmol_c\ kg^{-1}$) | *(top)* | 1.4 | 1.8 | 1.5 | 1.4 | 0.0 | 0.1 | 0.1 | 1.6 | 0.1 | 0.1 | 1.0 | 2.2 |
| | *(sub)* | 1.1 | 1.8 | 2.2 | 2.1 | 0.3 | 0.5 | 0.0 | 1.7 | 0.0 | 0.0 | 0.2 | 0.4 |
| $CEC_{pH7}$ ($cmol_c\ kg^{-1}$) | *(top)* | 8.8 | 5.4 | 9.9 | 6.4 | 22.7 | 16.2 | 19.3 | 13.8 | 6.6 | 5.5 | 20.0 | 36.0 |
| | *(sub)* | 4.0 | 3.0 | 6.9 | 5.6 | 25.0 | 19.5 | 14.9 | 12.0 | 3.2 | 3.6 | 4.7 | 6.0 |
| Base Saturation (%) | *(top)* | 19 | 16 | 26 | 23 | 87 | 17 | 49 | 54 | 44 | 35 | 18 | 17 |
| | *(sub)* | 19 | 26 | 12 | 15 | 67 | 38 | 52 | 43 | 39 | 43 | 43 | 36 |

$(\bar{x})$ = Mean *(top)* = topsoil 0–20 cm
$(s)$ = Standard deviation *(sub)* = subsoil 70–100 cm
FERR = Ferralsols (30), ACRI = Acrisols (33), LUVI = Luvisols (9), CAMB = Cambisols (30), AREN = Arenosols (5), PODZ = Podzols (6). The numbers between the brackets indicates the number of reference soils in ISIS database available for statistical analysis.

ralsols according to the revised legend (FAO, 1988). In view of the poor dispersing properties of tropical soils before measuring the silt and clay fractions it is recommended that further investigation of the applicability of the silt/clay ratio takes place.

*Bulk density, rootable pore volume and plant available moisture*

The results of bulk density, potential Rootable Pore Volume (RPV) and potential Plant Available Moisture (PAM) for the dominant soils are presented in table 2.

The mean bulk density of topsoils of the humid tropics is low (1.1–1.3 g $cm^{-3}$), but subsoils approach the average values found elsewhere. There are no large differences comparing the major soil groupings, but Ferralsols and Cambisols have the lowest and Acrisols, Arenosols and Podzols have the highest mean bulk density.

The potential Rootable Pore Volume (RPV) for topsoils is comparable for the dominant soils. Acrisols are lower with 13 % RPV, but this is considered not to be limiting for root development in their topsoils. Large differences are found in the subsoil, where Ferralsols and Cambisols have a high RPV and Acrisols and Luvisols have a lower RPV. Possible explanations for this are: i) the occurrence of clay accumulation and the formation of clay skins on structural elements in Acrisols and Luvisols; ii) Ferralsols have more stable micro-aggregates and biopores; and iii) Acrisols and Luvisols possibly have less bio-activity in the subsoil, but this could not be inferred from the field descriptions.

As expected, the potential Plant Available Moisture (PAM) is greater in the topsoil than in the subsoil due to a higher organic matter content and a better structure. Cambisols show the highest potential PAM and Ferralsols the lowest, with Acrisols and Luvisols taking intermediate positions. As dry periods are relatively short in the humid zone, if present at all, PAM should not form a constraint for crop production.

### 3.3.2 Chemical characteristics

*Soil reaction*

Soil reactions, pH-$H_2O$ and pH-KCl, of the dominant soils range from extremely acid to neutral (table 3). Ferralsols, Acrisols and Podzols have similar pH ranges and are predominantly strongly acid, Cambisols and Arenosols are acid and Luvisols are neutral. Comparing the soil reaction of topsoil and subsoil, except for the Luvisols, the dominant soils have a pH-$H_2O$ level of the topsoil similar or slightly lower than the subsoil.

A measure for the net charge status of the soil is delta pH, referring to the difference between pH-KCl and pH-$H_2O$. Ferralsols in ISIS have an average delta pH of –0.7 pH unit and the other soils around –1.0 pH unit (net negative charge: CEC). A frequency analysis of the delta pH for all soil samples shows that a positive delta pH is rare (see figure 2). Positive delta pH's, indicative for a net positive charge, are mainly observed in Geric Ferralsols (FAO, 1988).

*Organic Carbon*

Organic carbon and organic nitrogen content are widely used as a measure of organic matter quantity and quality in a soil, and as a crude measure of the fertility status (*Landon*, 1991). The interpretation of the content of organic carbon and organic nitrogen is complex, because

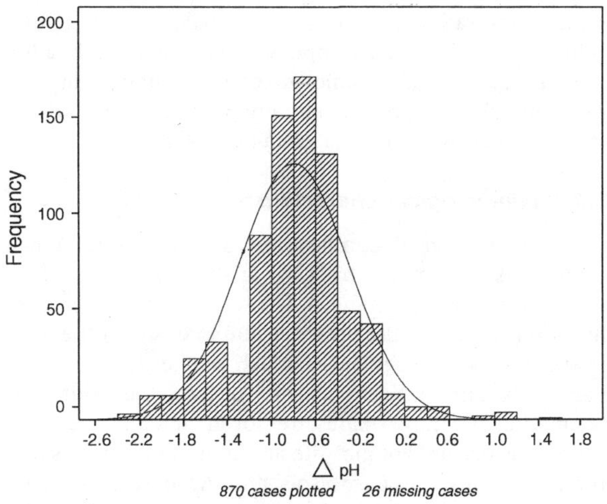


**Figure 2:** ΔpH histogram of 149 reference soils of the humid tropics. (ΔpH = pH-KCl minus pH-$H_2O$).

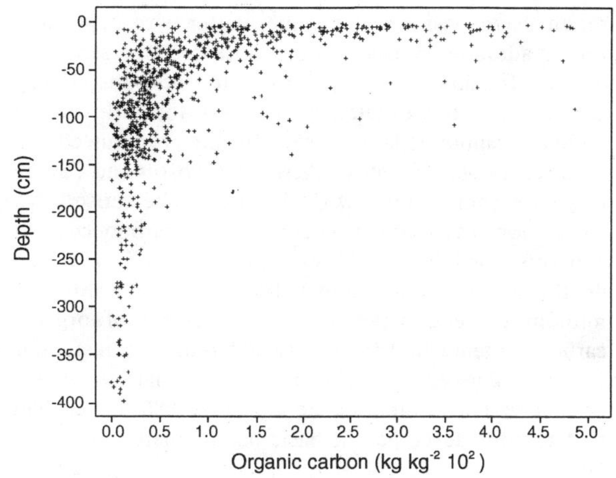


**Figure 3:** Scatter plot of organic carbon *vs.* depth of 149 reference soils from the humid tropics.

many other factors influence soil nutrient status and soil physical characteristics. Organic matter influences chemical and physical soil properties, especially in the tropics. In sandy soils and clay soils consisting dominantly of 1:1 lattice clays, a large part of the plant nutrients and about 90% of the capacity of the soil nutrient retention originates from organic matter. In unfertilized soils, organic matter is the source of 90–95 percent of nitrogen. Organic matter can be the major source of available phosphorus and available sulphur when humus is present in appreciable amounts (*Miller* and *Donahue*, 1990). Organic matter has also a strong positive influence on soil structure and the resistance of soils to erosion.

A plot of organic carbon (OC) content with depth of all samples not only shows the expected highest level of organic carbon in the upper 50 cm of the soil profiles, but also that substantial amounts can occur in the deeper subsoil of some soils (figure 3). Some soils have 1% OC at a depth of 50 cm, and 0.5% OC at a depth of 100 cm. Comparing field descriptions and analytical data of soils from Brazil and Indonesia, it is observed that such relatively high organic carbon levels in the subsoil are not matched by a darker soil matrix colour. These subsoils have a clear yellowish or reddish soil matrix, with Munsell colour codes of hue 5YR to 10YR and value/chroma combinations of 4/5 to 6/8, which outside tropical areas are indicative of a much lower content of organic carbon. The accumulation of organic carbon in the deeper subsoil and its importance for plant available water and nutrient retention is also reported from deeply weathered clay soils (Xanthic Ferralsols) in the states of Pará and Amazonia, Brazil (*Nepstad* et al., 1994). Therefore, further study of organic carbon in the deeper subsoil is advocated, irrespective of observed soil matric colour.

The organic carbon content and C/N ratio of topsoil and subsoil is given in table 3. Ferralsols, Acrisols, Luvisols and Cambisols have a comparable mean distribution of organic carbon within their profiles. Arenosols and Podzols clearly form a separate group with either a much lower or a higher organic carbon content. In addition, some deeper subsoil horizons of individual Podzol profiles have a much higher level, associated with the presence of a spodic horizon, i.e. an accumulation of organic material and iron in the subsoil, partly as an indurated layer ("ortstein"). In the humid tropics, the spodic horizon may be situated at a great depth ("giant Podzols").

The degree of decomposition (or mineralisation) of organic matter is characterized by the C/N ratio. High C/N ratios are indicative of slow decompostion and low values suggest a high degree of decomposition or a low organic nitrogen (N) content from vegetation with a low N content. The following interpretation is mainly used in temperate zones. Fresh organic matter may have a C/N ratio from around 15 (leguminous biomass) to about 80 (cereal straw). A C/N ratio of soil organic matter above 12 to 14 points to a shortage of nitrogen in the soil, resulting in a stagnation of the organic matter decompostion process. A C/N ratio from 8 to 10 is indicative for good quality 'stabilized' humus (*Miller* and *Donahue*, 1990).

The mean C/N ratios for the dominant soils are presented in table 3. Cambisols have the lowest C/N ratio in the topsoil, indicative of good quality organic matter. High C/N ratios in Podzols indicate that humified organic matter has a very low nitrogen content. It is a reflection of the poor parent material of the soil and poor quality of organic materials. The relatively high C/N ratio of the topsoils of Ferralsols and, remarkably, also of Luvisols points to a nitrogen shortage.

The overall difference of the C/N ratio between topsoil and subsoil is large. This may be explained by a more thorough decomposition of organic matter with depth and age. In addition, transport of the more decomposed, finer, organic fraction from the topsoil to the subsoil may contribute to this difference. However, the influence of the latter cannot be large, because of the pronounced bio-homogenisation. Ferralsols, Acrisols, Luvisols and Cambisols have comparably low C/N ratios in the subsoil. No correlation was found between organic carbon content and soil reaction for acid soils (pH-$H_2O$ < 5.5), at any depth, nor is a correlation with the exchangeable aluminium content apparent. The correlation of organic carbon content with parent material has also been studied. For soils derived from acid parent materials the mean organic carbon content of the topsoil is 0.8% lower than that of soils developed in basic parent materials.

*Nutrients*

In addition to the soil reaction and organic carbon content, the nutrient status of a soil is further characterized by the exchangeable bases (calcium, magnesium, potassium and sodium), the exchangeable acidity (dominantly exchangeable aluminium), and the measured cation exchange capacity (CEC) at pH 7. Mean values for the sum of exchangeable bases, cation exchange capacity and base saturation are low or very low in topsoil and subsoil of Ferralsols, Acrisols, Arenosols and Podzols (table 3). Cambisols and Luvisols form a separate group with moderate to high values.

There is no good correlation between the content of bases and the parent material classes, although soils developed in basic parent materials have a slightly higher CEC, sum of bases, and base saturation compared with acid, intermediate or old alluvial parent materials. This low correlation can be explained because the Leaching Rainfall (see section 3.2) is very high and therefore old, i.e. (very) deeply developed, soils in the humid tropics are strongly leached, irrespective of the original base content of the parent material. As observed before, physical characteristics, such as clay percentage and structure, are more strongly correlated with parent material. For example, basic parent material results in soils with a high clay content. However, it is noted that for a restricted area correlation between parent material and chemical characteristics of soils can be stronger.

A correlation of pH-KCl and aluminium saturation for all soil samples is presented in figure 4. The exchangeable aluminium content is negligibe when pH-KCl exceeds 4.8 (which corresponds approximately to a pH-$H_2O$ > 5.5), as free aluminium is neutralized. Soils with pH values below these levels can have a high level of exchangeable aluminium, inducing aluminium toxicity, which limits root growth and other biological activity.

The exchangeable aluminium content in Acrisols and Ferralsols is high, but nil or very low in the other dominant soils. The value of exchangeable aluminium is low in the subsoil of the Ferralsols. Nonetheless, the mean aluminium saturation expressed as percentage of the ECEC[3] is high, about 50% and comparable with Acrisols. In addition to the use of organic materials, liming is usually recommended to prevent aluminium toxicity by raising the pH value to about 5.5 (*Landon*, 1991).

### 3.3.3 Mineralogical characteristics

The mineralogical composition of the 149 reference profiles, as determined by X-ray analysis of the clay fraction is summarized in table 4. The semi-quantitative classification is based on the presence and intensity of the peak in the clay mineralogy diagram. All soils have a predominance of kaolinite. Nearly all soils have some goethite and haematite, but these oxides are not included in the table. Moderate amounts of gibbsite are found in Ferralsols and remarkably also in Cambisols and Arenosols. Acrisols, Cambisols and Luvisols have minor amounts of chlorite and illite. It should be noted that the amount of clay in Arenosols is low in comparison to the other soils.

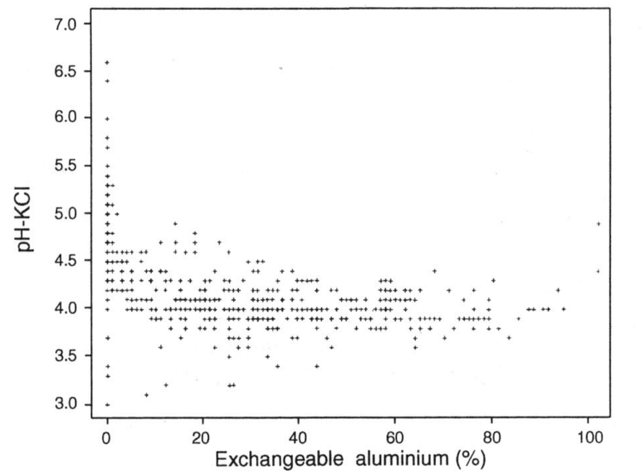


**Figure 4:** Scatter plot of pH-KCl *vs.* exchangeable aluminium saturation (%) using 149 reference soils of the humid tropics.

**Table 4:** Clay mineralogy of dominant soils in the humid tropics.

| | FERR | ACRI | LUVI | CAMB | AREN |
|---|---|---|---|---|---|
| Gibbsite | + | – | | + | + |
| Kaolinite | ++ | +++ | ++ | ++ | + |
| Halloysite | | ++ | | + | |
| Illite | | – | – | – | |
| Chlorite | | – | – | – | |
| Smectite | | | + | | – |
| Vermiculite | | – | | | |

+++ = dominant, ++ = sub-dominant, + = moderate and – = minor presence of a clay mineral
FERR = Ferralsols, ACRI = Acrisols, LUVI = Luvisols, CAMB = Cambisols and AREN = Arenosols

[3] The ECEC is calculated from the sum of exchangeable bases and exchangeable acidity and is considered to approximate the actual cation exchange capacity of the soil under field conditions.

### 3.4 Major agronomic constraints

#### 3.4.1 All dominant soils

For the purpose of highlighting major soil/land constraints for agriculture, a qualitative evaluation of 15 land qualities was made based on the Framework for Land Evaluation (FAO, 1976, 1983). The land quality assessment is generally based on 2 or more single land characteristics. Critical values and other class limits for single soil parameters are adapted from several publications (*Ilaco*, 1981 and *Landon*, 1991). To make the assessment applicable for all tropical regions, criteria have been developed for a 'standard' crop, i.e. a deep rooting annual crop, assuming low technology and low inputs. In 'Framework' terms this corresponds with a major kind of land use, viz. "rainfed agriculture, annual cropping". Each land quality is rated in 5 classes which indicate the degree of limitation for the specified use.

Rating procedures have been computerized with the Automated Land Evaluation System (ALES), (*Rossiter* and *Van Wambeke*, 1993), to assess the present status of the land using the reference soils (*Mantel* and *Kauffman*, 1995). Depending on the land quality, an assessment is made for a specific depth range. For topsoil characteristics the first soil horizon is evaluated; for nutrient status related characteristics it is assessed over 0—50 cm, and for others over the depth 0—120 cm. For the land quality 'erosion hazard' the erodibility of the soil is rated. Climate factors are not considered.

This, briefly described, qualitative assessment should not be regarded as an absolute judgement, but rather as a first identification of constraints for agriculture. The assessment was executed for all soils, and results, i.e. the frequencies of the five degree classes for all the dominant soils, are expressed in percentages in table 5. A high frequency in the classes moderate, serious and very serious are indicative for the presence of a constraint. The most common constraints are: a low level of plant nutrients, a low level of nutrient retention capacity, and toxicity caused by high exchangeable aluminium levels. Less frequently occurring constraints are the potential for mechanisation, erosion hazard and potential soil moisture content.

#### 3.4.2 Individual dominant soils

In the preceding section, the constraints for the dominant soils as one group are summarized. Here a similar assessment is presented for each dominant soil. For ease of comparison the results of the dominant soils are summarized in one table. For each dominant soil, the median of the land quality ratings (1 = no limitation to 5 = severely limiting) was calculated. In table 6, the six major limitations — those that were moderately or (very) severely limiting — are given for the dominant soils.

From this table it can be seen that of the five dominant soils, Luvisols have the most favourable properties for agriculture. Cambisols in the humid tropics have frequently severe limitations for availability and retention of nutrients and have a moderate toxic aluminium level. Ferralsols and Acrisols have as limitations (very) low nutrient retention and nutrient availability, severe aluminium toxicity and moderate soil moisture retention. Arenosols

**Table 5:** Frequency (%)[1] of constraints by land quality of all dominant soils.

| degree of constraints[2] > land quality | no | weak | moderate | serious | very serious |
|---|---|---|---|---|---|
| length of growing period | 70 | 30 | 0 | 0 | 0 |
| drought hazard | 100 | 0 | 0 | 0 | 0 |
| soil moisture availability | 5 | 39 | 36 | 13 | 6 |
| oxygen availability | 68 | 18 | 6 | 7 | 1 |
| nutrient availability | 2 | 9 | 13 | 33 | 43 |
| nutrient retention capacity | 7 | 5 | 12 | 38 | 38 |
| rootable volume | 58 | 29 | 10 | 3 | 0 |
| conditions for germination | 97 | 2 | 1 | 0 | 0 |
| salinity | 99 | 0 | 1 | 1 | 0 |
| sodicity | 79 | 7 | 7 | 1 | 6 |
| aluminium toxicity | 24 | 0 | 33 | 13 | 30 |
| workability | 21 | 57 | 15 | 7 | 0 |
| potential for mechanisation | 38 | 22 | 19 | 16 | 5 |
| erosion hazard | 41 | 28 | 10 | 11 | 9 |
| flooding hazard | 77 | 14 | 8 | 0 | 1 |

[1] Because of rounding, the sum of percentages for each land quality may sum to 99 or 101.
[2] The qualitative degree of limitation reads as follows:

no = no constraint or limitation, no yield reduction
weak = slight limitation, slight yield reduction
moderate = moderate limitation, moderate yield reduction
serious = severe limitation, clear yield reduction
very serious = very severe limitation, strong yield reduction or no yield.

**Table 6:** Major constraints of dominant soils of the humid tropics.

| land quality | FERR | ACRI | LUVI | CAMB | AREN/PODZ | ALL |
|---|---|---|---|---|---|---|
| soil moisture | x | x | x | | xxx | x |
| nutrient availability | xxx | xxx | | xx | xxx | xx |
| nutrient retention | xxx | xx | x | xx | xxx | xx |
| aluminium toxicity | xx | xx | | x | x | x |
| mechanisation | | | x | | | |
| erosion hazard | | | x | | | |

x = moderately limiting, xx = severely limiting, xxx = very severely limiting.
FERR = Ferralsols, ACRI = Acrisols, LUVI = Luvisols, CAMB = Cambisols, AREN = Arenosols, PODZ = Podzols and ALL = overall figure of the 6 dominant soils.

and Podzols have the least favourable properties for agriculture.

### 3.5 Landforms, soils and biodiversity

Soil properties strongly influence the characteristics of the forest growing on it. However, there are few strictly paired observations across the three continents concerned with soil conditions and the local structure of the natural vegetation: multistorey closed-canopy forest *vs* open-canopy forest *vs* savanna-forest, its total timber volume and biomass, including organic carbon storage, and its biodiversity.

In overview papers on these relationships for the Amazon region (*Sombroek*, 1990, 1992) it was concluded that the highest gross timber volumes (over 200 $m^3$ $ha^{-1}$) are found in those areas where the total annual rainfall is not excessively high (less than 3000 mm per year); where a short relatively dry season exists (less than four consecutive months with less than 50 mm rainfall each) and where moreover the soils are deep (more than 200 cm) and well drained, not very sandy and lacking a compact subsoil. Such soils, with a substantial effective soil moisture storage capacity, are mainly of the Xanthic Ferralsol type which predominate in the easter axial part of the Amazon basin. If the dry season is too long, and/or the soils are less deep or having somewhat compact subsoil (such as occurs in some Acrisols of the fringe areas of the Amazon region), an open-canopy forest prevails with a lower biomass and a stronger propensity to degrade under anthropogenic influences such as shifting cultivation into bamboo-, liana- or savanna vegetation. Arenosols, Podzols, Plinthosols, Dystric Gleysols and Histosols have a natural forest of low timber volume ("caatinga Amazonia"), tree savanna vegetation with bare white-sand patches ("campina"), scrubby to open grass savanna ("campo"), or a strong predominance of palm species. Such patterns are present in Zaire and on the island of Kalimantan in Indonesia.

The relationship between soil conditions and the biodiversity of the natural vegetation is more complex. This is linked with the various aspects of biodiversity: species diversity, endemism and speciation. The *richness in species* of a range of habitats in the geographic unit under consideration ("gamma" biodiversity) is bound to be highest in areas with short-distance strong variation in habitat, where the landscape is composed of land facets of strongly different topographic, soil hydrological and microclimatic conditions. These areas are to be found in uplands and hilly areas where geomorphological processes were active in several episodes of the Pleistocene or Early Holocene, especially where such activity resulted in exposure of fresh rocks or sediments of short-distance contrasting lithology and mineralogy. Hilly lands and uplands with convex slopes over cystalline basement or older sedimentary rocks are, therefore, likely to have the highest degree of species richness.

*Endemism*, or the restriction of species to specific areas within the biome is often linked to extreme climatic, hydrologic or soil conditions, where such species have a competitive advantage. For the Amazon region this applies, for instance, to the species Cedro (*Cedrela odorata*), Mahogany (*Swietenai macrophylla*) and Castanho-do-Pará (*Bertholletia excelsa*).

The "campina" and "caatinga" units on the Arenosols and Podzols of the Amazon have a well-documented low degree of species richness but a high number of endemic species. This is also reported for other areas with such kind of soils, such as parts of Kalimantan.

Even in areas with a seemingly monotonous cover of well drained, strongly weathered non-sandy soils such as Ferralsols and with uniform climatic conditions, phytogeographic differences exists. This holds true, for instance, for the occurrence of Angelim-pedra (*Hymenolobium excelsum*) and Pau-amarelo (*Euxilophora paraensis*) on only some landscape facets of eastern Amazonia; in both cases a specific micro-nutrient content of the substratum is considered to be the determining factor (*Sombroek*, 1990).

Large *intra-specific variation* and the evolution of new species, as forms of genetic diversity, are supposed to occur where past climatic, geomorphologic and soil conditions were not much different from those of "refuge" areas today. For the Amazon region, such areas have tentatively been identified as isolated sandstone table-lands, intact high-level sedimentary plains of the Late Tertiary – Early Pleistocene age (the Amazon "planalto"), and areas of relic valleys with long concave slopes.

## 4 Conclusions

The number of reference soil profiles in ISIS of each major soil group per ecological zone is limited (between 10 and 30) in comparison to the large areas involved (between 50 and 500 million hectares). On the other hand, the reference profiles have been carefully selected and are well distributed over the humid tropical zone. Uniform analytical procedures (one soil laboratory), makes correlation for the ISIS datasets justified at major soil grouping level. Results, presented in section 3 and the conclusions which follow should be seen as indicative of possible trends. Departures from these trends will be common in view of the natural spatial variability of soil characteristics, especially when looking at sub-group level. For most major soil groupings the number of reference profiles currently in ISIS is insufficient for statistical analyses at subgroup level. For such analysis, a database with a larger number of soil profiles by major soil grouping is needed. Arenosols and Podzols can easily be distinguished from the Ferralsols and Acrisols on the basis of their diagnostic properties. The overall data presented in this paper show that Ferralsols and Acrisols, covering 57% of the humid tropics, have rather similar key characteristics. The separation is based on a relatively small increase in clay content with depth, which does not affect the major vegetation type or crop productivity level. These conclusions tally with recent research in the Colombian Amazon (*Duivenvoorde* and *Lips*, 1995). On the other hand, variability of each key characteristic for both Ferralsols and Acrisols is large. Agricultural land assessment studies and soil vegetation/bio-diversity correlation studies should be based on measurements of key soil characteristics, reflecting the bio-physical functioning of the soil, and should not rely only on a soil taxonimic approach.

In view of the very low to low nutrient content of both topsoil and subsoil in the studied dominant soils, the decomposing organic materials covering the mineral soil as the litter layer are the main source of soil fertility and should be included as a potential key property. A clearer topsoil characterization is required in present international soil taxonomic systems, which usually put emphasis on stable subsoil parameters. Organic carbon content is remarkably constant in the subsoils of the low fertility soils with an average content of 0.4% for all major soil groupings (at a depth of 70 to 100 cm). The top/subsoil ratio of the organic carbon content is about 5 to 6. In low fertility soils, organic matter largely determines the soil nutrient level and the cation exchange capacity. It is therefore remarkable that this large top/subsoil ratio of organic carbon content is much less reflected in the nutrient level (expressed as sum of bases), with a top/subsoil ratio around 2 to 4. It becomes even less when considering the top/subsoil ratio of the effective cation exchange capacity with a ratio between 1 to 2. This means that the mineral part of the soil, including the subsoil, fulfils a more important role then one would expect from the general accepted dominant role of organic carbon.

A large extent of the humid tropics is covered by Ferralsols, Acrisols, Arenosols and Podzols of low fertility, all presenting various degrees of limitations when forests are cleared for (low input) arable farming. Major constraints are a low plant nutrient content, a low nutrient retention capacity and a high exchangeable aluminium content. The combination of these soil-related constraints with a very high Leaching Rainfall makes fertilization of these soils difficult. It is disputed whether high fertilizer input, low biomass turn-over, mono-cropping agro-systems are sustainable and whether it could be easily introduced to the farmers on these soils in the humid tropics (*Weischet* and *Caviedes*, 1994). Sustainable agro-ecosystems in the humid tropics aim at a closed nutrient cycle (thus minimalising nutrient losses to the deeper subsoil). Agroforestry-systems imitating the rainforest, thus mimicking a permanent vegetative coverage of the soil, crop/plant diversity and a large biomass turn-over, seem to be most promising from the point of view of sustainability. The required higher (manual) labour input and the lack of education are probably the main limitations to the adoption of these sustainable agro-forestry systems by farmers in the humid tropics. Some of the extremely infertile soils need a cropping system with long fallows. In reality this is often no longer feasible. Technically, a well-timed and well-spaced, moderate fertilizer application (leading to more biomass and thus more crop residue), combined with topsoil organic matter conservation may improve rooting and fertility conditions. In some of the very acid soils, liming or the use of aluminium tolerant crop varieties may be imperative. However, economic conditions will dictate whether such, generally costly, fertilizers and soil amendments are available for the average farmer.

The relatively fertile Cambisols and Luvisols have fewer limitations. In view of the persistent ecological constraints of the low fertile soils mentioned before, these fertile soils, covering a smaller area of the humid tropics, should attract the attention of governments and farmers for use in highly productive sustainable agricultural systems.

To accurately assess the value of high floristic biodiversity, multidisciplinary terrain studies are needed on the geographic relationships between floristic species occurrence, soil conditions, historic stability of landscapes, past climatic conditions and past anthropologic influences (the latter may have contributed to a higher bio-diversity as well!). Full or partial protection of forest areas of the upper part of hydrological catchments, of areas with unstable or otherwise fragile soils, and of areas of indigenous forest dwellers, has to go hand-in-hand with protection of proven high-biodiversity forests. This requires a process of holistic (agro-) ecologic and (socio-) economic zoning and subsequent land use or non-use planning, the latter in close cooperation with all the participants interested in the future of the tropical forest lands.

## Acknowledgements

The authors are grateful for comments received on the draft text from Hans van Baren, Niels Batjes, Mike Bridges, Roel Oldeman and Otto Spaargaren. We are indebted to Jacqueline Resink for map compilation and to Piet Tempel for his programming support, which facilitated selection of profiles and calculations using the information in the ISIS database.

## References

*Batjes, N. H., V. W. P. van Engelen, J. H. Kauffman* and *L. R. Oldeman* (1994). Development of global databases for environmental research. Transactions 15th World Congress of Soil Science (Mexico), Vol. 6a, 40–57.

CGIAR (1992). Review of CGIAR priorities and strategies. Report no. AGR/TAC: IAR/92/18 – Part I. TAC secretariat, FAO, Rome.

*Duivenvoorden, J. F.* and *J. M. Lips* (1995). A land-ecological study of soils, vegetation and plant diversity in Colombian Amazonia. Tropenbos series 12. The Tropenbos Foundation, Wageningen.

FAO-Unesco (1974). FAO-Unesco Soil Map of the World. Volume I Legend. Unesco, Paris.

FAO (1971–1981). FAO-Unesco Soil Map of the World. Volumes II – X. Unesco, Paris.

FAO (1976). A framework for land evaluation. Soils Bulletin 32. FAO, Rome.

FAO (1977). Guidelines for soil profile description. 2nd edition. FAO, Rome.

FAO (1978–1981). Report on the Agro-ecological Zones Project. Vols. 1–4. World Soil Resources Report 48 (1–4). FAO, Rome.

FAO (1983). Guidelines: land evaluation for rainfed agriculture. Soils Bulletin 52. FAO, Rome.

FAO (1988). FAO-Unesco Soil Map of the World. Revised Legend. World Soil Resources Report 60. FAO, Rome.

FAO (1989). FAO-ISRIC Soil Database (SDB). World Soil Resources Report 64. FAO, Rome.

FAO (1990). Guidelines for soil description. 3rd edition. FAO, Rome.

FAO (1992). Framework for characterization and classification of topsoils in the world. World Soil Resources Report, first draft. FAO, Rome.

FAO (1993). World Soil Resources. An explanatory note on the FAO world soil resources map at 1:25 000 000 scale. World Soil Resources Report 66 Rev. 1. FAO, Rome.

FAO (1995). Forest Resources assessment 1990, Global synthesis. Forestry Paper no. 124. FAO, Rome.

*Greenland, D. J., G. Bowen, H. Eswaran, R. Rhoades* and *C. Valentin* (1994). Soil, water and nutrient management research – a new agenda. IBSRAM Position Paper, Bangkok.

ILACO (1981). Agricultural compendium for rural development in the tropics and sub-tropics. Elsevier Scientific Publishing Company.

ISRIC (1994–1995). Country Reports – Field and analytical data of soil reference profiles. International Soil Reference and Information Centre, Wageningen. (Country reports are available for: Brazil, China, Colombia, Costa Rica, Cuba, Côte d'Ivoire, Ecuador, Indonesia, Nicaragua, Peru, Thailand, Turkey and Zambia).

ISRIC (1993). Bi-annual Report 1991–1992. International Soil Reference and Information Centre, Wageningen.

ISRIC (1995). Bi-annual Report 1993–1994. International Soil Reference and Information Centre, Wageningen.

*Kauffman, J. H., S. Mantel* and *O. C. Spaargaren* (1996). Soils of the humid and seasonally dry (sub) tropics: a correlation of reference soil data and their assessment for agriculture using the ISIS database. In: Proceedings National Soil reference and Databases Workshop, Vol. 2. Use of ISRIC's databases for the characterization of soils of major agro-ecological zones. International Soil Reference and Information Centre, Wageningen.

*Landon, J. R.* (Ed.), (1991). Booker tropical soil manual: a handbook for soil survey and agricultural land evaluation in the tropics and subtropics. Booker Agriculture International, London.

*Lal, R.* (1987). Need for, Approaches to, and Consequences of Land Clearing and Development in the Tropics. In: Lal, R., H. W. Scharpenseel and M. Sudjadi (Eds.), (1987). Tropical land clearing for sustainable agriculture, IBSRAM Proceedings, No. 3, Bangkok.

*Leemans, R.* and *W. Cramer* (1991). The IIASA database for mean monthly values of temperature, precipitation and cloudiness on a global terrestrial grid. Research Report RR-91-18, International Institute of Applied Systems Analyses, Laxenburg.

*Mantel, S.* and *J. H. Kauffman* (1995). Agricultural Suitability of Reference Soils, the Automated Land Evaluation System applied to ISRIC Soil Information System. Working Paper and Preprint 95/13. International Soil Reference and Information Centre, Wageningen.

*Nepstad, D. C., C. R. De Carvalho, E. A. Davidson, P. H. Jipp, P. A. Lefebre, G. H. Negreiros, E. D. Da Silva, T. A. Stone, S. E. Trumbore* and *S. Vielra* (1994). The role of deep roots in the hydrological and carbon cycles of Amazonian forest and pastures. Nature, 372: 668–669.

*Miller, R. W.* and *R. L. Donahue* (1990). Soils, an introduction to soils and plant growth. 6th edition. Prentice-Hall International.

*Rossiter, D. G.* and *A. R. Van Wambeke* (1993). Automated Land Evaluation System, ALES Version 4 User's Manual. Department of Soil, Crop and Atmospheric Sciences, Cornell University, Ithaca.

Soil Survey Staff (1992). Keys to Soil Taxonomy, fifth edition. SMSS Technical Monograph no. 19. Pocahontas Press, Blacksburg.

*Sombroek, W. G.* (1990). Amazon landforms and soils in relation to biological diversity. Proceedings of a workshop to determine priority areas for conservation areas in Amazonia, January 1990, Manaus, Brazil, INPA, Manaus and Conservation International, Washington (advance publication in Annual Report 1990 of ISRIC, Wageningen, 7–25).

*Sombroek, W. G.* (1992). Biomass and carbon storage in the Amazon ecosystems. Interciencia 17 (5).

*Spaargaren, O. C.* (Ed.) (1994). World Reference Base for Soil Resources. Draft. ISSS-ISRIC-FAO, Wageningen/Rome.

Statistix (1994). User's manual, Statistix, version 4.1. Analytical software. Tallahassee.

*Trewartha, G. T.* (1968). An introduction to Climate. Fourth Edition. McGraw-Hill Book Company, New York.

*Van Reeuwijk, L. P.* (1993). Procedures for soil analysis. Technical Paper 9 (4th edition). International Soil Reference and Information Centre, Wageningen.

*Van Ijssel, W. I.* and *W. G. Sombroek* (1987). Spatial variability of tropical rainforest and forestlands. In: Van Beusekom C. F., C. P. Van Goor and P. Schmidt (Eds.): Wise utilization of tropical rainforest lands. Scientific series 1, Tropenbos, Wageningen.

*Van Waveren, E. J.* and *A. B. Bos* (1988). ISRIC Soil Information System, user manual and technical manual, Technical Paper 15, International Soil Reference and Information Centre, Wageningen.

*Weischet, W.* and *C. N. Caviedes* (1993). The persistent ecological constraints of tropical agriculture. Longman.

*Whitten, D. G. A.* and *J. R. V. Brooks* (1974). A dictionary of geology. Penguin Books.

*Young, A.* (1976). Tropical soils and soil survey. Cambridge Geographical Studies no. 9. Cambridge University Press.

# The World Reference Base for Soil Resources

## An introduction with special reference to soils of tropical forest ecosystems


**Otto C. Spaargaren**[1] and **Jozef Deckers**[2]

[1]International Soil Reference and Information Centre, P.O. Box 353, 6700 AJ Wageningen, The Netherlands. e-mail: spaargaren@isric.nl
[2]Institute for Land and Water Management, Catholic University of Leuven, Vital Decosterstraat 102, B-3000 Leuven, Belgium. e-mail: seppe.deckers@agr.kuleuven.ac.be



### Abstract

The World Reference Base for Soil Resources (WRB) is the successor to the International Reference Base for Soil Classification (IRB). Its task is to apply the IRB principles of definitions and linkages to the existing classes of the Revised FAO-Unesco Soil Map of the World Legend (FAO, 1988). The main objective is to provide scientific depth and background to the Revised Legend, and to ensure that the latest knowledge relating to global soil resources and interrelationships is incorporated in a world-wide soil reference system.

At present, the WRB comprises 30 reference soil groups. Three new reference soil groups, i.e. Cryosols, Durisols and Umbrisols, are introduced compared with the Revised Legend, Greyzems are amalgamated with the Phaeozems, and Podzoluvisols renamed Albeluvisols. The concept and terminology of diagnostic horizons and properties is retained and expanded. In addition, diagnostic soil materials are defined. In the description and definition of the WRB diagnostic horizons, properties and materials emphasis is placed on field identification. Analytical criteria are given to help the identification. Some modifications are proposed to diagnostic horizons, properties and materials as defined in the Revised Legend and a number of new ones are formulated.

The basic philosophy of WRB is that the soil groups must represent a minimal geographical coverage and are the result of a major pedogenetic process. Soils are characterized by their morphological expression rather than by analytical data.

Ferralsols, Acrisols, Lixisols, Alisols and Nitisols constitute the main reference soil groups in tropical forest ecosystems occupying well drained upland positions. Other soil groups of local or regional importance are Andosols, Podzols, Umbrisols and Arenosols. The main soils occupying poorly drained positions in the tropics are Histosols, Fluvisols, Plinthosols, Gleysols and, to a lesser extent, Planosols. On one hand, the World Reference Base for Soil Resources, like the FAO-Unesco Legend, continues to build up on existing soil classification systems. Examples are the description, definition and subdivision of Anthrosols and Andosols. On the other hand, some proposals result from new ideas which are not yet reflected in published classification systems.

The final aim is a well described and defined World Reference Base for Soil Resources that is internationally accepted by the community of soil scientists. It also intends to facilitate the international use of pedological data, not only by soil scientists, but also by other users of soil and land.


## 1 Introduction

After some hundred years of modern soil science no system of soil classification has yet been adopted universally (*Dudal*, 1990). This situation arises partly from the fact that soils constitute a continuum which, unlike individual plants and animals, needs to be subdivided into classes by convention. To remedy this situation work has been going on during the past 15 years to arrive at what now is known as a *World Reference Base for Soil Resources*.

**History.** The World Reference Base for Soil Resources (WRB) is the successor to the International Reference Base for Soil Classification (IRB), an initiative taken by the International Society of Soil Science (ISSS) in the early 1980's. The intention of the IRB project was to arrive at a framework through which existing soil classification systems could be correlated and ongoing soil classification work harmonized. The objective was to reach an international agreement on major soil groupings to be recognized at a global scale and on the criteria and methodology to be applied for defining and separating them. Such an agreement was meant to facilitate the exchange of information and experience, to provide a common scientific language, to strengthen the applications of soil science, and to enhance the communication with other disciplines.

In 1992, it was therefore decided that the IRB adopt the Revised Legend of the Soil Map of the World (FAO, 1988) as the frame for its further work. It would become IRB's task to apply its principles of definitions and linkages to the defined soil units to give them more depth and validation. The merger of the two efforts was launched under the name: 'World Reference Base for Soil Resources', an ISSS/FAO/Unesco/ISRIC undertaking. During the 15th World Congress of Soil Science in Acapulco, Mexico, a Draft World Reference Base for Soil Resources (ISSS-FAO-ISRIC, 1994) was presented for further comments and scrutiny.

**Objectives and principles.** The main objective of the World Reference Base for Soil Resources is to provide scientific depth and background to the Revised Legend, so that it incorporates the latest knowledge relating to global soil resources and interrelationships. More specifically, the objectives are:

- to develop an internationally acceptable framework for delineating soil resources to which national classification systems can be related and through which the national systems can be linked, using the Revised Legend as a guideline;
- to provide this framework with a sound scientific basis so that it facilitates the international use of pedological data, not only by soil scientists, but also by other users of soil and land, such as geologists, botanists, agronomists, hydrologists, ecologists, farmers, foresters, civil engineers, architects, etc.;
- to acknowledge in the framework important lateral relationships of soils and soil horizons as characterized by topo- and chronosequences; and
- to emphasize the morphological characterization of soils rather than to follow an approach purely based upon laboratory analyses.

Consequently, the basic framework of the FAO-Unesco Legend, with its two categoric levels and guidelines for developing classes at a third level (*Nachtergaele* et al., 1994), was adopted. Subsequent developments have led to a system of 30 reference soil groups, each with uniquely defined modifiers, thus giving the World Reference Base for Soil Resources a maximum flexibility and broadest possible scope with respect to national classification systems which it, amongst others, is supposed to link.

**Elements of the World Reference Base for Soil Resources.** For describing and defining the reference soil groups and soil units of the WRB, use is made of soil characteristics, properties and horizons, which in combination will define soils and their interrelationships.

*Soil characteristics* are single parameters which are observable or measurable in the field, in the laboratory, or can be analyzed by using microscope techniques. They include characteristics such as colour, texture and structure of the soil, features of biological activity, arrangement of voids and pedogenetic concentrations (mottles, cutans, nodules, ...) as well as analytical determinations (soil reaction, particle-size distribution, cation exchange capacity, exchangeable cations, amount and nature of soluble salts, ...).

*Soil properties* are combinations ("assemblages") of soil characteristics which are known to occur in soils and which are considered to be indicative of present or past soil-forming processes (e.g. vertic properties are a combination of heavy texture, smectitic mineralogy, gilgai, slickensides, hard consistence when dry, sticky consistence when wet, shrinking when dry and swelling when wet).

*Soil horizons* are three-dimensional pedological bodies which are more or less parallel to the earth's surface. Each horizon is characterized by one or more properties, occurring over a certain depth, with a certain degree of expression. The thickness varies from a few centimetres to several metres; most commonly it is about a few decimeters. The upper and lower limits ("boundaries") are diffuse, gradual, clear or abrupt. Laterally, the extension of a soil horizon varies greatly, from a metre to several kilometres. However, a soil horizon is never infinite. Laterally, it disappears or grades into another horizon.

*Reference soil groups* are defined by a vertical combination of horizons within a defined depth, and by the lateral organization of these horizons, or by the lack of them.

Soil horizons and properties are intended to reflect the expression of genetic processes which are widely recognized as occuring in soils. They can therefore be used to describe and define soil classes. They are considered to be "diagnostic" when they reach a minimum degree of expression, which is determined by visibility, prominence, measurability, importance and relevance for soil formation and soil use, and quantitative criteria. To be considered diagnostic, soil horizons also require a minimum thickness, which must be appraised in relation to bioclimatic factors (e.g. a podzolic B horizon in boreal regions is expected to be less thick than one in the tropics.).

Table 1 illustrates the interrelationships between soil characteristics, diagnostic properties, diagnostic horizons and reference soil groups, with Ferralsols as an example.

It is recognized that a number of the reference soil groups may occur under different climatic conditions. It was decided, however, not to introduce separations on account of climatic characteristics in order to keep the units within manageable limits and in order not to subordinate the classification of soils to the availability of climatic data.

## 2 The reference soil groups in the WRB

**The WRB reference soil groups.** The 30 reference soil groups presently proposed are Histosols, Cryosols, Anthrosols, Leptosols, Vertisols, Fluvisols, Solonchaks, Gleysols, Andosols, Podzols, Plinthosols, Ferralsols, Planosols, Solonetz, Chernozems, Kastanozems, Phaeozems, Gypsisols, Durisols, Calcisols, Albeluvisols, Alisols, Nitisols, Acrisols, Lixisols, Luvisols, Umbrisols, Cambisols, Arenosols and Regosols.

**Table 1:** Relationships for Ferralsols and ferralic soil units between soil characteristics, diagnostic properties, thickness, diagnostic horizons and reference soil groups.

| Soil characteristics | Limiting value | Diagnostic property | Thickness | Diagnostic horizon | Reference soil group |
|---|---|---|---|---|---|
| CEC | $\leq$ 24 $cmol_c$ $kg^{-1}$ clay | Ferralic (protic) | ($\geq$15 cm) | (Cambic) | Ferralic Cambisols |
| | $\leq$ 16 $cmol_c$ $kg^{-1}$ clay | Ferralic (orthic) | $\geq$30 cm | Ferralic | Other Ferralsols |
| ECEC* | $<$ 12 $cmol_c$ $kg^{-1}$ clay | Ferralic (orthic) | $\geq$30 cm | Ferralic | Other Ferralsols |
| | $<$ 1.5 $cmol_c$ $kg^{-1}$ clay | Geric | $\geq$30 cm | Ferralic | Geric Ferralsols |
| Soil reaction | $pH_{KCl}$ - $pH_{water}$ $\geq$ 0 | Geric | $\geq$30 cm | Ferralic | Geric Ferralsols |
| Water-dispersible clay* | $<$ 10% | Ferralic (orthic) | $\geq$30 cm | Ferralic | Other Ferralsols |
| Gravel content** | $<$ 90% | Ferralic (orthic) | $\geq$30 cm | Ferralic | Other Ferralsols |
| Weatherable minerals * | $<$ 10% | Ferralic (orthic) | $\geq$30 cm | Ferralic | Other Ferralsols |
| Texture*** | sandy loam or finer | Ferralic (orthic) | $\geq$30 cm | Ferralic | Other Ferralsols |
| Structure | weak to moderate | Ferralic (orthic) | $\geq$30 cm | Ferralic | Other Ferralsols |

* Failing one or more of these, the soil classifies as Ferralic Cambisols.
** Failing this the soil classifies as Leptosol.
*** Failing this the soil classifies as Ferralic Cambisol.

**New groups and shifts at the highest level.** The newly introduced reference soil groups are **Cryosols, Durisols** and **Umbrisols**. Cryosols are introduced at the highest level to identify a group of soils which occur under the unique environmental conditions of alternating thawing and freezing. These soils have a "*cryic horizon*" (i.e. a horizon which shows evidence of perennial ice segregation and/or cryoturbation and/or macro- and microstructures resulting from ice segregation, cryoturbation, freeze-thaw processes, frost heave, cryogenic sorting, and/or thermal cracking).

Durisols are soils occurring in semi-arid to arid environments which have, within 100 cm of the surface a silica-cemented layer ("*duripan*").

Umbrisols are proposed to cover the soils which have an umbric horizon. They are a logical counterpart of the Chernozems, Kastanozems and Phaeozems.

The Podzoluvisols as known in the FAO-Unesco Revised Legend have been renamed **Albeluvisols** in the WRB. The name Podzoluvisol suggests that in these soils both the processes of cheluviation (leading to Podzols) and subsurface accumulation of clay (resulting in Luvisols) take place, while in fact the dominant process consists of removal of clay and iron/manganese along preferential zones (ped faces, cracks) in the argic horizon (resulting in "*albeluvic tonguing*"). The name Albeluvisol is therefore thought to be more appropriate.

For the World Reference Base it was decided to exclude from the Leptosols soils which have pedogenetic horizons such as indurated calcic or gypsic horizons or hardened plinthite. This necessitated inclusion of these soils in the appropriate reference soil group. Included in the **Plinthosols, Gypsisols** and **Calcisols** are soils which have a petroplinthic, petrogypsic or petrocalcic layer at shallow depth. In the Revised Legend the latter soils belong to the Leptosols.

## 3 Well drained soils of tropical forest ecosystems

Five reference soil groups constitute the main soils in well drained upland positions in the tropics, viz. Ferralsols, Acrisols, Lixisols, Alisols and Nitisols. Diagnostic for these soil groups are the presence of a *ferralic* horizon, an *argic* horizon dominated by low activity clays accompanied by a low (in Acrisols) or high (in Lixisols) base saturation, *alic* properties, and a *nitic* horizon, respectively. Other soil groups of local or regional importance are Andosols (Southeast Asia, Central America), Podzols (South America, Southeast Asia). Umbrisols (tropical highlands) and Arenosols (South America, Africa, Southeast Asia).

The main characteristic of **Ferralsols** is the dominance of low activity clays (mainly kaolinite) resulting from strong weathering and mineral alteration. They can usually be recognized by their rather uniform appearance, their weak to moderately developed macrostructure and friable consistence, and gradual to diffuse horizon boundaries. Microstructure, however, is normally strongly developed, resulting in "pseudosand" or "pseudosilt" structures. The dominance of low activity clays is expressed by a low CEC, in Ferralsols of 16 $cmol_c$ $kg^{-1}$ clay or less (corrected for organic matter). In addition, they have a low content of water-dispersible clay. Some Ferralsols exhibit a significant clay increase somewhere in the solum, but this is not considered as an argic horizon if the water-dispersible clay content is low (less than 10%). These characteristics form together the requirements for the *ferralic* horizon, which is diagnostic for the Ferralsols.

Ferralsols have low nutrient reserves and available aluminium and manganese may reach toxic levels. Phosphate fixing capacity is usually high, more than 85%. The available water content is low (normally less than 100 mm per meter soil). The favourable physical properties ensure that these soils are easy to work and hardly prone to erosion. However, the surface is liable to compaction and crusting if heavy machinery is used to clear forest. Ferralsols occur mainly in South America and Central Africa, and to a lesser extent in South and East Asia.

**Acrisols** and **Lixisols** are characterized by having an increase in clay content sufficient to meet the requirements for an argic horizon and by a rather advanced stage of weathering reflected in a CEC of less than 24 $cmol_c$ $kg^{-1}$ clay. Distinction between Acrisols and Lixisols is made on the basis of their base saturation; Acrisols having a base saturation $< 50\%$, and Lixisols $\geq 50\%$.

They often show a clearly developed horizonation, particularly in the upper part of the soil. Many Acrisols and Lixisols have topsoils of sandy loam or sandy clay loam textures with clear boundaries to an argic horizon having a sandy clay or clay texture. Soil structure is normally only weakly developed. Surface horizons are generally thin, with a low amount of organic matter, especially in regions with pronounced dry seasons. Only in tropical highlands, under fairly humid conditions and/or low temperatures, organic matter accumulation may be considerable.

Acrisols and Lixisols have poor chemical and physical properties. Adverse characteristics are the low nutrient retention capacity and nutrient reserves, the high aluminium saturation in Acrisols, and a low structural stability of the topsoil. They are easily erodible. Careful soil management, including use of adapted varieties, is required to make these soils productive. Amongst the measures to be taken are minimum tillage (to prevent slaking and caking), preservation of the topsoil and its organic matter status, erosion control, split application of fertilizers and minimal disturbance when clearing the land. Agroforestry has shown to be beneficial to these soils, in particular by improving their organic matter status, and as soil erosion control measure.

Main areas with Acrisols and Lixisols in the world are Southeast Asia, West Africa and Latin America. Generally, Acrisols predominate in the more humid parts of the tropical regions, while Lixisols mainly occur in the drier parts of the tropics, further away from the equator.

**Alisols** form a relatively new group in the system. The name Alisols was coined in the Revised Legend for soils having an argic B horizon, a CEC of 24 $cmol_c$ $kg^{-1}$ clay or more and a base saturation (by *1M* $NH_4OAc$) of less than 50%. This group of soils were separated from the Acrisols as originally defined in the 1974 FAO-Unesco Legend because they generally have a high total exchangeable aluminium content (FAO, 1988).

Conceptually, Alisols are soils with 2:1 clays (chlorite, smectite, vermiculite) in which a process of mineral alteration is taking place in a silica and base depleted environment. This group of soils is characterized by an intermediate to advanced stage of weathering, and, having lost most of their primary minerals, are dominated by high activity clays. Through weathering of the secondary 2:1 clay minerals, large amounts of aluminium and magnesium are released, giving rise to the strongly acid conditions. Figure 1 shows the relationship between *1M* KCl extractable Al and CEC (both in $cmol_c$ $kg^{-1}$ clay) in subsurface horizons of Ferralsols, Acrisols, Alisols, Luvisols and so-called "red montmorillinitic soils", all having a pH KCl of 4.0 or less, and illustrates that the magnitude of KCl extractable aluminium is an important diagnostic characteristic to distinguish between Alisols and other "luvic" soils.

The total reserve of bases (i.e. the content of total basic cations (Ca, Mg, K and Na), including both exchangeable

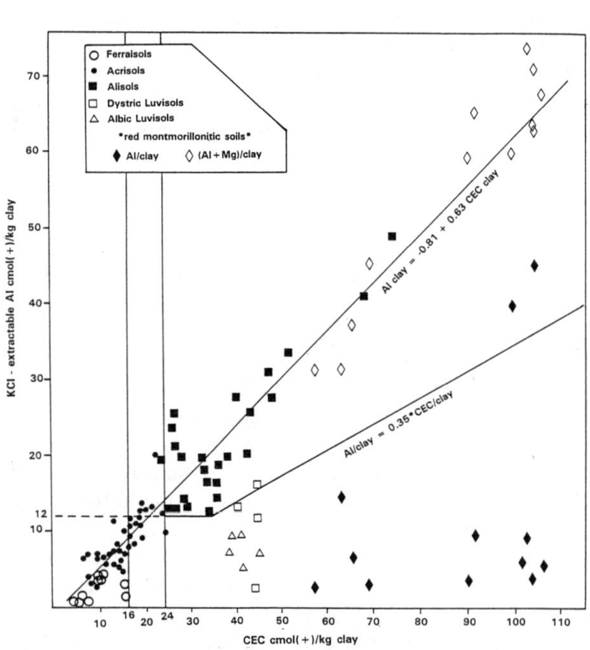


**Figure 1:** Relationship between 1*M* extractable Al and CEC (both in $cmol_c$ $kg^{-1}$ clay) in subsurface horizons of Ferralsols, Acrisols, Alisols, Luvisols and "red montmorillonitic soils", all having pH KCl values of 4.0 or less (from ISSS-FAO-ISRIC, 1994).

and non exchangeable pools, and expressed in $cmol_c\ kg^{-1}$ soil) is concentrated mainly in the clay fraction and constitutes usually 80% or more of the total reserve of bases in the fine earth fraction. In the field, many Alisols have a dark coloured surface horizon with massive or weakly developed structures. Subsurface horizons normally have angular blocky or prismatic structures and develop cracks upon drying, unlike Acrisols, Lixisols and Ferralsols. They usually have a medium to high clay content, a low content in silt (silt-clay ratio less than 0.6) and are often derived from, or associated with, basic rocks.

Besides the specific chemical problems, Alisols have unfavourable physical properties in the surface horizon(s). The low structural stability of the surface horizon results in slaking and a reduced permeability and internal drainage (FAO, 1991). Alisols are known to be erosion-prone.

The importance of recognizing Alisols as reference soil group is that these soils with high activity clay have a very high aluminium saturation, which is toxic to most plants. However, unlike the Acrisols and Lixisols, these soils may be made very productive after heavy fertilization and liming. Minimum tillage is recommended to preserve the surface soil.

The extent of Alisols is not precisely known, but major occurrences are recorded from Southeast USA, Latin America, West Indies, West and East Africa, Indonesia and China.

Throughout the tropics and subtropics well drained soils occur which consist of dusky red to dark brown clays with a moderately strong to strong angular blocky structure. The peds are "polyhedric or nutty" elements and have shiny faces resulting from thin clay and/or manganese coatings or pressure faces. These soils, called **Nitisols**, have a high aggregate stability, friable consistence, high porosity, fair to good moisture storage capacity and an easy rooting. Horizon boundaries between the surface layer(s) and the underlying horizons are usually gradual to diffuse. The upper horizons contain variable amounts of organic matter and may be acid or neutral in reaction. They have in common that they are predominantly composed of low activity clay minerals. P-adsorption capacity is high but this does not, however, result in acute P deficiencies.

Characteristic for the Nitisols is the occurrence of a *nitic horizon*. Besides the typical structure, nitic horizons have a clayey texture although in the field the material feels loamy. Most Nitisols show a gradual clay increase in the subsoil. Silt content is low (silt-clay ratios of 0.20–0.35) as is the water-dispersible clay content (less than 10% of the total clay). Usually no gravel or stone-sized concretions are present, but some fine Fe/Mn concretions ('shot') may occur.

The CEC (by *1M* $NH_4OAc$) in humus-poor nitic layers is between 8 and 24 $cmol_c\ kg^{-1}$ clay, sometimes higher. These values can increase substantially with an increase in organic matter. ECEC is about half of the CEC. This points to a predominance of 1:1 lattice type clay minerals (kaolinite, halloysite).

Typical for nitic horizons is the high amount of sesquioxidic minerals in the clay fraction, especially those of iron (total iron content may be as high as 15 percent). Hematite, goethite and gibbsite are the most common. Magnetite and maghemite are present in smaller but not insignificant amounts, making the soil material magnetic in dry conditions. Also crystalline titanium- and manganese oxides are present.

Main regions with Nitisols are East Africa, South Brazil, Central America and Southeast Asia (Java, Philippines). Depending on the topographic/hydrologic position, the age of the landscape elements, and the admixture of airborne materials, especially volcanic ash, Nitisols occur in lateral relation with Vertisols, Andosols, Ferralsols, Luvisols and Cambisols. These relationships are illustrated in Figure 2.

Nitisols are much sought after for sustained smallholders' farming and plantation crops such as cocoa and coffee, despite the low CEC and frequently low base saturation. The fact that these soils are less prone to erosion than most other upland soils in the (sub)tropics, their good tilth, easy workability and other favourable physical attributes have contributed to the success of sustainable low-input agriculture on these soils.

An important change in the concept of Nitisols with respect to the Revised Legend is that these soils are not necessarily characterized by the presence of an argic horizon. Moreover, to separate Ferralsols from Nitisols in which all requirements for a ferralic horizon are met, Ferralsols are not permitted to have a nitic horizon.

## 4 Poorly drained soils of tropical forest ecosystems

Four reference soil groups constitute the main soils in poorly drained positions in the tropics, viz. Histosols, Fluvisols, Plinthosols and Gleysols. Diagnostic for these soil

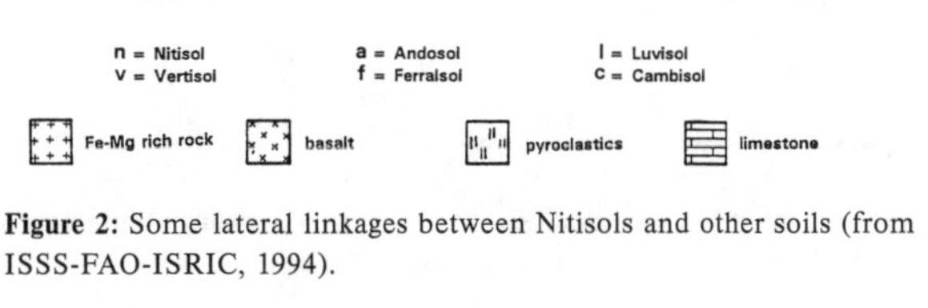


**Figure 2:** Some lateral linkages between Nitisols and other soils (from ISSS-FAO-ISRIC, 1994).

groups are the presence of a thick *histic* horizon, *fluvic* properties, a *plinthic* or *petroplinthic* horizon, and *gleyic* properties, respectively. Soils of less importance in tropical forest ecosystems are Planosols, and gleyic or stagnic soil units of other reference soil groups.

**Histosols** differ from other soils in that they are formed of 'organic soil material' with physical, chemical and mechanical properties that are very different from those of mineral soil materials. They develop in conditions where organic material is produced by an adapted (climax) vegetation, and where biochemical decomposition of plant debris is retarded by low temperatures, persistent waterlogging, extreme acidity, oligotrophy and/or the presence of high levels of electrolytes or organic toxins.

**Fluvisols** occur on materials deposited in aqueous sedimentary environments. *Driessen* and *Dudal* (1991) suggest that there are three situations where fresh material is continually added by sedimentation from water. These are (1) the inland fluvial and lacustrine fresh-water environments, (2) the marine environment and (3) the coastal salty or brackish marsh environment, of which deltas are a special case.

The concept of **Plinthosols** is one of soils affected, at present or in the past, by groundwater or stagnating surface water in which iron has been segregated to such an extent that a humus-poor, mottled layer has been formed, which irreversibly hardens when exposed to the air and sunshine. Included in the concept are those soils which have such a hardened layer at shallow depth. Although it is realized that soils with shallow petroplinthic layers and soils having plinthite usually occupy different positions in the landscape, it was felt appropriate to group them together as they are genetically related. This genetic relation is illustrated in Figure 3, showing the development of petroplinthite (*"cuirasses"*) as a result of landscape inversion.

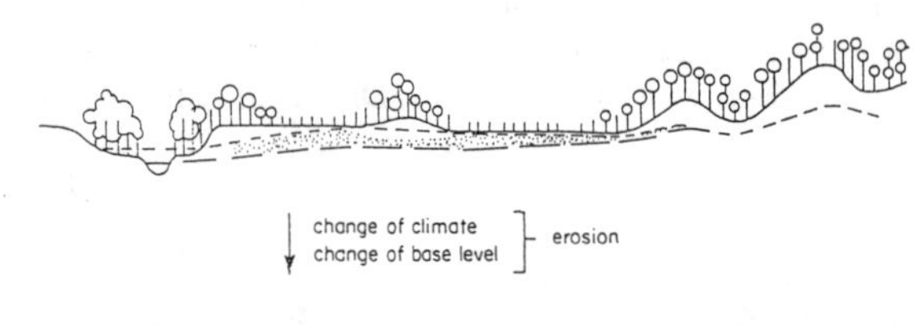


**Figure 3:** Historic inversion of relief in landscapes with plinthite and petroplinthite (from ISSS-FAO-ISRIC, 1994).

**Gleysols** are soils influenced by groundwater, unlike stagnic (sub)units, which are conditioned by stagnating surface water. The subsoil of Gleysols is permanently wet and reduced (unless drained), contains free $Fe^{2+}$ and/or has an rH (the negative logarithm of hydrogen pressure) of 19 or less. The topsoil may have features associated with oxidation, either by alternating moist/wet conditions, or by permanent aeration. This depends on variations of the groundwater table, but also on rainfall and evapotranspiration.

## 5 Correlation of soils with tropical forest ecosystems

**Well drained tropical forest ecosystems.** Perhumid evergreen forest is the most dominant natural vegetation type in regions around the equator. These forests occur in areas without a prolonged dry period, while precipitation exceeds 1600 mm, often 2000 mm annually (FAO-Unesco, 1971; 1977; 1979). Temperatures are high throughout the year with little variation between the seasons. In well drained sites these conditions favour strong weathering and leaching. Consequently, soils associated with these forests are mainly Ferralsols, Acrisols and Lixisols, and to a lesser extent, Alisols and Nitisols.

Where prolonged dry seasons occur or where soil moisture storage capacity in subhumid regions is too low to supply sufficient water during short dry periods, tropical deciduous and semi-deciduous forests or open woodlands occur. Dominant soils are Acrisols and Lixisols, associated with Nitisols and Ferralsols with minor occurrences of Alisols, or sandy soils (Podzols or Arenosols).

Semi-arid ("dry subhumid") areas usually have a shrub vegetation cover with isolated larger trees, some of which do have economic value. Soils associated with this vegetation generally are Lixisols and Arenosols.

In mountain regions with tropical evergreen and semi-deciduous forests Acrisols and Umbrisols prevail, associated with Leptosols (shallow soils overlying continuous hard rock) and Cambisols (moderately deep soils characterized by slight to moderate soil development). To a minor extent also Nitisols, Alisols and Histosols may be found.

**Poorly drained tropical forest ecosystems.** These tropical forest ecosystems are found in low-lying areas which are daily, seasonally or permanently inundated with fresh, brackish or salt waters. Daily inundations, in particular with salt or brackish water, occur along the coastlines in many tropical regions, and give rise to vast mangrove forests. Dominant soils are Fluvisols, many of which contain sulphidic materials in the subsoil. When drained these soils become extremely acid (pH water less than 3.5).

In deltaic regions and other low-lying alluvial plains prolonged seasonal inundations may occur giving rise to fresh water swamp forests of which the different types of vegetation are usually closely related to the duration and depth of flooding. Main soils found here are Fluvisols and Gleysols, associated with Plinthosols.

Areas permanently inundated by fresh water have swamp or bog forests with Histosols and Gleysols as dominant soils.

Poorly drained areas influenced by groundwater at shallow depth often have a savanna woodland vegetation. Gleysols are the main soil type. Where the groundwater level fluctuates, Plinthosols may be dominant.

**Importance of forest cover for maintenance of fragile soil systems.** Many soils in tropical upland regions have in common a surface layer which is highly susceptible to degradation. The organic matter accumulation in the top parts of Ferralsols, Lixisols and Alisols is usually restricted to the first 10 centimetres. Organic matter in these soils contributes significantly to the nutrient holding capacity, water storage, and structural stability. Indiscriminate cutting will expose these fragile topsoils to sun and rain, and the content of organic matter will subsequently be reduced rapidly. This reduction often results in structural decline (compaction, crusting), loss of fertility and difficult regeneration of the vegetation.

The vegetation cover is also of great importance for the soil moisture balance and runoff protection. Under tropical forests the soil is kept almost permanently moist and a rather constant temperature is maintained. The foliage protects the surface from erosion. This is particularly important in areas with Plinthosols. These soils will harden irreversibly if they are not protected from desiccation or erosion. When hardened, Plinthosols are rendered practically useless.

Similarly, desiccation, either by cutting of the vegetation or by drainage, causes Thionic Fluvisols – common soils under mangrove forests – to become extremely acid through oxidation of the sulphide materials present.

## 6 Conclusion

**Correlation with existing classification systems.** The World Reference Base for Soil Resources (WRB), having taken the framework of the FAO-Unesco Revised Legend as guide, obviously bears many similarities to it. The nomenclature of the Revised Legend has been adopted and, where necessary, adapted. Its concepts of diagnostic horizons and properties, supplemented by diagnostic soil materials, have been taken up. The original Legend (FAO, 1974) itself has been built on the knowledge and experience of many soil scientists from all over the world, and as such reflects a consensus of a number of classification systems. For instance, the separation of Greyzems, Chernozems and Kastanozems is related to the older Russian classification of Grey Forest soils, Chernozems and Chestnut soils. Similarly, Cambisols coincide largely with the German "Braunerde" and French "Sols bruns", while Ferralsols closely follow the concept of the US Oxisols and Brazilian Latosols.

Many of the definitions in both the original and revised FAO-Unesco legends were adopted from the USDA-SCS Soil Taxonomy (Soil Survey Staff, 1975; 1987). These definitions are in most cases summarized and simplified, in accordance with the requirements of the legend.

The WRB also continues to build on existing classification systems. The proposals for the Anthrosols contain many elements from the recently published Chinese Soil Taxonomic Classification System (Gong, 1994). The description and definition of Andosols is closely related to that of the Andisols in the Référentiel Pédologique Français (RPF) (AFES, 1990); similarly the definition of the Podzols has been influenced by the RPF criteria for Podzosols, but to a lesser extent than for the Andosols. On the other hand, some of the proposals result from new ideas not yet reflected in published classification systems. The Durisols are a good example of this.

Definitions of diagnostic horizons, properties and materials derived from existing classification systems which are proposed for the WRB, but which are not used in the Revised Legend, concern anthric horizons, the anthraquic horizon sequence (*Gong*, 1994), and tephric soil material (*Hewitt*, 1992).

**Latest developments.** The draft World Reference Base for Soil Resources has been undergoing a period of testing and refining. This exercise showed the strengths and weaknesses of the system. Particular emphasis was placed on determining sound field criteria. Limits between the reference soil groups and soil units were checked for usefulness and pragmatism. Gaps were identified and removed. "Floating" soils, i.e. soils that are either difficult to classify or that are unsatisfactorily classified, needed to be re-assessed and, consequently, some proposed WRB classes were adapted. Significant work has been done on developing the "modifier level", in fact amalgamating the soil unit level of the Revised Legend and the proposals for distinguishing soil subunits (*Nachtergaele* et al., 1994).

It is hoped that the final result may become a well described and defined World Reference Base for Soil Resources, which is internationally acceptable by the soil science community and which facilitates the international use of pedological data, not only by soil scientists but also by other users of soil and land, such as geologists, agronomists, hydrologists, ecologists, farmers, foresters, botanists, civil engineers, architects, etc. Funds permitting, it may be supplemented by a publication on "Benchmark Soils of the World", in which the WRB soil units are briefly described and defined, their potential and problems highlighted and are illustrated by a typical pedon with relevant analytical data, as well as a "World Atlas of Soils", presenting the distribution of the main soil groups and derived variables in more detail.

## 7 References

*AFES* (Association Française pour l'Etude du Sol). 1990. Référentiel Pédologique Français. 3ème proposition, avril 1990. Plaisir. 279 p.

*Delvaux B.* and *A. Herbillon*. 1993. Alisols, searching for significant diagnostic criteria, WRB meeting, November 1993. Leuven. 16 p.

*Driessen P. M.* and *R. Dudal* (Eds.). 1991. The major soils of the world. Lecture notes on their geography, formation, properties and use. Wageningen/Leuven. 310 p.

*Dudal R.* 1990. An International Reference Base for Soil Classification (IRB). "Ernst Schlichting" Memorial Workshop, Stuttgart-Hohenheim, Germany, 21–23 April, 1990, p 54–57.

*FAO*. 1988. Soil Map of the World. Revised Legend. Reprinted with corrections. World Soil Resources Report no. 60. Food and Agriculture Organization of the United Nations, Rome. 119 p.

*FAO*. 1991. World Soil Resources. An explanatory note on the FAO World Soil Resources Map at scale 1:25 000 000. World Soil Resources Report no. 66. Food and Agriculture Organization of the United Nations, Rome. 58 p.

*FAO-Unesco*. 1971. Soil Map of the World 1:5 000 000. Vol. IV. South America. Unesco, Paris. xiv + 193 p.

*FAO-Unesco*. 1974. Soil Map of the World 1:5 000 000. Vol. I. Legend. Unesco, Paris. vii + 59 p.

*FAO-Unesco*. 1977. Soil Map of the World 1:5 000 000. Vol. VI. Africa. Unesco. Paris. xiv + 299 p.

*FAO-Unesco*. 1979. Soil Map of the World 1:5 000 000. Vol. IX. Southeast Asia. Unesco, Paris. xiv + 149 p.

*Gong Zitong* (Ed.). 1994. Chinese Soil Taxonomic Classification System (1st Proposal). Institute of Soil Science, Academia Sinica, Nanjing. viii + 93 p.

*Hewitt A. E.* 1992. New Zealand Soil Classification. DSIR Land Resources Scientific Report no. 19. Lower Hutt. 133 p.

*ISSS-FAO-ISRIC*. 1994. World Reference Base for Soil Resources. Draft. ISSS-ISRIC-FAO, Wageningen/Rome. 161 p.

*Nachtergaele F., A. Remmelzwaal, J. Hof, J. van Wambeke, A. Souirji* and *R. Brinkman*, 1994. Guidelines for distinguishing soil subunits. Transactions 15th World Congress of Soil Science. Volume 6a, p. 818–833.

*Soil Survey Staff.* 1975. Soil Taxonomy. A basic system of soil classification for making and interpreting soil surveys. Soil Conservation Service, U. S. Dept. Agric., Agriculture Handbook no. 436. U. S. Govt. Pr. Off., Washington, DC. vii + 754 p.

*Soil Survey Staff.* 1987. Keys to Soil Taxonomy (third printing). SMSS Technical Monograph no. 6. Soil Management Support Services, Ithaca, New York. 280 p.

# Soils of the tropical forests of Leyte, Philippines I: Weathering, soil characteristics, classification and site qualities

**Reinhold Jahn**[1] and **Victor B. Asio**[2]

[1]Institute of Soil Science (310), University of Hohenheim, 70593 Stuttgart, Germany
[2]Dept. of Agronomy and Soil Science, Visayas State College of Agriculture, ViSCA, Baybay, Leyte, 6521-A, Philippines

## Abstract

Two forest soils representative of two research sites in Leyte, Philippines, are described as to their physical, chemical and mineralogical properties. The soils, an Andosol and an Alisol, which have developed from intermediate quaternary and pleistocene to pliocene volcanoclastics, respectively, represent a chronosequence of soils. The course of weathering is explained in terms of intensity and extent of the turnover of elements, loss of basic elements, desilification and development of pedogenic minerals. The site properties are described quantitatively in terms of rootability, availability of air, water and nutrients, as well as nutrient reserves. Both soils are highly weathered and show more or less the same percentage of desilification. The loss of basic elements is slightly lower in the Andosol than in the Alisol. In addition, there is a distinctly smaller loss of P and Mn in the Andosol. By calculating the turnover and losses of elements in kg per $m^2$, the much higher extent of weathering in the Alisol can be illustrated. The clay mineral association is dominated by imogolite in the Andosol and by halloysite and kaolinite in the Alisol. The progress of mineral alteration can also be observed from the Fe-mineralogy. Both soils have excellent physical qualitites regarding rootability and aeration and water supply to plants. The limiting factor for plant nutrition is the very low amount of available P in both soils, which is aggravated further by high P-retention in the Andosol.

## 1 Introduction

Site qualities are in general a very complex aggregation of observable and measurable site characteristics and which have to be evaluated for specific land uses (FAO, 1976). With the aim of ecological land evaluation in mind, the large number of site characteristics should be summarized in terms of possible rooting depth and rootability as well as water, air, energy and nutrient budgets (*Schlichting* et al., 1995). Furthermore, statements about elasticity and stability of a site under a specific use are needed in order to establish a sustainable land use. Since many site characteristics are the result of pedogenesis, the development of qualities is therefore closely related to the development of soils which in turn is time dependent. Land evaluation usually accepts site characteristics more or less only as time dependent for the surveyable period of a particular use. This is relatively short compared to the formation of soils. Thus, in order to figure out the long term behavior of soils, it is the objective of this paper to combine statements about soil formation with statements about site qualities formation on a long time scale.

The bigger part of the Island of Leyte (7955 $km^2$) is made up of volcanic rocks from different phases of volcanic activity during the tertiary and quaternary age. The NNW-SSE running central cordillera (max. height 1333 m) consists at the center of andesitic, basaltic and dacitic flows and breccia of miocene age which are covered by younger lava flows (mostly of intermediate nature) and deposits of volcanoclastics. Two research areas (see also part II by *Asio* et al., this book, Fig. 1) were chosen at the western slope of the cordillera with hilly landscapes about 12 km NE of Ormoc (700 m asl) and 8 km N of Baybay (100 m asl).

## 2 Material and methods

The two soils, which are discussed here, have been chosen from a set of similar soils (compare *Asio* et al. (part II), Fig. 2) identified in each area, being as representative of more or less undisturbed conditions and maximum soil development in both areas.

Analyses of granulometry, porosity, bulk density, extractions with $NH_4$-oxalate, NaOH, Na-pyrophosphate and Na-dithionite-citrate and pH were carried out using the methods of *Schlichting* et al. (1995). $C_t$ was determined by the Walkley-Black method (*Nelson* and *Sommers*, 1982), $N_t$ by the Kjeldahl method (*ISRIC*, 1986), cation exchange capacity (CEC) and exchangeable cations by the $NH_4$-acetate method (*Black*, 1965), available P by Bray No. 2 (*PCARR*, 1980) and P-retention by the method of *Blake-*

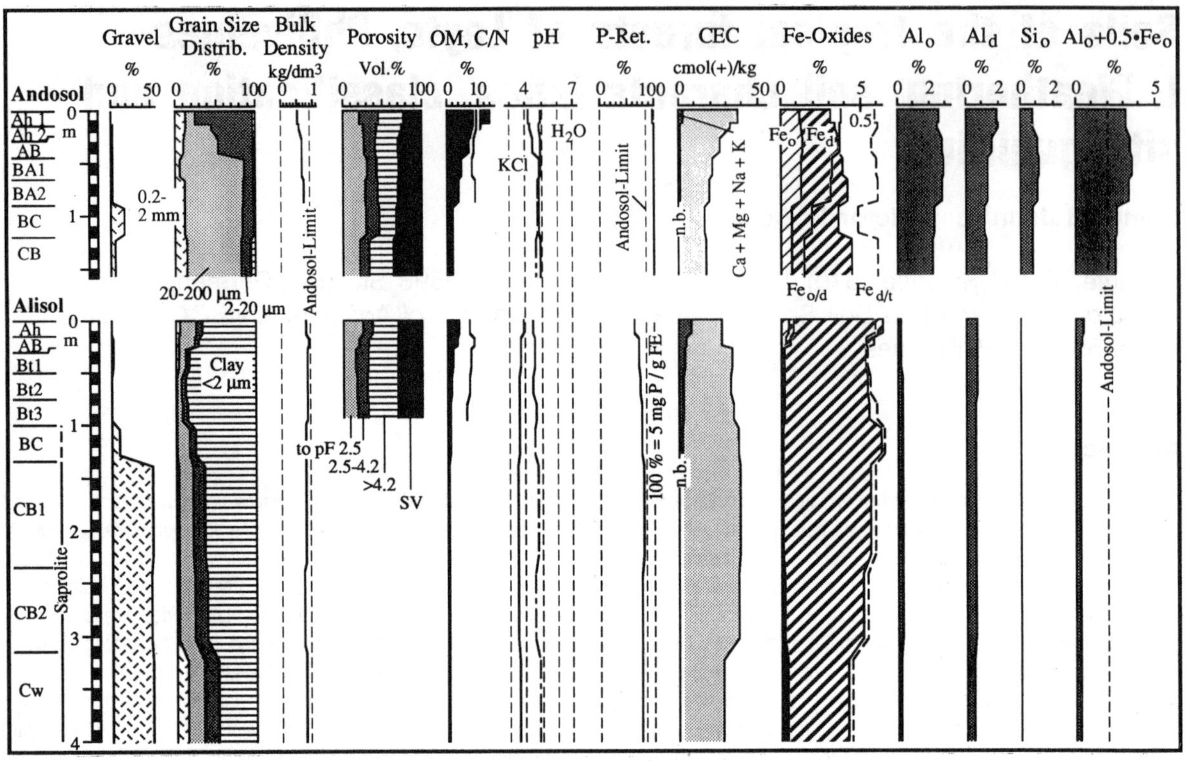

**Figure 1:** Depth functions of important soil characteristics for an Andosol and an Alisol from Leyte (Philippines).

*more* et al. (1981). Minerals were determined using a Siemens X-ray equipment on powder- (sand and silt) and texture specimens (clay) and total element contents were determined by X-ray fluorescence spectrometry. Soil forming processes were quantified using the methods discussed by *Barshad* (1964), *Alaily* (1984), *Jahn* (1988) and *Schlichting* et al. (1995).

## 3 Results and discussion

### 3.1 Soil forming factors

**Climate:** The present climate is chracterized as a humid tropical monsoon climate ("tropical rainforest climate, monsoon type" after *Köppen*, 1923) with an annual rainfall of about 3000 mm and an average annual air temperature of 24,2 (Ormoc) and 27,4°C (Baybay). The rainfall distribution generally shows a more pronounced peak rain period (up to nearly 500 mm per month) on the eastern side of the central cordillera than on the western side. Only for a very short period does the precipitation in the western side drop below 100 mm per month which corresponds to the average monthly potential evapotranspiration. Typhoons regularly cross the Island during the times of the SW-monsoon (June to October) and NW-monsoon (November to February). The difference between the coldest and the warmest month is between 2 and 3°C. From these data (*Barrera* et al., 1954, VISCA-Meteorological Station) the soil moisture regime can be classified to be udic and the soil temperature regime to be isohyperthermic.

**Geology and Geomorphology:** The site near Ormoc is situated in a hilly landscape with volcanic cones which are rounded due to erosion. The remnants of the cones are a clear indication of the quaternary age of the landscape. The parent material of the investigated site is intermediate (55% $SiO_2$) volcanoclastics, classified as local trachytic basalt-andesite. The parent material of the Baybay site is more basic (50% $SiO_2$) and is classified as porphyritic basalt. The generally very steep western slope of Mt. Pangasugan, where the site is located, shows some plain areas at different heights which are strongly dissected by rivers. This indicates that they were probably the remnants of one or more old surfaces (peneplaines). The association of the strongly eroded sites in the valleys and the deep weathered saprolites in the plains, is a further argument for old (Pleistocene and/or Pliocene) erosion surfaces.

**Vegetation:** Both areas belongs to the natural rainforest type of dipterocarp forest. The Ormoc site actually is at the edge of a residual forest which has a mixture of shrubs and trees such as *R. rosaefolius, L. capitellata, A. stipulaua, M. philippica* and some dipterocarps; herbs such as *H. capita, E. tomentosus;* grasses (*S. spontaneum)* and some ferns. The Baybay site is located in a secondary forest which is presently dominated by tree species like *A. lebbeck, P. tinc-*

| Minerals in Sand and Silt (identified by RBA) (quantitative numbers, 1=1000 RBA-counts) | | | | | | | | | | | |
|---|---|---|---|---|---|---|---|---|---|---|---|
| Gibbsite | | Cristobalite | | Quartz | | Feldspars | | Hornblende | | Magnetite | |
| Sand | Silt | Sand | Silt | Sand | Silt | Sand | Silt | Sand | Silt | Sand | Silt |
| 2 | 2 | 0.7 | 4 | 2 | 5 | 8 | 0.5 | 5 | 0.5 | 3 | 0.5 |
| 1 | 3 | 0.4 | 4 | 1 | 5 | 6 | 0.5 | 5 | 0.5 | 2 | 0.5 |
| 5 | 4 | 2 | 7 | 3 | 7 | 3 | 4 | 4 | 1 | 1 | 0.5 |
| 4 | 4 | 1 | 7 | 2 | 7 | 20 | 1 | 4 | 1 | 0.7 | 0.7 |
| 0 | 0 | 0 | 1 | 1 | 2 | 0.5 | 0.2 | 2 | 0.4 | 1 | 1 |
| 0 | 0 | 0 | 1 | 2 | 3 | 0 | 0.7 | 6 | 0,2 | 2 | 1 |
| 0 | 0 | 0 | 1 | 3 | 1 | 0.5 | 0 | 0.7 | 0.2 | 3 | 1 |
| 0 | 0 | 0.4 | 1 | 1 | 0.5 | 0 | 0 | 0 | 0 | 3 | 1 |
| 0 | 0 | 2 | 1 | 0 | 0.4 | 0.4 | 0 | 0 | 0 | 2 | 2 |

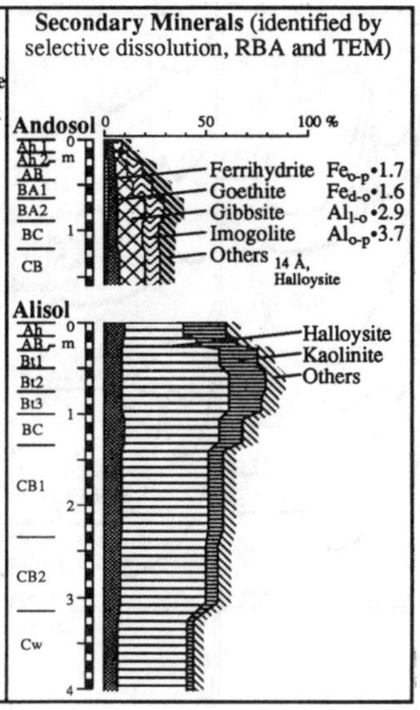


**Figure 2:** Mineral content of sand (2000-63 μm) and silt (63-2 μm) identified by X-ray analysis (peak height of powder specimens, 1 = 1000 counts) and content of secondary minerals identified by selective dissolution, X-ray analysis and electron microscopy.

*torium, A. blancoi, B. racemosa* and *G. album*. Many non-tree plant species are also abundant in the area.

### 3.2 Soil morphology and classification

**Ormoc:** A deep (about 160 cm) weakly developed loamy soil, characterized by very high friability, good granular structure, non-plasticity and non-stickness over the whole soil depth. The color ranges from dark yellowish brown (10 YR 3/4) on the surface, to yellowish brown (10 YR 5/5) in the lowest horizon. It has a horizon sequence of Ah-AB-BA-BC-CB. The soil meets all requirements of an umbric epipedon and andic properties (compare Fig. 1). A cambic B-horizon, rich in organic matter is developed. Therefore, this pedon is classified as an Umbric Andosol (FAO, 1988) or a Typic Hapludand (USDA, 1992) respectively.

**Baybay:** A very deep (> 400 cm) well-developed heavy clay soil. It has a granular structure in the Ah-horizon and moderate to strong angular blocky structure in the Bt-horizon. It has high amounts of fine pores and a color ranging from dark brown (7.5 YR 3/4) on the surface to yellowish red (5 YR 4/6) in the saprolite layer which starts at a depth of about 100 cm. It is plastic and sticky and has a horizon sequence of Ah-AB-Bt-BC-CB-Cw, the Cw still being a strongly weathered saprolite. The unweathered rock may be at a depth of > 600 cm. The epipedon can be classified as ochric due to the colour requirements. The B-horizon is to be classified as an Argic horizon with low base saturation. The CEC, analyzed from fine earth, can be calculated for the clay and yield a value > 24 cmol(+)/kg. Therefore, this pedon is classified as a Haplic Alisol (FAO, 1988) or a Typic Hapludult (USDA, 1992). Because of the decrease in clay below 100 cm, the pedon does not meet the requirements of a Nitisol or a Paleudult in the FAO and USDA systems, respectively.

### 3.3 Physical properties

Both pedons are low in rock fragments, but the Alisol shows considerable amounts of saprolitized rock at a depth > 200 cm (Fig. 1), whose structure is easy to destroy by hand. The texture in the two pedons is very different, being silty loam to sandy loam in the Andosol, and heavy clay in the Alisol. The measured low clay content of the Andosol (< 5%) is too low to be realistic and is in contradiction to the water content at 1500 kPa (25–32%) and the CEC (17–38 cmol(+)/kg). The bulk density is generally low for both pedons (< 0.85 kg/dm$^3$), with the Andosol having the lower value (0.51 kg/dm$^3$) in the A-horizon and the Alisol (0.65 kg/dm$^3$) in the saprolite. As a result, both pedons have high porosity of 69 to 77%.

### 3.4 Chemical properties

The pH (KCl) shows that the Alisol (pH 3,5–3.9) is more acidified than the Andosol (pH 4.2–4.9). The pH measured in $H_2O$ gives smaller variations in the Andosol compared to that measured in KCl. In the lower horizons of the Andosol, a positive ΔpH occurs, indicating a positive net charge (*Mekaru* and *Uehara*, 1972). In contrast to the differences in the particle size distribution, a relatively similar CEC can be observed in both pedons which decreases with depth from 38 to 17 cmol(+)/kg in the Andosol but increases with depth from 27 to 39 cmol(+)/kg in the Alisol. The base saturation is lower in the Andosol (< 10%) than in the Alisol (< 30%) and decreases with depth in both pedons. The organic carbon stored in the first meter in the Andosol is twice that of the Alisol. As an indicator of the anion exchange capacity, P-retention was measured. This gives absorption rates of > 96% in the Andosol and 60 to 77% in the Alisol (100% = 5 g P/kg fine earth).

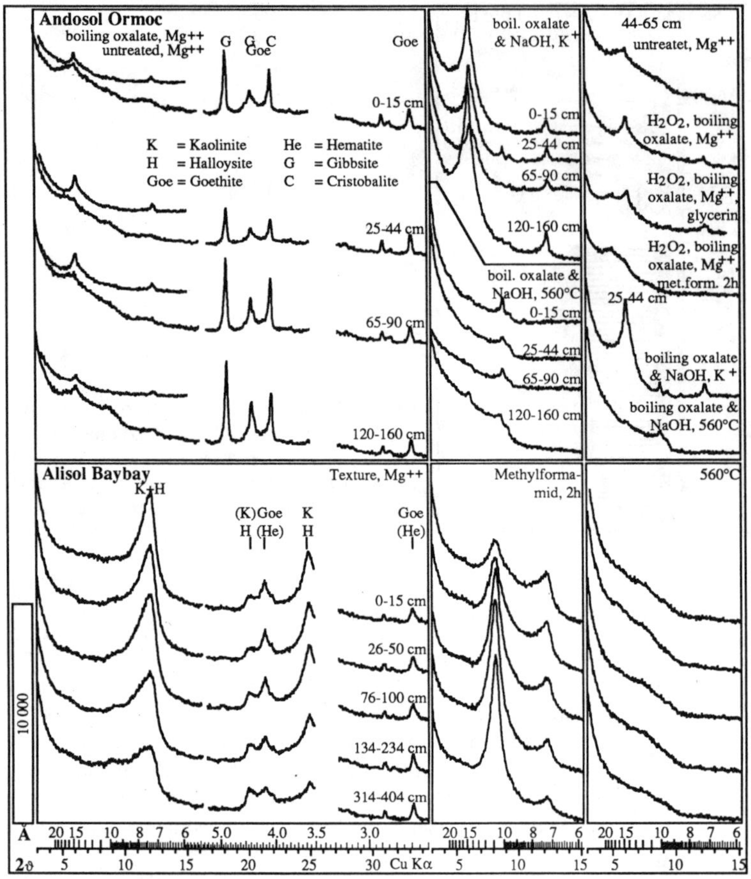


**Figure 3:** X-ray diffraction diagrams of oriented clay samples for an Andosol and an Alisol from Leyte (Philippines).

### 3.5 Mineralogical properties

As a general feature, the Andosol has more weatherable primary minerals (feldspars and hornblende) in the silt and sand fraction than the Alisol (Fig. 2). Hornblende was found in reasonable amounts in the Alisol only in the topsoil, thus giving an indication that the pedon was probably covered by younger (or less weathered) material in the past. The distribution of feldspars in the profile, however, shows no significant differences with depth. This is also true for more stable minerals like quartz, cristobalite and magnetite. Hence, it is believed that both pedons have been weathered from a parent material which is mineralogically (and geochemically) comparable to the material which is actually found below the soil. But the age of the parent material which actually formed the different parts of the soil may be different from the underlying rock. The new formations of pedogenic minerals are by weight much bigger in the Alisol (≈ 1800 kg clay per $m^2$, 404 cm depth) than in the Andosol (≈ 50 kg measured clay per $m^2$, 160 cm depth) but the variety is higher in the Andosol. The identification of layer silicates is difficult in the Andosol. But after treatments with boiling oxalate or NaOH, expanding and nonexpanding 14 Å-minerals as well as 7 Å-minerals can be found (Fig. 3). Gibbsite is an important constituent in all particle size fractions (Fig. 2 and Fig. 3). In addition, other pedogenic minerals can be found in the Andosol by electron microscopy (imogolite, allophane) or estimated by selective dissolution treatments (imogolite, allophane, ferrihyrite, goethite and gibbsite). From the different dissolution treatments ($X_o$ = oxalate, $X_d$ = dithionite, $X_p$ = pyrophosphate, $X_l$ = NaOH) based on fine earth, 1–2% ferrihydrite ($Fe_{o-p} \cdot 1.7$),3–5% goethite ($Fe_{d-o} \cdot 1.6$), 4–9% imogolite ($Al_{o-p} \cdot 3.7$) (more imogolite was identified by electron microscopy than allophane) and gibbsite ($Al_{l-o} \cdot 2.9$) (which increases from 0% in the topsoil to 13% in the lower horizons) can be estimated. In the Alisol, the clay mineralogy consists almost entirely of halloysite and kaolinite with the halloysite/kaolinite ratio changing from 6/4 in the topsoil to 9/1 at 400 cm depth (Fig. 3). A side from these layer silicates, goethite, and in much smaller amounts hematite, can be identified by X-ray analyses.

### 3.6 The weathering process

Early stages of weathering are indicated by the loss of bases and the turnover of Fe from primary minerals (or

**Table 1:** Losses and gains of elements, pedogenic formations from weathering of volcanoclastic parent rock for an Andosol and an Alisol and rock remnants (in kg).

| Original Rock Mass | Element Losses $(\ldots_xO_x)$ | Element Gains (H, O) | Pedogenic New formations $Fe_{o-p}\cdot 1,7$ (Ferrihydrite) | $Fe_{d-o}\cdot 1,6$ (Goethite) | $Al_{o-p}\cdot 4,5$ (Imog.+Alloph.) | $Al_{l-o}\cdot 2,9$ (Gibbsite) | Layer Silicates | Weathered Rock Remnants |
|---|---|---|---|---|---|---|---|---|
| Andosol Ormoc (0–160 cm) derived from originally 1776 kg/m² andesitic volcanoclastics | | | | | | | | |
| 1776 | 808 | 177 | 17 | 54 | 106 | 111 | ≈100 | 757 |
| 1000 (Ø) | 455 | 99 | 9 | 30 | 60 | 63 | ≈ 56 | 426 |
| Alisol Baybay (0–404 cm) derived from originally 4439 kg/m² basaltic volcanoclastics | | | | | | | | |
| 4439 | 1828 | 367 | 6 | 254 | – | – | 1555 | 1163 |
| 1000 (Ø) | 412 | 83 | 1 | 57 | – | – | 350 | 262 |

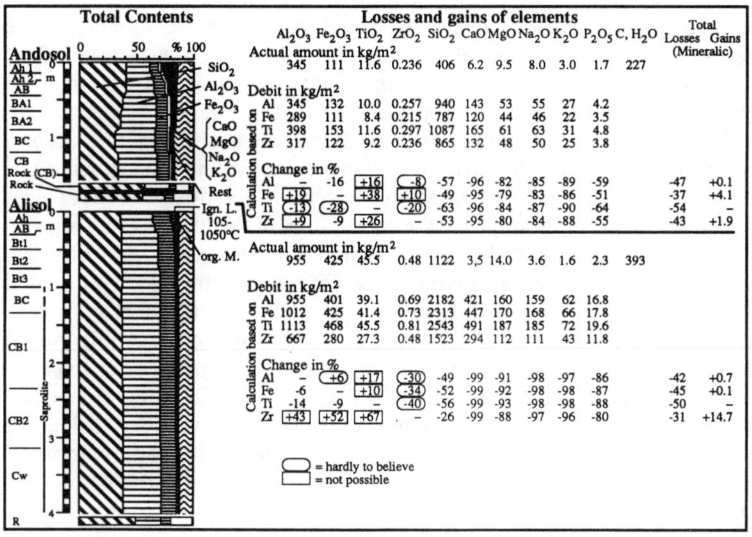


**Figure 4:** Depth functions of total element contents and losses and gains of elements.

volcanic glass) to secondary Fe-oxides. The intensity of Fe-oxide formation seems to be nearly equal in both soils as the $Fe_{d/t}$ ratio indicates (Fig. 1). However, the amout of $Fe_d$ in the Alisol (162 kg per m², 404 cm depth) is much higher than the Andosol (45 kg per m², 160 cm depth). A qualitative difference is given by the crystallinity of Fe-oxides, shown by the $Fe_{o/d}$ ratio (*Blume* and *Schwertmann*, 1969), which for the whole pedon is 0.25 in the Andosol and 0.02 in the Alisol (compare Tab. 1).

From the distribution of total element content in the profile (Fig. 4), the large losses of Si, Ca, Mg, Na, K and P, the surplus of structural water (ignition loss) and the residual increase of Al and Fe due to weathering of the rock into soil, can be observed. Under the assumption of a homogeneous parent material and relative stability of index elements, the changes in the other elements can be calculated based on the ratio of each element to an index element in the rock and in the soil. As index (stable) elements Al, Fe, Ti and Zr have been used, but the most accurate results can be obtained by using Al (Fig. 4). The loss of Si amounts to about 50% of its original content (Tab. 2). The %-loss in both pedons are in the same range but the amount in the Alisol (0–404 cm) is two times higher than in the Andosol (0–160 cm). This means, that the intensity of desilification is about equal in both pedons, but the absolute amount is much greater in the Alisol. The loss of basic cations (Ca, Mg, Na, K) from the original mass is about 90% in the Andosol and 97% in the Alisol. This shows a more pronounced loss of basic cations than of silicon in the more intensively weathered Alisol. The loss of Mn and P is clearly lower in the Andosol than in the Alisol. The changes in the amounts of Fe (loss in the Andosol, surplus in the Alisol) are in the range of measurement errors (related to the homogenuity of parent material) and should therefore not be interpreted.

There is a large difference in the amount of clay in the two pedons, it being 40 times higher in the Alisol than in the Andosol (based on analyzed grain size distribution). Because of the already strong desilification in the Andosol (molar Al/Si ratio in the uppermost horizons 0.6 to 0.7, in the lower horizons 0.9 to 1.3) it seems not possible for more imogolite to be formed in the future. Therefore, it is estimated that further weathering favors the formation of halloysite through lower leaching rates of Si (*Parfitt* and *Wilson*, 1985). This pathway was obviously followed in the formation of the Alisol, too, as indicated by its high

**Table 2:** Original amount (Or.a.) and loss of elements, based on an Al-balance, for the first meter and the total soil depth.

| Si | | Fe | | Mn | | Ca | | Mg | | Na | | K | | P | |
|---|---|---|---|---|---|---|---|---|---|---|---|---|---|---|---|
| Or.a kg/m² | Loss % | Or.a. kg/m² | Loss % | Or.a. kg/m² | Loss % | Or.a. kg/m² | Loss % | Or.a. kg/m² | Loss % | Or.a. kg/m² | Loss % | Or.a. kg/m² | Loss % | Or.a. kg/m² | Loss % |
| Andosol Ormoc 0–100 cm | | | | | | | | | | | | | | | |
| 232 | 50 | 49 | (13) | 1.1 | 32 | 54 | 95 | 17 | 80 | 21 | 82 | 12 | 86 | 1.0 | 43 |
| Andosol Ormoc 0–160 cm | | | | | | | | | | | | | | | |
| 439 | 57 | 92 | (16) | 2.1 | 35 | 102 | 96 | 32 | 82 | 41 | 85 | 22 | 89 | 1.9 | 59 |
| Alisol Baybay 0–100 cm | | | | | | | | | | | | | | | |
| 272 | 48 | 75 | (–9) | 1.4 | 66 | 80 | 99 | 26 | 92 | 31 | 97 | 14 | 97 | 2.0 | 82 |
| Alisol Baybay 0–404 cm | | | | | | | | | | | | | | | |
| 1020 | 49 | 281 | (–6) | 5.3 | 50 | 301 | 99 | 97 | 91 | 118 | 98 | 52 | 97 | 7.4 | 86 |

amount of halloysite which in turn is now partly altered to kaolinite (higher in the topsoil than in the subsoil). For a more complete kaolinitization of primarily formed halloysite, it seems necessary to have land surfaces with a minimum age of 5 million years, as shown by examples from the moist parts of the Canary Island. Indications for this course of clay formation are given in pedons described by *Caldas* et al. (1982) and *Salguero* et al. (1985) for example. Thus, it is deduced from the Si-balance that the large part of the halloysite in the Alisol was not formed via imogolite and/or allophane.

As often observed (*Wada*, 1986; *Miehlich*, 1991) gibbsite is also found in the Andosol, occurring in remarkable amounts in all grain size fractions. Since smectite is also found in the pedon, the formation of gibbsite seems to be problematic as Al-chlorite should be formed first (*Pa Ho Hsu*, 1989). In the case of the Andosol described here, the formation of gibbsite seems to be limited to feldspar-pseudomorphoses on the surface of gravels. Very high contents of gibbsite were found by X-ray analysis on selected surface materials of feldspars. The same conclusion was reached by analyzing gravels with electron microscope and EDAX.

The great diversity of pedogenic minerals (ranging from smectite to gibbsite) in the Andosol leads to the assumption of a very difficult movement of the soil solution with different velocities and concentrations of elements. The extremes are to be found on the one hand, at the surface of grains where fast moving water with very low element concentrations percolates, making the formation of gibbsite possible and on the other hand, in voids of the vesicular grains in Andosols in which penetrating water remains for a long time. This results in an increase in the concentration of elements including bases, thereby enhancing the formation of smectite.

### 3.7 Site qualities

Land evaluation should be done for a specific purpose (FAO, 1976). One example is the evaluation for the suitability of a site for agricultural use. For that purpose, site qualities evaluated from soil characteristics can be given in qualitative ratings, based on a given set of quantitative criteria which are related to the plant requirements, site stability and elasticity. For this paper, site qualities of the soils studied were evaluated using the method of *Schlichting* et al., 1995).

Table 3 shows the different site qualities of the Andosol and the Alisol with their corresponding ratings. As can be seen, rooting depth which is the potential amount of available space for the root system of the plant, is very high for both soils. Rootability indicates the ease or difficulty with which plant roots are able to penetrate the soil and is dependent upon soil porosity, aggregate stability and amount of stones or rock fragments in the profile. As the results show, the Andosol has a higher rating than the Alisol due to the fact that the former is more porous than the latter. Erodibility, which was evaluated using the method of *Wischmeier* and *Smith* (1978), ranges from low to moderate in the Andosol and is low in the Alisol. This rating is based on the computed erodibility factor K which is low (<0,05) for both soils. However, it is believed that the Andosol has a high erodibility as evidenced, for instance, by the presence of heavily eroded portions of the site. This unrealistic low value for the erodibility factor can be due to the very high organic matter content of the Andosol or the very high clay content of the Alisol. This indicates one problem to be encountered when using this method for soils with the above-mentioned characteristics, especially for the humid tropics. With regard to air capacity and drainage (very important site qualities in humid areas like Leyte), the results indicate high ratings for both soils although the Andosol received a higher rating than the Alisol. Moreover, available water capacity is very high in both soils, indicating that water supply is not a constraint under the given climatic conditions for the specified use. In terms of nutrient availability, available N is high in both soils but available P is very low which can be attributed to their high P retention capacity (due to mineralogy). This results indicate that P is a limiting nutrient for the growth of plants in both sites. Available K is moderate in both soils while available Ca is deficient in the Andosol only. Reserve nutrients (N, P, K, Ca, Mg) which indicates the possible long term supply of these nutrients to the plant, ranges from high to very high in both soils, except the reserve of K, which appears low in the Alisol.

**Table 3:** Site qualities of an Andosol and an Alisol from Leyte, Philippines.

| | Andosol Ormoc | | Alisol Baybay | |
|---|---|---|---|---|
| Rooting depth | >100 cm | +++ | >100 cm | +++ |
| Rootability | | +++ | | ++ |
| Erodibility | | ± | | —— |
| Air capacity (Ø) | 28 % | +++ | 18 % | ++ |
| Available watercapacity | 160 l/m² | +++ | 140 l/m² | +++ |
| Drainage | | +++ | ? | ++ |
| Organic Matter | 42 kg/m² | | 18 kg/m² | |
| Available N | 27 g/m² | ++ | 34 g/m² | ++ |
| Available P | 0.7 g/m² | —— | 0.3 g/m² | —— |
| P-retention (Ø) | 98 % | +++ | 70 % | ++ |
| Available K | 45 g/m² | ± | 36 g/m² | ± |
| Available Ca | 7 g/m² | —— | 198 g/m² | + |
| Available Mg | 15 kg/m² | ± | 194 g/m² | +++ |
| Reserves of N | 2.7 kg/m² | +++ | 1.4 kg/m² | ++ |
| Reserves of P | 0.5 kg/m² | ++ | 0.3 kg/m² | + |
| Reserves of K | 1.6 kg/m² | ++ | 0.3 kg/m² | — |
| Reserves of Ca | 2.9 kg/m² | +++ | 1.1 kg/m² | ++ |
| Reserves of Mg | 3.4 kg/m² | +++ | 2.1 kg/m² | ++ |

—— = very low, — = low, ± = moderate, + = moderately high, ++ = high, +++ = very high

As these values are evaluated from total element contents, strong bonding forces can be assumed for P, K, Ca and Mg.

## 4 Conclusions

The described soils can be seen as a chronosequence. The high degree of desilification and loss of basic elements marks a final stage for the Andosol. Desilification as well as loss of bases due to further soil development, will probably not result in an increase in intensity but in an increase in the absolute amount (through deep weathering). The pedogenic mineral association in the Andosol is of a greater variety (ferrihydrite, goethite, imogolite, allophane, smectite/vermiculite, halloysite, gibbsite) than in the Alisol (halloysite, kaolinite, goethite, hematite). The high amount of halloysite in the Alisol is formed preferentially from preweathered material and not necessarily from imogolite. The kaolinite is a product of the aging of halloysite. Gibbsite in the Andosol occurs as a feldspar-pseudomorphoses on locations with high leaching rates. Both soils have excellent physical qualitites regarding rootability and air and water supply to plants. The minimum factor for plant nutrition is given by the very low amount of available P in both soils, aggravated further in the Andosol by high P-retention. With the exception of K in the Alisol, both soils show reasonable amounts of nutrient reserves with the Andosol having higher amounts than the Alisol. Except for N, strong bonding forces are assumed for the nutrient reserves.

## 5 Acknowledgement

This research was supported by the German Agency for Technical Corporation (GTZ) and the German Academic Exchange Service (DAAD).

## 6 References

*Alaily, F.* (1984): Heterogene Ausgangsgesteine von Böden: Die Rekonstruierung und Bilanzierung von Böden aus heterogenen Ausgangsgesteinen. Landschaftsentwicklung und Umweltforschung, Schriftenr. d. FB Landschaftsentwicklung d. TU Berlin 25. 236 pp.

*Barrera, A., Arisorenas, I.* and *Tingzon, J.* (1954): Soil Survey of Leyte Province, Philippines. Departm. of Agric. and Natural Resources, Manila, 103 pp.

*Barshad, I.* (1964): Chemistry of soil development. In: Bear (Editor): Chemistry of the soil. 2. Aufl. Reinhold Publ. Corps. New York, 515 pp.

*Black, C. A.* (1965) (editor): Methods of Soil Analysis. Part 2. Chemical and Microbiological Properties. Amer. Soc. Agron. Monograph No. 9. Madison Wisconsin, 1572 pp.

*Blakemore, Searle* and *Daly* (1981): Methods for chemical analysis of soils. New Zealand Soil Bureau Scientific Report 10 A.

*Blume, H.-P.* and *Schwertmann, U.* (1969): Genetic evaluation of profile distribution of aluminum, iron and manganese oxides. Soil Sci. Soc. Amer. Proc. 33, p. 438–444. Madison, Wisconsin.

*Caldas Fernandez, E., Tejedor Salguero, M. L.* and *Quantin, P.* (1982): Suelos de Regiones Volcanicas – Tenerife – Islas Canarias. Coleccion Viera y Clavijo IV, Santa Cruz de Tenerife. 250 pp.

*FAO* (1976): A Framework for Land Evaluation. Soils Bulletin 32. Rome, 72 pp.

*FAO-UNESCO* (1988): Soil map of the world – Revised Legend. World Soil Resources Report 60, Rome, 119 pp.

*ISRIC* (1986): Procedures for Soil Analyses (van Reenwijk, L. P., editor). Wageningen, 106 pp.

*Jahn, R.* (1988): Böden Lanzarotes – Vorkommen, Genese und Eigenschaften von Böden aus Vulkaniten im semiariden Klima Lanzarotes (Kanarische Inseln). Hohenheimer Arbeiten. Ulmer, Stuttgart, 257 pp.

*Köppen, W.* (1923): Die Klimate der Erde, Grundriss der Klimakunde. W. de Gruyter & Co., Berlin und Leipzig, 369 pp.

*Mekaru, T.* and *Uehara, G.* (1972): Anion adsorption in ferruginous tropical soils. Soil Sci. Soc. Amer. Proc., Vol. 36, p. 296–300.

*Miehlich, G.* (1991): Chronosequences of volcanic ash soils. Hamburger Bodenkundliche Arbeiten 15. Hamburg, 207 pp.

*Nelson, D. W.* and *Sommers*, L. E. (1982): Total Carbon, Organic Carbon and Organic Matter. In Page, A. L. (editor): Methods of Soil Analysis (2. edit.). Part II. Chemical and Microbiological Properties. Amer.

Soc. Agron. Inc. and Soil Sci. Soc. Amer. Madison Wisconsin, p. 539–594.

*Pa Ho Hsu* (1989): Aluminium hydroxides and oxihydroxides. In: Dixon, J. B. and Weed, S. B. (editors): Minerals in soil environments. SSSA, Madison, Wisconsin, p. 331–378.

*Parfitt, R. L.* and *Wilson, A. D.* (1985): Estimation of allophane and halloysite in three sequences of volcanic soils, New Zealand. CATENA SUPPLEMENT 7, Braunschweig, p. 1–8.

*PCARR* 1980: Standard methods for soil, plant tissue, water and fertilizer. Philippine Council for Agriculture and Resources Research. Los Banos, Laguna. 194 pp.

*Salguero Tejedor, M. L., Mendoza, J. C., Rodriguez, R. A.* and *Caldas, F. E.* (1985): Polygenesis on deeply weathered pliocene basalt, Gomera (Canary Islands): From ferrallitization to salinization. CATENA SUPPLEMENT 7, Braunschweig, p. 131–151.

*Schlichting, E., Blume, H.-P.* und *Stahr, K.* (1995): Bodenkundliches Praktikum. Blackwell, 295 pp.

*USDA* (1992): Keys to Soil Taxonomy, 5th Edition. SMSS Technical Monograph No. 19. Pocahontas Pr., Blacksburg, Virg., 541 pp.

*Wada, K.* (1986) (Editor): Ando Soils in Japan. Kyushu University Press, Japan, 276 pp.

*Wischmeier, W. H.* and *Smith, D. D.* (1978): Predicting Rainfall Erosion Losses – a Guide to Conservation Planning. Agriculture Handbook No. 537, U.S. Department of Agriculture, Washington D.C., 58 pp.

# Soils of the tropical forests of Leyte, Philippines II: Impact of different land uses on status of organic matter and nutrient availability

**Victor B. Asio**[1,3], **Reinhold Jahn**[2], **Karl Stahr**[2] and **Josef Margraf**[3]

[1]Soil Science Section, Department of Agronomy and Soil Science, Visayas State College of Agriculture, ViSCA, Baybay, Leyte 6521-A, Philippines
[2]Institute of Soil Science (310), University of Hohenheim, 70593 Stuttgart, Germany
[3]Tropical Ecology Program, Philippine-German Environmental Research and Development Program, ViSCA, Baybay, Leyte 6521-A, Philippines

## Abstract

In order to verify the hypothesis that conversion of forest to agriculture and other secondary land uses does not necessarily cause OM and nutrient depletion in the soil, we conducted this research at two locations, one an Andosol (Andisol) area and the other an Alisol (Ultisol) in Leyte, Philippines. In each location, adjacent sites having different land uses but with the same soil type, physiography and geology, were chosen as sampling sites. These included: forest, grassland, pasture and bushland for the Andosol in Ormoc, and forest, reforestation, coconut + bushes, shifting cultivation and coconut + kudzu, for the Alisol in Baybay. Our results indicate that a change in land use or cultivation decreased the soil OM, total and available N, available K but tended to increase pH and available Ca and Mg. Available P, which was very low in both soils, showed no distinct differences measurable. The soil changes in the Andosol were greater and more distinct than in the Alisol. Land use history helps explain soil changes due to land use.

## 1 Introduction

Tropical forests are being converted to agriculture at the rate of at least 10 million hectares per annum (*Theng*, 1991). For the Philippines, the annual rate of forest conversion in the 1980s was estimated to be 379 000 hectares (*Villavicencio*, 1987) and in 1990, the country had only about 21% forest remaining (*DENR*, 1990). For Leyte island, we estimate the present forest cover to be only 10%. Moreover, our analysis of old and new land use maps revealed that most of the former forest areas are now dominated by grassland, scrubland and coconut (Fig. 1). Forest conversion has received increasing attention in recent years because of its impact on soil productivity and contribution to global warming. Starting with the classic works on the effects of clearing tropical forests for agriculture on the soil in the early 1960s (*Nye & Greenland*, 1960 and 1964), this topic has been the subject of many studies, often with conflicting results, particularly in South America and Africa but very little in Southeast Asia. While it is the opinion of many researchers that this land use conversion results in depletion of organic matter (OM) (e.g. *Nye & Greenland*, 1964; *Sanchez*, 1976; *Sombroek* et al., 1993), this is generally true only during the first few years after conversion since the effect could vary depending upon the type of land use, soil, physiography and geology. *Stevenson* (1986) reported that OM levels usually, but not always, decline when soils are first placed under cultivation. *Lugo* and *Brown* (1993) proposed that changes of soil OM in response to changes in land use follow different pathways depending upon the type of land use. *Theng* (1991) believed that the extent and severity of soil degradation upon clearing of tropical forests for agriculture would depend on the cultivation system being imposed and how the soil is being managed. It has also been argued that shifting cultivation seldom results in substantial OM depletion (*Sanchez*, 1976).

Forest conversion to agricultural land uses is believed to be the major cause for the widespread occurrence of degraded lands in the Philippines (*Asio*, 1996). However, the authors of this paper are not aware of any published data on the impact of different secondary land uses on the status of soil OM and available nutrients under Philippine conditions. This research was therefore conducted to verify the hypothesis that a change in land use from forest to agriculture and other secondary land uses does not necessarily cause OM and nutrient depletion in the soil but depends on the type of land use, management and site characteristics.

## 2 Methodology

### 2.1 Research area

The research area is the island of Leyte in southeastern Visayas which lies approximately between 124°

17′ and 125° 18′ east longitude and between 9° 55′ and 11° 48′ north latitude. The island has a total area of 799500 hectares, about 70% of which is hilly and mountainous.

Two research locations at the western slope of the NNW-SSE running central cordillera were chosen. The first one is located about 12 km NE of Ormoc (700 m asl) and the other 8 km N of Baybay (100 m asl) (Fig. 1). The climate is characterized by humid tropical monsoon with an annual rainfall of 3,391 mm (Ormoc) and 2,621 mm (Baybay) and average air temperature of 24.2°C (Ormoc) and 27.4°C (Baybay). The soil in Ormoc is Andosol (Andisol) developed from Quaternary volcanics of intermediate chemistry (55% $SiO_2$) while that in Baybay is an Alisol (Ultisol) developed from late Pleistocene to Miocene volcanics of intermediate chemistry (50% $SiO_2$). The two soils are considered to represent the younger and older stages of a chronosequence, respectively. Details of the site and soil characteristics in both locations are discussed in Part I by *Jahn* and *Asio* (this book) and in *Asio* (1996). Likewise, a comprehensive review of the ecology of the area is found in *Margraf* and *Milan* (1996).

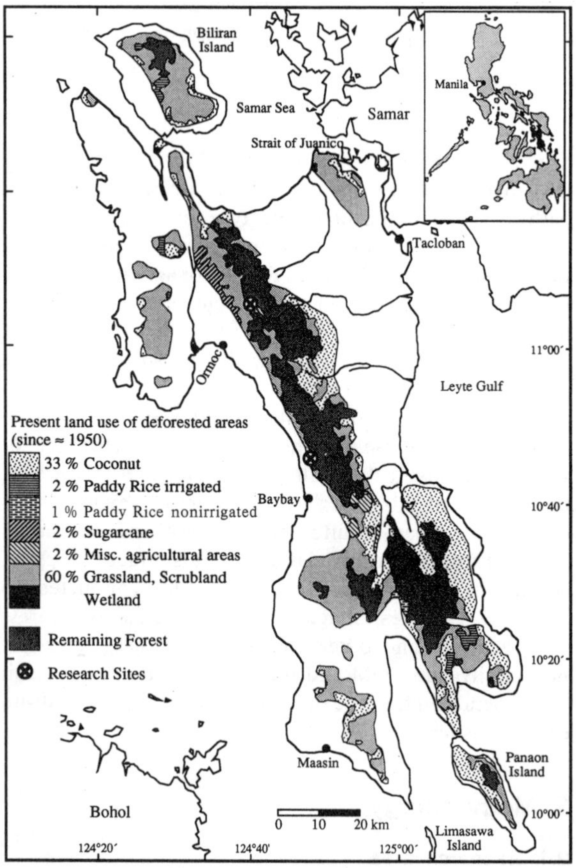


**Figure 1:** Location of research sites and the present land uses of the deforested areas in Leyte, Philippines.

## 2.2 Research approach and selection of sampling sites

The adjacent area approach or space-for-time approach for soil change studies (*Robarge* and *Johnson*, 1992) was used. This approach is commonly used to assess the effect of different land uses or different vegetations on soil properties and has been recently used by *Djegui* et al. (1992), *Shiddieq* and *Badayos* (1993) and *Maggs* and *Hewett* (1993).

In each research location, adjacent sites having comparable physiography, geology and soils but with different land uses were selected. All sites were under forest until the 1950s as revealed in old maps, aerial photographs and from actual interviews with land owners and farmers cultivating the lands. For comparison purposes, forest land use was included. However, only a residual forest was present in Ormoc and an undisturbed site at its edge was chosen to represent the forest land use. The land uses (sampling sites) selected represent major land uses in Leyte and in many parts of the country. These include: forest, grassland, pasture and bushland for the Andosol in Ormoc, and forest, reforestation area, coconut + bushes, shifting cultivation and coconut + kudzu for the Alisol in Baybay. Details of the vegetation and land use history of the sampling sites are given in Table 1. Comparability of the soils under the land uses studied is indicated by some profile morphological characteristics shown in Fig. 2.

## 2.3 Soil sampling and analysis

Soil samples were collected from every horizon of the soil profile in each land use (sampling site) quantitatively

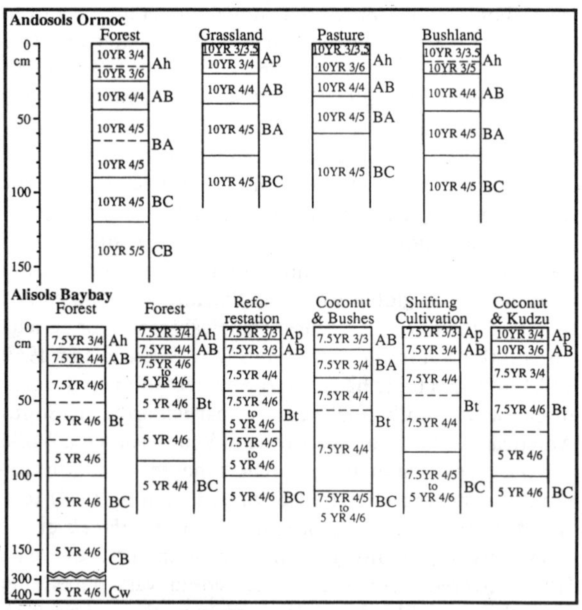


**Figure 2:** Some morphological characteristics of the soil profiles under various land uses studied.

**Table 1:** Vegetation and land use history of sampling sites in Ormoc and Baybay, Leyte

| Sampling Site | Description |
|---|---|
| **Ormoc site** | |
| **Forest** | Located at the edge of a residual forest. Vegetation includes shrubs & trees (*Rubus rosaefolius, Leucosyke capitellata, Acalypha stipulacea, Mussaenda philippica & a few dipterocarps*) herbs (*Hyptis capitata, Elephantopus tomentosus*) and grasses (*Saccharum spontaneum, Imperata cylindrica, Paspalum conjugatum*). The exact profile is actually at the boundary between the residual forest and the bushland. No evidence of any cultivation can be found. |
| **Grassland** | Grass fallow for about 8 years following 25 years of intensive sugarcane production. The dominant grass is *Imperata cylindrica.* |
| **Pasture** | Grazingland for 30 years now following a few years of cultivation. Dominant vegetation are grasses (*Cyrtococcum patens, Ottochloa nodosa, Paspalum conjugatum, Imperata cylindrica, Axonopus compressus, Cyrtococcum accrescens*), herbs (*Elephantopus tomentosus, Stachytarpheta jamaicensis, Ageratum conyzoides, Hyptis capitata, Crassocephalum crepidioidis), shrubs (Mimosa pudica, Melastoma offine, Psidium guajava*), vines (*Mikania cordata, Desmodium trifolium*). |
| **Bushland** | Cultivated for a few years in mid 1960s and followed by grassland until early 1980s. It is now dominated by a mixture of herbs (*Elephantopus tomentosus, Polygala japonica*), ferns (*Nephrolepis hirsutula, Sphaerostephanus unitus, Pteridium aquilinum*), shrubs and trees (*Mussaenda philippica, Melastoma offine, Cratoxylum sumatranum, Acalypha stipulacea*) and some grass like *Saccharum spontaneum, Imperata cylindrica, Paspalum conjugatum.* |
| **Baybay site** | |
| **Forest** | Secondary forest dominated by tree species like *Albizia lebbeck, Pterocymbium tinctorium, Artocarpus blancoi, Barringtonia racemosa, Glochidion album.* About 15 other tree species and many non-tree species are found. |
| **Reforestation** | Three-year old reforestation project of ViSCA-GTZ Ecology Program. It experienced several years of cultivation in the 1960s, as bushland in 1970s and cultivated in the 1980s. Trees include *Artocarpus blancoi, V. parviflora, T. microcarpa*, etc. (see *Margraf* and *Milan*, 1996 for details). |
| **Coconut + bushes** | Abandoned shifting cultivation field now dominated by *Cocos nucifera*, and several species of small trees, grasses and vines. The area was under shifting cultivation from the 1950s to early 1980s with a few years fallow in the 1960s. |
| **Shifting cultivation** | Five-year old shifting cultivation after 10 years fallow. It is said to have been cultivated in the late 1960s followed by short grassland period. The field is now cultivated to *Ipomoea batatas, Manihot esculenta, Zea mays*, and some vegetables. |
| **Coconut + kudzu** | Cleared in the 1950s, the area was cultivated for many years until about 1970s. After abandonement, it has been dominated by grass (*Imperata cylindrica*) and coconut. In the early 1980s new coconut varieties replaced the old ones. The legume cover crop kudzu (*Pueraria phaseoloides*) replaced the grass. |

by taking a continuous and uniform slice from the surface horizon down to the lowest. Three mixed surface samples (first and second horizons) were also collected from each sampling site with the use of an auger about two meters away from the soil profile at a triangular formation and with the soil profile at the center. Each mixed surface sample was the composite of five subsamples. Samples were air dried (except those for pH in $CaCl_2$), weighed (after separating the stone fragments) sieved to pass 2 mm and stored in covered glass bottles. Organic C was determined by Walkley-Black method (*Nelson* and *Sommers*, 1982) and then multiplied by 1.72 to obtain percent OM, total N by Kjeldahl method (ISRIC, 1986), available N by steam distillation after replacement of $NH_4^+$ and $NO_3^-$ in the exchange complex by 1N KCl (*Bremner*, 1965), available P using Bray 2 solution for extraction and by the method of *Murphy* and *Riley* (1962) for color development. Available K, Ca and Mg were analyzed by atomic absorption spectrophotometry and flame photometry after extraction with ammonium lactate solution, soil pH (0.01 M $CaCl_2$) was determined potentiometrically on fresh soil samples using a ratio of 1:2.5, and bulk density was by core method according to the methods of *Schlichting* et al. (1995). $CO_2$ evolution was measured by the soda-lime method of *Edwards* (1982).

## 3 Results and discussions

### 3.1 Organic Matter

Results showed that a change in land use or cultivation decreased the OM status of both Andosol and Alisol (Fig. 3 and Table 2). For the Andosol, which has a natural OM content twice higher than the Alisol, this effect was observed down to 80 cm depth and indicated the following trend: grassland < pasture = bushland < forest. The soil under grassland showed the lowest OM content. The soils under pasture and bushland had comparable OM contents although the former tended to have more OM on the surface while the latter had more on the subsurface. For the Alisol, the changes in OM status occurred largely in the upper 50 cm of the soil. In general, the soil under forest had the highest OM content followed by the soil under coconut + kudzu. The soils subjected to long period of human disturbance until recent years such as in the coconut + bushes area and reforestation area, showed the highest decrease in OM down to a depth of about 80 cm.

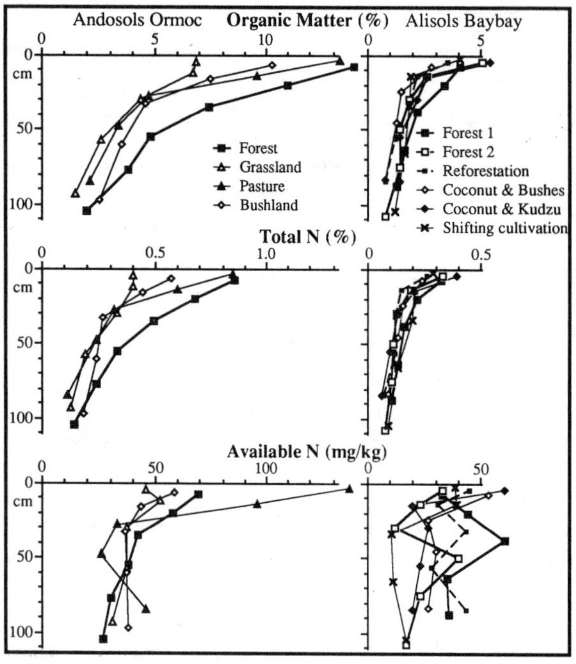

**Figure 3:** Organic matter, total N and available N of Andosols and Alisols under various land uses in Leyte, Philippines.

**Table 2:** Total amount of humus and N in the soil profile computed based on constant soil profile mass as affected by the different land uses*

| Land Use | Depth (cm) | Total Humus (kg) | Total N (kg) |
|---|---|---|---|
| Andosol (Ormoc), soil mass 669 kg | | | |
| Forest | 100 | 42.1 | 2.7 |
| Grassland | 86 | 26.6 | 1.8 |
| Pasture | 93 | 32.3 | 2.1 |
| Bushland | 92 | 32.4 | 2.0 |
| Alisol (Baybay), soil mass 825 kg | | | |
| Forest | 100 | 18.5 | 1.4 |
| Reforestation | 93 | 13.4 | 1.0 |
| Coconut + bushes | 68 | 13.1 | 1.2 |
| Shifting cultivation | 69 | 15.8 | 1.5 |
| Coconut + kudzu | 92 | 17.5 | 1.2 |

* The constant profile mass is used to correct differences due to variations in bulk density. *Nye* and *Greenland* (1964) suggested that since soil density may change due to cultivation, it is essential that comparison be based on the same soil mass in estimating changes in organic matter. The soil mass and total humus represent the sum of the values in all horizons within the considered depth. Values for each horizon was computed using its corresponding bulk density value.

The soil under shifting cultivation showed a decrease in OM particularly in the upper horizon but the total amount in the profile was still relatively high. The second forest soil profile showed a low and relatively irregular trend in OM distribution suggesting possible past local disturbance in the site where the profile was located and therefore should not be included in the comparisons. Values from surface soil samples (Table 3) indicate moderate spatial variability of OM on the surface horizon but still indicate a trend closely similar to the profile values.

The observed decrease in OM status of both soils agrees with the results of *Djequi* et al. (1992) and *Shiddieq* and *Badayos* (1992) and can be attributed to increased OM oxidation, losses from erosion and decreased quantity of plant residues returned to the soil. These could also explain the tendency of soils with long history of disturbance such as grassland, reforestation and coconut + bushes, to contain low OM. *Eghball* et al. (1994) concluded that tillage systems that do not greatly disturb the soil surface tend to have greater amounts of OM. The low OM status of the soil under grassland is due to lower productivity (less litter production) and better mineralization. This agrees with the results of *Herrmann* et al. (1994) who found a lower OM accumulation under grass vegetation than under trees. The low litter production of the grassland is partly due to the periodic burning and partly to the previous land use which was 25 years of sugarcane cultivation. On the other hand, the relatively high OM under pasture is due to the fact that pastures are not tilled hence, the loss of carbon is less compared to cropland soils (*Dale* et al., 1993). Moreover, the periodic addition of organic material as animal waste could also have contributed to it. The higher OM of the soil under bushland (which has evolved from grassland) than under the grassland indicates improvement in its OM status as a result of fallow and vegetational succession. The relatively high OM of soils under coconut + kudzu particularly in the upper horizons, could be more the effect of the legume cover crop (kudzu) than of coconut and could also be due to lesser soil disturbance. Moreover, this result implies that this land use can improve the OM status of degraded soils. This area was cultivated from the 1950s to the 1970s and was degraded as indicated by the dominance of *Imperata cylindrica* after its abandonement in the early 1970s. This grass requires low nutrients especially P which enables it to develop in soils of low fertility (*Gerold*, 1991). Hence, it is considered an indicator of land or soil degradation in the tropics (*Theng*, 1991).

### 3.2 Total and available N

Figure 3 and Table 2 show that the total N generally decreased in the secondary land uses with the greatest decrease occurring in soils under grassland for the Ormoc Andosol, and in soils with long history of human disturbance particularly coconut + bushes and reforestation, in the case of the Baybay Alisol. In contrast to the trend in OM, that of total N is not so clearly defined especially in the Alisol. Values from mixed surface samples (Table 3) show some degree of spatial variability on the surface horizon but the average values still show the general decrease indicated by the profile samples.

**Table 3:** Ranges and average values of organic matter and total N of the surface soil samples as a function of land uses.

| Land Use | Depth (cm) | OM (%) Range[2] | OM (%) Average[1] | Total N (%) Range[2] | Total N (%) Average[1] |
|---|---|---|---|---|---|
| Andosol (Ormoc) | | | | | |
| Forest | a 0–25 | 14.0–14.5 | 14.2 | 0.82–0.88 | 0.84 |
| | b 25–44 | 12.5–13.5 | 13.2 | 0.73–0.84 | 0.80 |
| Grassland | a 0–20 | 6.6– 7.4 | 6.9 | 0.35–0.40 | 0.38 |
| | b 20–40 | 6.2– 6.7 | 6.5 | 0.32–0.40 | 0.37 |
| Pasture | a 0–20 | 11.5–11.6 | 11.6 | 0.54–0.70 | 0.64 |
| | b 20–35 | 8.2– 8.9 | 8.5 | 0.50–0.56 | 0.53 |
| Bushland | a 0–20 | 10.8–10.1 | 11.0 | 0.58–0.60 | 0.59 |
| | b 20–45 | 8.5– 9.0 | 8.7 | 0.41–0.51 | 0.48 |
| Alisol (Baybay) | | | | | |
| Forest | a 0–20 | 6.2– 7.5 | 7.0 | 0.38–0.43 | 0.41 |
| | b 20–40 | 0.8– 3.3 | 1.9 | 0.20–0.37 | 0.27 |
| Reforestation | a 0– 8 | 2.6– 3.8 | 3.2 | 0.29–0.35 | 0.32 |
| | b 8–20 | 0.5– 2.2 | 1.6 | 0.11–0.27 | 0.19 |
| Coconut + bushes | a 0–15 | 3.5– 4.6 | 4.4 | 0.28–0.59 | 0.40 |
| | b 15–34 | 2.9– 3.3 | 3.1 | 0.20–0.28 | 0.24 |
| Shifting cultivation | a 0– 5 | 4.5– 4.6 | 4.8 | 0.29–0.36 | 0.33 |
| | b 5–22 | 3.1– 3.4 | 3.2 | 0.19–0.24 | 0.22 |
| Coconut + kudzu | a 0– 9 | 5.0– 5.5 | 5.3 | 0.26–0.45 | 0.35 |
| | b 9–20 | 3.3– 3.5 | 3.4 | 0.17–0.31 | 0.22 |

[1]Average from three composite surface samples.
[2]The minimum and maximum values of the three composite surface samples.
a = first horizon, b = second horizon

The less defined differences in total N among secondary land uses in the Alisol can be attributed to the complex history of most of the land uses. For example, the inclusion of legume crops during a particular cropping season which is a practice by upland farmers in the study area, the cutting and regrowth of leguminous trees, and burning of the plant residues, all could have an effect on the amount of total N in the soil profile. *Pagel* et al. (1982) reported that the over-all N balance in tropical soils is greatly dependent on the type and intensity of land use. Likewise, *Jenny* (1950) explained the high N content of Colombian forest soils as due to the contribution of leguminous tree species.

Figure 3 shows the behavior of available N ($NO_3^-$ and $NH_4^+$) in the soil profile under different land uses. Although the trend is not as defined as that of OM and total N, there is nevertheless a general tendency for the soils under forest to contain more available N than that under the other land uses. This is related to the amount of OM and total N in the soil profile which in this case are high in soils under forest. *Pagel* et al. (1982) reported that N mineralization in tropical soils is enhanced by high amount of easily decomposable material. Consequently, soils under land uses which accumulate lesser OM would have also lower available N. *Vitousek* et al. (1989) also indicated that the long-term effect of disturbance is a decline in N mineralization. The low level of available N in soil under grassland agrees with the results of *Maggs* and *Hewett* (1993). The high level of available N on the soil surface under pasture could be due to OM addition as animal wastes. The complex behavior of available N in the Alisol profiles subjected to long periods of human disturbance can also be attributed to the management practices employed by the farmers. *Eghball* et al. (1994) presented conclusive evidence that tillage methods and crop sequences greatly affect available N distribution (especially $NO_3^-$) in the soil profile.

### 3.3 Soil pH

The behavior of pH in the soil profile under the different land uses is shown in Figure 4. As can be seen, there is a tendency for pH to be slightly higher in the disturbed or cultivated sites than in the undisturbed or forest sites. *Nye* and *Greenland* (1964) found a marked rise in soil pH as a result of clearing tropical forests. This may be the effect of ash in the case of the cultivated sites since burning is an integral part of the farmers practices, or of animal wastes in the case of pasture.

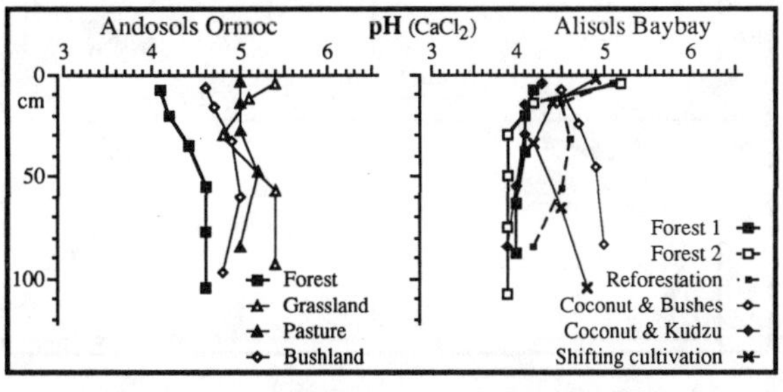


**Figure 4:** Soil pH in 0.01 MCaCl₂ of Andosols and Alisols under various land uses in Leyte, Philippines (using fresh samples).

### 3.4 Available P, K, Ca, and Mg

The results revealed that regardless of land use, the available P in Andosol was very low between 2–4 mg/kg in the surface horizon and less than 1 mg/kg in the lower horizons. The same trend was found in the Alisol. Regardless of land use, available P was below 1 mg/kg (Fig. 5). These results can be expected considering the mineralogical characteristics of both soils studied. The Andosol has high amounts of gibbsite, imogolite, and amorphous compounds and the Alisol is dominated by kaolinite and halloysite and contains significant amounts of goethite and hematite. The very high P retention capacity of both soils amount to more than 90% for the Andosol and more than 60% for the Alisol (*Asio*, 1996). Please refer also to Part I by *Jahn* and *Asio* in this book.

Available K, Ca, Mg (extracted by ammonium lactate solution) in the soil profile under different land uses is presented in Figure 5. The principle behind this organic acid extractant is that it is produced by the roots of the plants (*Schlichting* et al., 1995) and therefore the fractions are closely related to plant available nutrients.

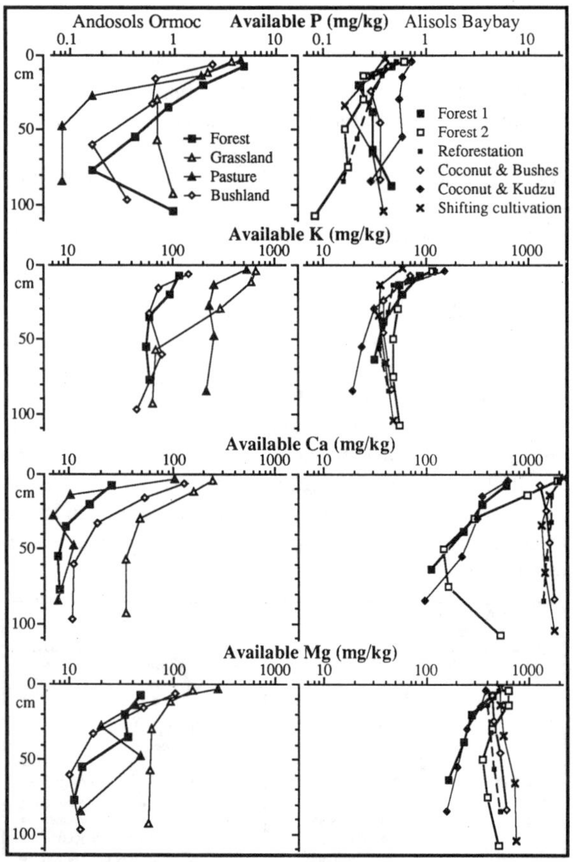


**Figure 5:** Available P, K, Ca and Mg of Andosols and Alisols under various land uses in Leyte, Philippines.

For the Andosol, available K was highest in the soil under grassland followed by the soil under pasture. The soils under forest and bushland gave the lowest available K. For the Alisol, a generally higher available K was found in the soil under forest than under the disturbed or cultivated sites. The high available K under grassland can be attributed to the ash from the periodic burning of the grass vegetation and the decomposition of the sugarcane residues during the previous land use. *Hebel* (1995) reported a very high amount of exchangeable K coming from the decomposition of crop residues. The high available K under pasture could have been contributed by animal waste. For the Alisol, the generally lower available K in the cultivated or disturbed sites can be attributed to losses by leaching and erosion. Average values from the surface samples (Table 4) revealed a trend generally similar to the profile values indicating that the variability of available K on the surface soil could have been caused by the land uses.

As regards available Ca and Mg, results revealed that a change in land use tended to increase their amounts in the soil profile. This is contrary to what is commonly reported in the literature (e.g. *Maggs* and *Hewett*, 1993). This could be attributed to several reasons. First, we hypothesize the possible role of soil rejuvenation through soil erosion in the accumulation of Ca and Mg. This means that erosion of the base-poor soil surface would expose the relatively less weathered subsoil containing more bases. Second is the contribution of the ash produced by burning practices. In an Alfisol in Ghana, the ash contained 1.5 to 3 tons Ca/ha and about 180 kg Mg/ha (*Sanchez*, 1976). Third is the possible release of Ca and Mg from the enhanced weathering of rock and mineral fragments due to burning and exposure to the sun when the soil surface is bare. It has been reported for instance, that the air temperature over a burning tropical forest may reach 450 to 650°C at 2 cm above the soil surface which decreases in the top 5 cm of the soil at a rate of 100°C/cm (*Sanchez*, 1976). This condition, coupled by high rainfall and the mechanical disturbance during cultivation, could have possibly enhanced mineral dissolution.

Average values of surface samples indicate the same trend for Andosol but not for Alisol. This implies that spatial variability in available Ca and Mg on the surface horizon is high especially as a result of human disturbance which is much greater in the latter than in the former.

### 3.5 Carbon Dioxide Evolution

Carbon dioxide in soils comes mainly from the decomposition of OM. To check if the OM data obtained under different land uses were reliable, $CO_2$ evolution measurements were made under field conditions.

Table 5 shows that it is difficult to compare the absolute amounts of $CO_2$ evolved because of generally high variability of values. However, generalizations can be made

**Table 4:** Average values of available K, Ca and Mg of the surface soil samples (a = first horizon and b = second horizon) as a function of land use.

| Land Use | Depth (cm) | Available K (mg/kg) | Available Ca (mg/kg) | Available Mg (mg/kg) |
|---|---|---|---|---|
| Andosol (Ormoc) | | | | |
| Forest | a 0–25 | 182 | 22 | 52 |
| | b 25–44 | 145 | 16 | 33 |
| Grassland | a 0–20 | 659 | 217 | 127 |
| | b 20–40 | 558 | 174 | 87 |
| Pasture | a 0–20 | 367 | 93 | 203 |
| | b 20–35 | 201 | 15 | 51 |
| Bushland | a 0–20 | 365 | 105 | 106 |
| | b 20–45 | 210 | 44 | 47 |
| Alisol (Baybay) | | | | |
| Forest | a 0–20 | 170 | 3096 | 820 |
| | b 20–40 | 72 | 1655 | 666 |
| Reforestation | a 0– 8 | 154 | 2457 | 580 |
| | b 8–20 | 60 | 1965 | 461 |
| Coconut + bushes | a 0–15 | 88 | 1632 | 483 |
| | b 15–34 | 40 | 1530 | 435 |
| Shifting cultivation | a 0– 5 | 73 | 2605 | 614 |
| | b 5–22 | 46 | 2057 | 608 |
| Coconut + kudzu | a 0– 9 | 189 | 686 | 375 |
| | b 9–20 | 100 | 413 | 298 |

All values are the average of the three composite surface samples collected in each sampling site.

**Table 5:** Carbon dioxide evolution under different land uses in Ormoc and Baybay sites

| Land Use | July 1994 $CO_2$ ($g/m^2/day$) | July 1994 SD | July 1994 CV (%) | January 1995 $CO_2$ ($g/m^2/day$) | January 1995 SD | January 1995 CV (%) |
|---|---|---|---|---|---|---|
| Andosol (Ormoc) | | | | | | |
| Forest | 14.3 | 2.97 | 21 | 16.8 | 4.18 | 25 |
| Grassland | 12.3 | 5.63 | 46 | 9.7 | 3.00 | 31 |
| Pasture | 15.3 | 5.66 | 37 | 14.2 | 5.80 | 41 |
| Bushland | 10.9 | 2.95 | 27 | 5.8 | 3.10 | 53 |
| Alisol (Baybay) | | | | | | |
| Forest | 16.6 | 3.44 | 21 | 14.9 | 4.60 | 31 |
| Reforestation | 13.2 | 2.94 | 22 | 14.2 | 4.20 | 29 |
| Coconut + bushes | 14.8 | 3.16 | 21 | 10.5 | 3.60 | 34 |
| Shifting cult | 13.5 | 3.83 | 28 | 13.1 | 4.80 | 37 |
| Coconut + kudzu | 16.5 | 2.56 | 15 | 17.0 | 4.40 | 26 |

The $CO_2$ value for each land use is the average of 10 determinations.
SD = standard deviation
CV = coefficient of variation

based on the trend of the means. In Ormoc site, there is a good tendency for $CO_2$ evolution to be higher in soils under forest and pasture than under grassland and bushland. Similarly, soils under forest and coconut + kudzu in Baybay site evolved more $CO_2$ than the other land uses. This general tendency is related to the fact that these land uses in both sites which produce higher $CO_2$, have also higher OM. Thus it supports the hypothesis that higher OM generally results in higher $CO_2$ evolution. An exception occurs in soils with high clay-OM complexes like Ormoc soil where the decomposition of OM is inhibited. Thus, Ormoc and Baybay soils show relatively the same range of $CO_2$ evolution values despite the much higher OM of the former than the latter.

## 4 Conclusions

We conclude that forest conversion to secondary land uses does not necessarily decrease the nutrient status of the soil. Moreover, the behavior of OM and nutrient status depends upon the type of land use and soil. Specifically, we conclude that:

1. Forest conversion to other land uses decreased the OM, total and available N, available K but tended to increase pH, available Ca and Mg in both Andosol and Alisol. However, in terms of available P which was very low in the soils studied, no distinct changes were measurable.
2. The type of land use determined the OM and nutrient status. Further, soils subjected for many years to a mixture of management practices tended to result in poorly defined distribution of OM and nutrients in the soil profile. In general, the changes in OM and nutrients due to land use were greater and more defined in the Andosol than in the Alisol.
3. Phosphorus is the most limiting nutrient for the growth of plants in both soils studied. Consequently, it may have contributed indirectly to the accumulation in the soil of Ca and Mg coming from burning, weathering and other sources.
4. For researches like this, land use history is an important key in explaining OM and nutrient status under different land uses.

## Acknowledgement

We are thankful to the German Agency for Technical Cooperation (GTZ) and the German Academic Exchange Service (DAAD) for financial and scholarship supports, respectively, particularly to the senior author. We also thank Karin Breuer in Hohenheim as well as Cynthia Godoy, Jovie Modina and Allan Modina in ViSCA, Leyte, for their laboratory assistance. We are grateful to Dr. J. Quimio for his help in plant identification.

## References

*Asio, V. B.* (1996): Characteristics, Weathering, Formation and Degradation of Soils from Volcanic Rocks in Leyte, Philippines. Hohenheimer Bodenkundliche Hefte 33, Stuttgart, 209 pp.

*Bremner, J. M.* (1965): Inorganic forms of nitrogen, in C. A. Black: Methods of Soil Analysis. Part II. Chemical and Microbiological Properties. Agron. Soc. Amer., Madison, Wisconsin, p. 1179–1232

*Dale, V. H., R. A. Houghton, A. Grainger, A. Lugo* and *S. Brown* (1993): Emissions of greenhouse gases from tropical deforestation and subsequent uses of the land, in *National Research Council:* Sustainable Agriculture and the Environment in the Humid Tropics, National Academy Press, Washington, D.C., 702 pp.

*Department of Environment and Natural Resources* (1990): The Philippine Environment in the Eighties. EMB, Diliman, Q. C., Philippines.

*Djegui, N., P. de Boissezon* and *E. Gavinelli* (1992): Statut organique d'un sol ferrallitique du Sud-Benin sous foret et differents systemes de cultures. Cah. Orstom. ser. Pedol. XXVII: 5–22

*Edwards, N. T.* (1982): The use of soda-lime for measuring respiration rates in terrestrial systems. Pedobiologia 23: 321–330

*Eghball, B., L. N. Mielke, D. L. McCallister* and *J. W. Doran* (1994): Distribution of organic carbon and inorganic nitrogen in a soil under various tillage and crop sequences. J. Soil and Water Cons. 49: 201–205

*Gerold, G.* (1991): Human impact on forest ecosystems and soil deterioration in tropical Bolivia, in Erdelen, W. et al.: Tropical Ecosystems (Symposium Proceedings), Margraf Scientific Books, Weikersheim, Germany, p. 107–119

*Hebel, A.* (1995): Einfluss der organischen Substanz auf die räumliche und zeitliche Variabilität des Perlhirse-Wachstums auf Luvic Arenosol des Sahel (Sadore/Niger). Hohenheimer Bodenkundliche Hefte 24, Stuttgart, 170 pp.

*Herrmann, L., A. Hebel* and *K. Stahr* (1994): Influence of microvariability in sandy Saharian soils on Millet growth. Z. Pflanzenernähr. Bodenk. 157: 111–115

*International Soil Reference and Information Center* (1986): Procedures for Soil Analysis (*van Reeuwijk, L. P.* ed). ISRIC, Wageningen, The Netherlands, 106 pp.

*Jenny, H.* (1950): Causes of the high nitrogen and organic matter content of certain tropical forest soils: Soil Science 69: 63–69

*Lugo, A. E.* and *S. Brown* (1993): Management of tropical soils as sinks or sources of atmospheric carbon. Plant and Soil 149: 27–41

*Maggs, J.* and *B. Hewett* (1993): Organic C and nutrients in surface soils from some primary rainforests, derived grasslands and secondary rainforests on the Atherton Tableland in North East Queensland. Aust. J. Soil Res. 31: 343–350

*Margraf, J.* and *P. P. Milan* (1996): Ecology of Dipterocarp Forests of Leyte, Philippines, in A. Schulte and D. Schöne: Dipterocarp Forest Ecosystems. Towards sustainable management. World Scientific, Singapore p. 124–154.

*Murphy, J.* and *J. P. Riley* (1962): A modified single solution for the determination of phosphate in natural waters. Anal. Chim. Acta 27: 31–36

*Nelson, D. W.* and *L. E. Sommers* (1982): Total carbon, organic carbon and organic matter, in A. L. Page: Methods of Soil Analysis. Part II. Chemical and microbiological Properties. Second Edition. Amer. Soc. Agron. Inc. and Soil Sci. Soc. Amer., Madison, Wisconsin, p. 539–594

*Nye, P. H.* and *D. J. Greenland* (1960): The soil under shifting cultivation. Technical Communication No. 51. Commonwealth Bureau of Soils, Harpenden, 156 pp.

*Nye, P. H.* and *D. J. Greenland* (1964): Changes in the soil after clearing tropical forest. Plant and Soil 21: 101–112

*Pagel, H., J. Enzmann* and *H. Mutscher* (1982): Pflanzennährstoffe in tropischen Böden – ihre Bestimmung und Bewertung. VEB Deutscher Landwirtschaftsverlag Berlin, 272 pp.

*Robarge, W. P.* and *D. W. Johnson* (1992): The effects of acid deposition on forested soils. Adv. Agron. 47: 1–83

*Sanchez, P. A.* (1976): Properties and Management of Soils in the Tropics. Wiley and Sons, New York, 618 pp.

*Schlichting, E., H.-P. Blume* and *K. Stahr* (1995): Bodenkundliches Praktikum (2. Aufl.), Blackwell, Berlin, 295 pp.

*Shiddieq, D.* and *R. B. Badayos* (1993): Effect of tropical rainforest conversion to agricultural farm on Ultisols in Subanjeriji, Sumatra, Indonesia. The Phil. Agric. 76: 207–230

*Sombroek, W. G., F. O. Nachtergaele* and *A. Hebel* (1993): Amounts, dynamics and sequestering of carbon in tropical and subtropical soils. Ambio 22: 417–426

*Stevenson, F. J.* (1986): Cycles of Soil. Carbon, Nitrogen, Phosphorus, Sulfur, Micronutrients. A Wiley-Interscience Publication. John Wiley and Sons, New York, 380 pp.

*Theng, B. K. G.* (1991): Soil Science in the tropics- the next 75 years. Soil Science 151: 76–89

*Villavicencio, V.* (1987): Philippines, in C. Lin Sien: Environmental Management in Southeast Asia. Faculty of Science, National University of Singapore, p. 77–107

*Vitousek, P. M., P. A. Matson* and *K. van Cleve* (1989): Nitrogen availability and nitrification during succession: primary, secondary, and old field seres. Plant and Soil 115: 229–239

# Soil chemical changes after tropical forest disturbance and conversion: The hydrological perspective

**L. A. (Sampurno) Bruijnzeel**

Faculty of Earth Sciences, Vrije Universiteit, De Boelelaan 1085, 1081 HV Amsterdam, The Netherlands

## Abstract

The soil chemical and hydrochemical changes accompanying tropical forest disturbances of varying intensity are reviewed. Various methodological problems are discussed with respect to the establishment of ecosystem nutrient balances and the determination of changes therein due to forest exploitation or conversion to other land uses. It is shown that amounts of nutrients removed from the forest ecosystem in harvested timber generally exceed those associated with enhanced leaching, both after selective logging and clearcutting operations. Losses of base cations through volatilization and ash dispersal upon burning residual biomass are substantial, contrary to common belief. Approximate periods required for the replenishment of lost nutrients at the ecosystem level are estimated by comparing the losses with inputs via bulk precipitation and/or mineral weathering. The results suggest typical recovery periods of 30–60 years, depending on harvesting intensity, rainfall regime and soil nutrient retention capacity. Issues requiring further research include: (i) the evapotranspiration from young regenerating and planted vegetation; (ii) nutrient losses via rapid flow through macropores during rainfall (as opposed to slow matrix flow); (iii) nutrient losses via erosion after mechanized tree harvesting and burning of slash (particularly in the context of short-rotation plantation forestry); and (iv) the possible contributions to the overall ecosystem nutrient budget by mineral weathering. Particular emphasis is placed on the integration of hydrological, pedological and plant ecological process research at various levels of scale within an overall catchment ecosystem context in order to facilitate the integration of different disciplines. A plea is made to concentrate future research efforts at a relatively small number of carefully selected and well-researched locations, possibly joined together in a network capturing the chief environmental variability encountered in the humid tropics.

## 1 Introduction

The classical view of a tropical rain forest as having a relatively rich internal nutrient economy perched on a nutrient-poor substrate has been shown by *Proctor* (1987) to be an overgeneralization. In addition, the areal extent of extremely infertile soils in the tropics appears to be much smaller than previously assumed (*Richter* and *Babbar*, 1991). Nevertheless, the notion that tropical forest disturbance or conversion to other land-use types will invariably lead to environmental havoc (including dramatically reduced soil fertility) – aptly summarized by the phrase 'from green hell to red desert' – is still widespread. As suggested by *Hamilton* and *King* (1983) and *Jordan* (1985) it is important in this respect to recognize the intensity of a specific disturbance rather than lump everything under such meaningless terms as 'deforestation'.

The hydrological and soil physical impacts of various kinds of tropical forest disturbance and conversion have been discussed in detail in a series of earlier review papers by the author (*Bruijnzeel*, 1990, 1992a, 1993, 1996, 1997). The present contribution focuses on the accompanying changes in soil chemical characteristics. In doing so, a specific attempt will be made to highlight the close link between hydrology and soil science in this respect. The paper starts with a brief discussion of the solute losses from undisturbed (lowland) tropical rain forest, against which any changes due to forest exploitation or clearance can be compared. Next, the respective soils impacts of various degrees of forest disturbance are reviewed in order of increasing intensity. After identifying a number of issues requiring further research, the paper concludes with the suggestion that future work should preferably be integrating hydrological and biogeochemical aspects at various levels of scale within an overall catchment ecosystem context.

## 2 Solute losses from undisturbed lowland forests

Tropical forests growing on very nutrient-poor substrates are only able to maintain a high biomass level through an array of nutrient conserving mechanisms, together producing a relatively 'tight' or 'closed' nutrient cycle with only small amounts of nutrients leaking from the system (*Herrera* et al., 1978; *Brinkmann*, 1985). Conversely, forests on more fertile substrates will exhibit a more 'open' type of nutrient cycle (*Baillie*, 1989). It is important to make a distinction here between primary and old secondary forests, which have essentially attained a state of dynamic equilibrium (*Uhl*, 1982; *Lieberman* et al., 1985), and young secondary forests or plantations which generally represent rapidly aggrading ecosystems. In the former the rate of nutrient uptake to maintain overall forest biomass will be covered by and large by the amounts of nutrients released from decomposing litter and logs, supplemented by contributions via atmospheric deposition (*Vitousek* and *Sanford*, 1986; Proctor, 1987). Net uptake rates in young secondary or plantation forests, however, normally exceed the amounts of nutrients cycled via litterfall and crown wash, especially during the first decade (*Ewel*, 1976; *Bruijnzeel*, 1983; *Jaffré*, 1985; *Gholz* et al., 1985). The remainder will have to be supplied by the soil and this may present difficulties under certain conditions. Examples include subsequent rotations of fast-growing exotics planted on very infertile soils (*Russell*, 1983; *Spangenberg* et al., 1996) or the repeated harvesting of nutrient-demanding hardwood species on fertile substrates (*Hase* and *Fölster*, 1983; *Bruijnzeel* and *Wiersum*, 1985). Although in the short run, nutrient availability is governed by the balance between processes releasing nutrients into available forms and those removing them (*Proctor*, 1987), in the long run the nutrient status of a forest ecosystem depends on the balance between overall nutrient inputs (atmospheric deposition, mineral weathering, gas absorption/fixation) and outputs (mainly leaching and volatilization). Because the ratio of solid matter (soil and weathered rock) over the liquid phase (percolating water) is very large, small changes in the chemical composition of the former may show up as quite detectable changes in the chemistry of the latter (*Verstraten*, 1980; *Bruijnzeel*, 1983). As such, amounts of nutrients carried in solution by low-order streams can be regarded as a useful, albeit relative, indicator of an area's nutrient status, both before and after a change in land use (*Likens* et al., 1977).

*Bruijnzeel* (1991) examined the results of 20 nutrient input-output budget studies for tropical forest ecosystems. Scatter plots of annual streamflow totals *vs.* annual calcium, magnesium and potassium losses revealed four groups with characteristic nutrient export patterns that corresponded closely with soil fertility levels. The results are shown in Figure 1 together with those obtained more recently for eight catchments in Malaysia (*Grip* et al., 1994; *Zulkifli* and *Abdul Rahim*, 1994), Central Amazonia (*Lesack*, 1993ab), Ivory Coast (*Stoorvogel*, 1993), Fiji (*Waterloo*, 1994), and Guyana (*Brouwer*, 1996).

As shown in Figure 1, calcium and magnesium losses from areas with spodosols and highly leached oxisols are distinctly lower than for the overall oxisol/ultisol group. In turn, losses from sites with inceptisols exceed those from the latter group whereas still higher exports have been reported for forested catchments underlain by rocks that are particularly rich in calcium or magnesium, such as limestones, calcareous shales and slates, or ultramafic rocks (see *Bruijnzeel* (1991) for details). Potassium exports exhibit a similarly upward trend over the first three groups, with some overlap between groups. Potassium losses from the fourth group (soils of high fertility) are low, reflecting the scarcity of the element in the corresponding bedrock types (Figure 1c).

Interestingly, exports of base cations in streamflow from areas with highly infertile soils (and presumably tight nutrient cycles) were nevertheless quite high for streams draining catchments larger than, say, 10 $km^2$. Solute losses from such areas were much smaller when evaluated for small (several hectares) catchments and/or plots (*Bruijnzeel*, 1991). Such discrepancies may be interpreted in terms of the relative depths of (i) the weathering front (where solutes are released), (ii) the degree of river incision into the substrate, and (iii) the fine root network (*Baillie*, 1989; *Burnham*, 1989; *Bruijnzeel*, 1991; *Eernisse*, 1993). This finding not only has implications for the experimental evaluation of solute losses from tropical forest ecosystems but also for the overall availability of nutrients to the vegetation. The role of deep roots in tropical rain forests has received insufficient attention (cf. *Eernisse*, 1993; *Nepstad* et al., 1994).

## 3 Chemical responses to forest disturbance

Site fertility is potentially threatened upon forest exploitation or conversion by the removal of nutrients in harvested timber, through surface erosion, and via enhanced leaching as a result of (i) the larger volumes of water percolating through the soil after opening up of the canopy; (ii) a temporarily reduced capacity of damaged vegetation to take up nutrients; and (iii) the sudden addition of large amounts of fresh organic debris to the forest floor that are either left to decompose or burned to facilitate future access for planting (*Martin*, 1970; *Ewel* et al., 1981; *Uhl* and *Jordan*, 1984; *Poels*, 1987; *Bruijnzeel*, 1990; *Sim* and *Nykvist*, 1991). Naturally, infertile sites will be particularly vulnerable because much of the ecosystem's nutrient capital will be stored in the vegetation rather than in the soil and the disturbance may bring about the partial or complete disruption of the chief nutrient conservation mechanisms (especially the surface root mat; *Jordan*, 1985). In the following we will examine the hydrochemical and soil chemical responses to disturbances of (i) low to intermediate intensity (treefalls, hurricanes, logging, forest fires, and shifting cultivation); and (ii) high intensity (mechanized clearfelling followed by burning and conversion to other land uses, primarily forest plantations).

Basically, two approaches have been followed when trying to evaluate such responses, viz. the study of changes in (i) (top)soil nutrient concentrations over time (e.g. *Gillman* et al., 1985; *Amir* et al., 1990), and (ii) the chemical composition of the drainage water or streamflow (e.g. *Parker*, 1985; *Zulkifli*, 1989; *Malmer*, 1993). Each method has its specific difficulties and limitations. The soils approach suffers primarily from problems related to

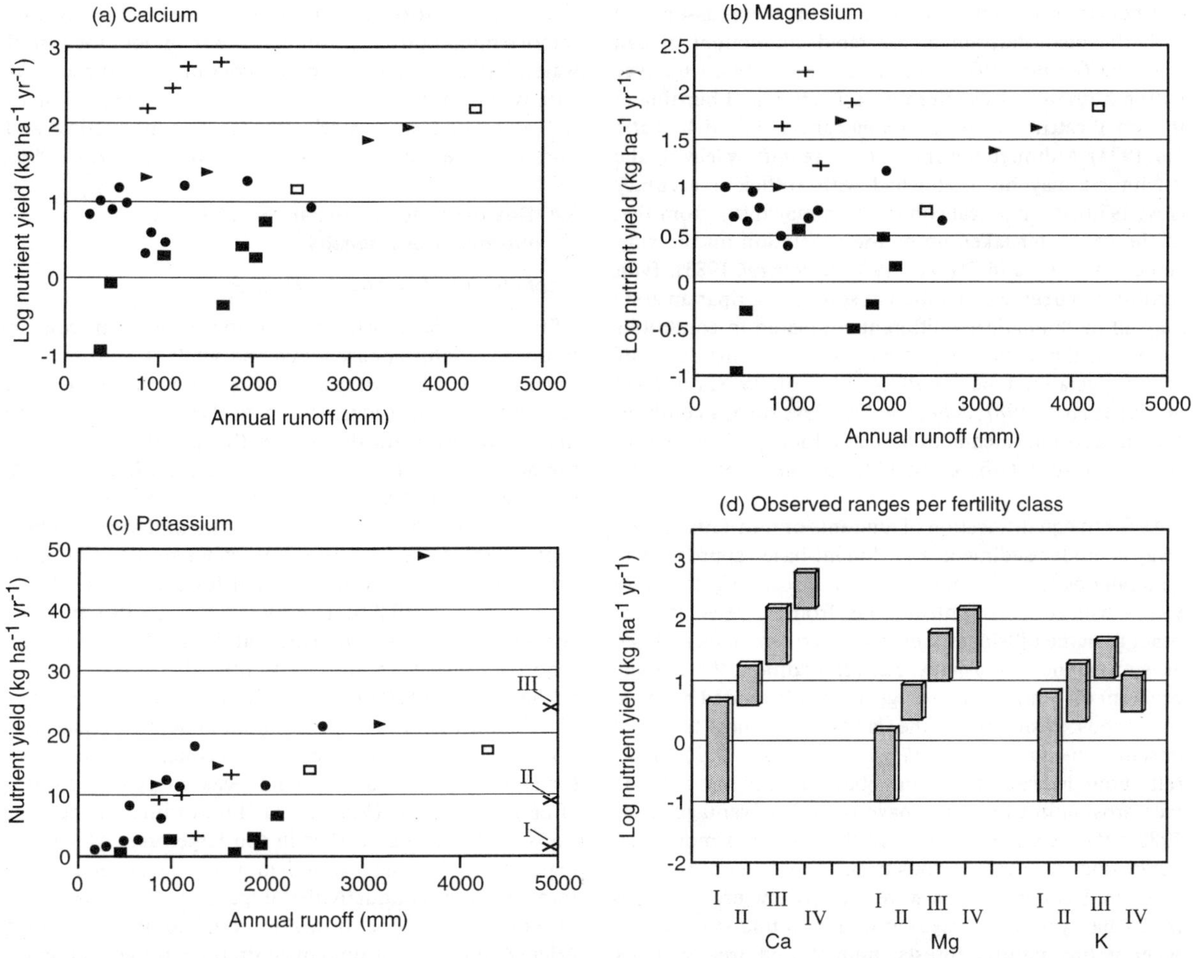


**Figure 1:** Solute losses of (a) calcium, (b) magnesium and (c) potassium (kg ha$^{-1}$ yr$^{-1}$) vs. amounts of streamflow and the ranges in output per soil fertility class (I–IV) for 28 tropical forest ecosystems (updated from Bruijnzeel, 1991). Legenda: ■ Group I, soils of very low fertility (spodosols, oxisols); ● Group II, moderately infertile soils (ultisols, some oxisols); ▶ Group III, moderately fertile soils (inceptisols); and + Group IV, soils of high fertility (mollisols, vertic soils); □ large catchment; × group average (potassium only, Figure 1c).

high spatial and/or temporal variability, particularly in such diverse ecosystems as tropical rain forests (*Proctor* et al., 1983; *Burghouts*, 1993) but also in even-aged plantations (*Lundgren*, 1978; *Hase* and *Fölster*, 1983). In addition, when dealing with long-term changes in soil nutrient reserves (e.g. during the respective phases of the shifting cultivation cycle or a rotation of plantation forest), the sampling of a single site repeatedly through time becomes often impractical and one will have to resort to sampling a 'false time series' representing the respective stages of vegetation development or the number of previous rotations. Needless to say, high spatial variability may easily confound the results obtained with this method (cf. *Hase* and *Fölster*, 1983; *Buschbacher*, 1984; *Spangenberg* et al., 1996).

*Bruijnzeel* (1991) discussed the numerous pitfalls associated with the hydrological approach at various levels of scale, whereas *Sollins* and *Radulovich* (1988) and *Lesack* (1993a) stressed the importance of the proper quantification of nutrient losses in macropore flow c.q. stormflow (cf. *Russell* and *Ewel*, 1985). Indeed, reliable estimates of nutrient losses via leaching are difficult to obtain. Within the present context, various investigators have used zero-tension or suction lysimetry to study the chemical composition of free-draining and matrix soil water, respectively, after experimentally cutting or burning tropical forest (e.g. *Toky* and *Ramakrishnan*, 1981; *Uhl* et al., 1982; *Russell*, 1983; *Brouwer, 1996 & this volume*). In the absence of sound hydrological measurements, however, the corresponding solute losses obtained by several of these studies must be considered doubtful (*Jordan*, 1989; *Bruijnzeel*, 1990, 1991). There is a dearth of reliable information on the water use of young regenerating vegetation in the humid tropics (*Hölscher*, 1995; *Roberts* et al., 1996).

Others have attempted to avoid the problems associated with the quantification of the rapid component of soil drainage (*Cooper*, 1979; *Russell* and *Ewel*, 1985) by evaluating nutrient losses after forest clearing and burning at the small catchment scale (*Malmer* and *Grip*, 1994; *Waterloo*, 1994). Although the amount of streamflow leaving the catchment may be established with sufficient accuracy (*Lee*, 1970), its chemical composition may differ from that of the soil water taken up by the vegetation on the side-slopes (*Nortcliff* and *Thornes*, 1978; *Bruijnzeel*, 1983). Type, fertility and degree of aeration of soils in the riparian zone may all be dramatically different compared to conditions found on the slopes and often result in a distinctly different vegetation type as well (*Brünig* et al., 1978; *Proctor* et al., 1983; *Poels*, 1987; *Johnston*, 1992). As such, a combination of plot- and catchment-based methods is likely to give the best results (*Bruijnzeel*, 1991; *Elsenbeer* et al., 1994; *Grip*, 1994).

As for the quantification of amounts of nutrients carried away in eroded sediment, a catchment-based approach will generally be less suitable because the resulting sediment yields will contain contributions from various sources other than the hillslopes themselves (stream bank and bed erosion; gullies; landslides, etc.). In addition, there is the problem of (temporary) storage of already eroded material in various positions in the landscape which tends to obscure on-site losses (*Walling*, 1983; *Rose*, 1993). Generally, bounded runoff plots have been used to measure surface erosion but these too have their disadvantages (*Lal*, 1988): *Wiersum* (1984) reviewed the results of more than 80 erosion studies in various tropical forest and tree crop systems. His conclusions were that (i) low erosion rates ($< 1$ t $ha^{-1}$ $yr^{-1}$ on average) prevailed in undisturbed and regenerating natural forests, multistoried tree gardens, and in forest plantations and tree crops with either a well-developed litter layer or some form of cover crop/mulch; (ii) intermediate levels of erosion (2–10 t $ha^{-1}$ $yr^{-1}$ on average; range 0.4–70) were observed during the cropping phase of the shifting cultivation cycle or during the inter-cropping stage of plantation forest establishment; and (iii) truly high erosion rates (c. 50 t $ha^{-1}$ $yr^{-1}$ on average; range 1–183) occurred in the case of clean-weeded tree crops or forests where the protective litter layer was destroyed by fire or harvesting. These conclusions have been confirmed by later work (*Fritsch* and *Sarrailh*, 1986; *Young*, 1989; *Nortcliff* et al., 1990; *Malmer*, 1993; *Ross* and *Dykes*, 1996). The nutrient content of either the surface runoff or the eroded material was not discussed by *Wiersum* (1984). Indeed, the associated nutrient losses have usually been quantified within an agricultural rather than a forest context (*Stocking*, 1984), although there are exceptions (e.g. *Kang* and *Lal*, 1981; *Roose*, 1981; *Hudson* et al., 1983; *Lima*, 1988; *Ross* et al., 1990; *Ramakrishnan*, 1992). The results of these studies are rather variable (if not contradictory at times), however, and no clear patterns emerge other than the obvious conclusion that more erosion implies more nutrient loss. Also, eroded sediment does not necessarily remove more nutrients from the site than does the runoff water. There is a need for more work on this aspect, particularly with respect to the nutrient losses in eroded material associated with mechanized forest harvesting and burning large volumes of slash (cf. *Malmer*, 1993, 1996).

### 3.1 Response to disturbances of low to intermediate intensity

#### *Treefalls, logging and hurricanes*

Studies of spatial variations in topsoil nutrient concentrations in mature-phase forest and small (47–164 $m^2$) *treefall gaps* in Costa Rica and Venezuela (on volcanic and old sedimentary substrates, respectively) were unable to detect any systematic differences that could be ascribed to the addition of organic matter in places where the bole and the crown of the fallen tree were downed (*Vitousek* and *Denslow*, 1986; *Uhl* et al., 1988b). Similarly, no significant increases in the nutrient concentrations of free-draining soil water were recorded for such gaps (*Parker*, 1985; *Uhl* et al., 1988b). However, a distinct rise in solute concentrations in soil moisture at 70 cm depth has been reported to occur about four months after the creation of artificial gaps of 500 $m^2$ (10 m x 50 m) and 2500 $m^2$ (50 m x 50 m) in lowland rain forest on an andept in Costa Rica (*Parker*, 1985). The pulse of elevated concentrations lasted for about a year and reached a peak some 6–8 months after gap creation (Figure 2a). Interestingly, concentrations started to rise earlier in the larger gap, which also experienced the highest soil moisture levels and, therefore, hydraulic conductivities (Figure 2b). Concentrations of nitrate showed a three- to fourfold increase while calcium and magnesium concentrations roughly doubled. Neither potassium, nor phosphate or ammonium concentrations responded markedly after clearing, possibly because of vigorous uptake by pioneer vegetation and nitrification of the ammonium (*Parker*, 1985; cf. *Robertson*, 1984). Apparently, enhanced leaching losses may occur if the size of the gap reaches a certain threshold value which is possibly located somewhere between 200 and 500 $m^2$ (*Parker*, 1985; *Uhl* et al., 1988b). Estimated leaching losses of base cations during the first 13 months after cutting were such that they could be replenished by nutrient inputs via rainfall within a few years (Table 1).

No material was removed from the site in the above studies nor was there any disturbance of the soil surface as would be the case with *mechanized removal of timber* (*Bruijnzeel*, 1992a). *Uhl* et al. (1982) presented an interesting comparison in this regard for an area underlain by waterlogged spodosols in the Amazon Territory of southern Venezuela. A 50 m x 100 m forest plot was (manually) cut and left to regenerate whilst a nearby patch of 4 ha of forest was cleared by bulldozing. The effects of the two disturbances on vegetation structure and composition as well as the chemical composition of the topsoil were evaluated after three years. Despite significantly increased nutrient concentrations in free-draining soil water during the first two-and-a-half years after forest cutting the nutrient concentrations in the top 10 cm of the soil were significantly higher than those in the original forest soil. The intensity of leaching was greatest for potassium, magnesium and nitrate but after three years concentrations in the leachate had started to approach the levels observed under undisturbed forest again. Unfortunately, the corresponding amounts of leached nutrients were not given. Biomass accretion in the manually cut site was mainly by sprouting and had attained the rather low value of 1290 g $m^{-2}$ after three years (*Uhl* et al., 1982). Leachate concentrations were not monitored at the bulldozed site but nutrient concentrations in the topsoil (and biomass) were much lower than before disturbance, mainly as a result of topsoil removal. Forest regeneration at this severely disrupted site was also slow, with only 77 g $m^{-2}$ of biomass having accumulated after three years. Interestingly, woody regeneration was rapid around the margins of the bulldozed plot where topsoil and organic debris were piled up (*Uhl* et al., 1982). Another effect of the drastic treatment was the gradual dying of all the trees in the surrounding forest downslope of the bulldozed plot, presumably due to a rise in the groundwater table as a result of the greatly diminished water uptake by the vegetation in the bulldozed area (*Uhl* et al., 1982).

*Gillman* et al. (1985) investigated the impact of mechanized logging on soil nutrient concentrations in upland Queensland. The harvesting of 33 % of the basal area associated with trees of 20 cm dbh and larger caused an initial reduction in canopy cover of 18 %, which recovered to 91 % of the original value within four years after logging. Statistically significant decreases in concentrations of organic

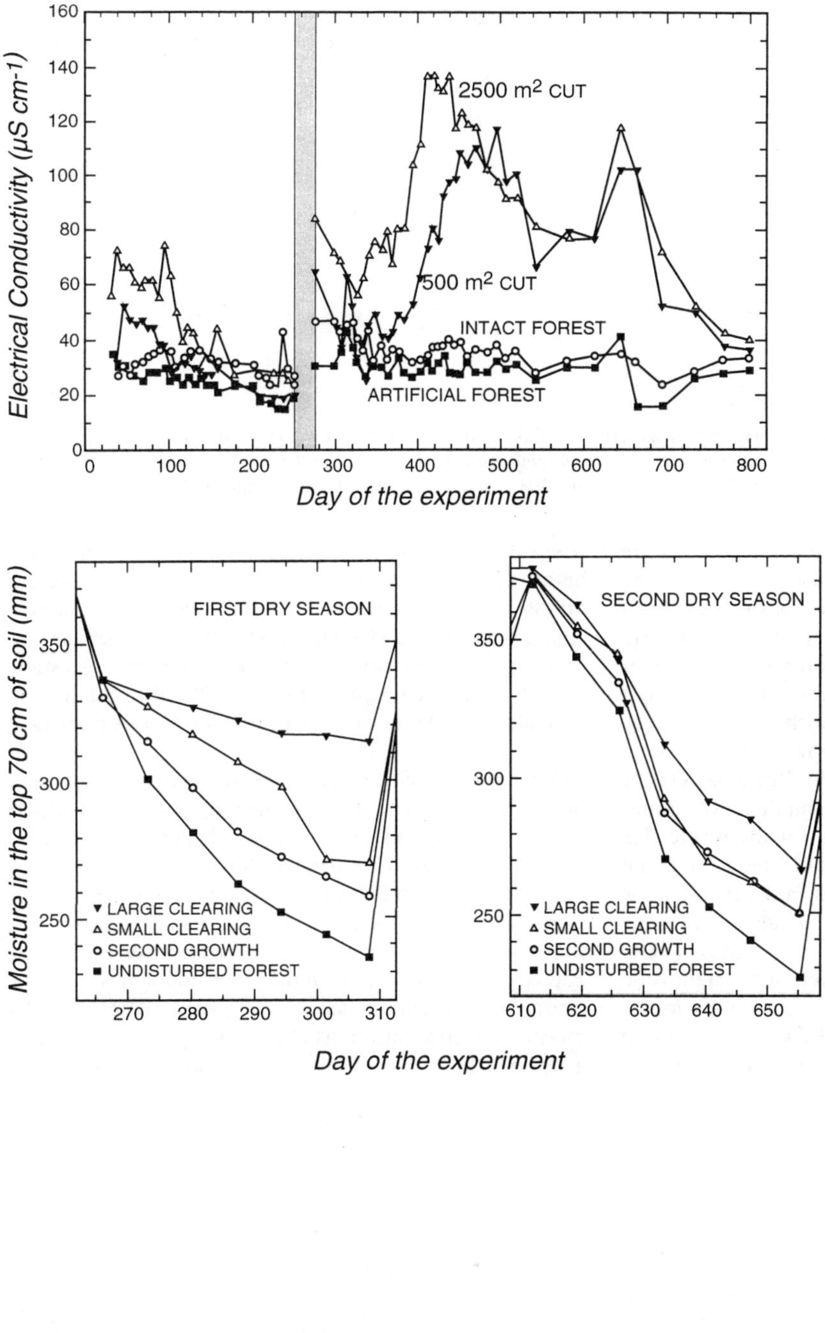


**Figure 2:** (a) Changes in the electrical conductivity and (b) amount of soil water at 70 cm depth in undisturbed forest, small artificial gaps and two clearings of contrasting size, Costa Rica (after Parker, 1985).

**Table 1:** (a) Nutrient losses in harvested timber and (b) increased leaching (kg $ha^{-1}$) after gap creation in tropical forest and the approximate time span (yr) required to compensate the losses via nutrient inputs from (c) atmospheric sources, and (d) *idem* plus weathering. Values above 10 have been rounded off to the nearest 5.

| Location | Volume of extracted timber ($m^3$ $ha^{-1}$) | | Ca | Mg (kg $ha^{-1}$) | K | Period for which leaching loss has been computed (yr) |
|---|---|---|---|---|---|---|
| La Selva, | – | (b) | 15 | 15 | 2 | 1.1 |
| Costa Rica[1, 2] | | (c) | *5* | *5* | *1* | |
| Mabura Hill, | ca. 200++ | (a) | 75 | 30 | 75 | |
| Guyana[3, 4] | large gap | (b) | 25 | 30 | 40 | 2.8 |
| | small gap | | 20 | 15 | 15 | |
| | large gap | (c) | *35* | *50* | *35* | |
| | small gap | | *35* | *35* | *30* | |
| Bukit Berembun, | 60 | (a) | 200 | 45 | 20 | |
| Malaysia[5, 6*] | | (b) | 30 | 15 | 75 | 5 |
| | | (c) | *60* | *60* | *20* | |
| Kabo, Suriname[6, 7*] | 20 | (a) | 80 | 7 | 35 | |
| | | (b) | 3 | 2 | 2 | 2 |
| | | (c) | *30* | *5* | *10* | |
| Sipitang, Malaysia[8, 9*] | 145 | (a) | 140 | 40 | 70 | |
| | | (b) | 25 | 8 | 105 | 2.8 |
| | | (c) | *55* | *45* | *35* | |
| | | (d)# | *10* | *4* | *20* | |
| Windsor Tableland, | 80 | (a) | 95 | 20 | 10 | |
| Northern Australia[10] | | (c) | *35* | *20* | *5* | |

[1]Parker (1985) (solute concentrations in soil water; nutrient input via precipitation);
[2]Bruijnzeel (1990) (water balance; his Table 4);
[3, 4]Brouwer (1996), *this volume;*
[5]Zulkifli (1989);
[6]Bruijnzeel (1992a);
[7]Poels (1987);
[8]Malmer (1993);
[9]Sim and Nykvist (1991);
[10]Gillman et al. (1985);
*basin study;
++65 and 16 $m^3$ of timber removed from the large and small gap, respectively (L. C. Brouwer, personal communication);
#estimated via nutrient outputs from undisturbed catchments with ultisols and spodosols (Malmer 1993), combined with proportions of catchment underlain by the two soil types (Sim and Nykvist, 1991); nutrient inputs via bulk precipitation according to Burghouts (1993).

matter, total nitrogen, pH, and exchangeable bases were found in skidder track depressions (occupying 13% of the total sample area of 5.6 ha) where topsoil had been removed by bulldozing (cf. *Uhl* et al., 1982). Similarly, increases were found for the mounds formed at the sides of the skidder tracks (15%), but concentrations were not significantly different in areas for which the only disturbance consisted of the addition of crown debris (40% of the area) or increased irradiation (16% of the area). On the basis of their well-planned stratified sampling programme, *Gillman* et al. (1985) concluded that, for the whole site, contents of nitrogen and exchangeable bases had not altered significantly after four years, although organic carbon had declined by about 15%. It would seem, therefore, that the only serious nutrient losses from this particular site would be those associated with stemwood removal and that losses via increased leaching were minor. Based on nutrient concentrations in bulk precipitation in the area (*Brasell* and *Gilmour*, 1980), it would take anywhere between four (potassium) and 34 (calcium) years before these losses would be compensated by atmospheric inputs alone (cf. Table 1). Therefore, the rotation period applied in the area (40–50 years) is probably sufficiently long for the ecosystem to recover from logging, suggesting that the Queensland polycyclic silvicultural system is sustainable from the nutrient point of view in this particular area.

Other studies of changes in soil chemistry over periods of one to two years after logging tropical forest include those by *Enright* (1978) in upland Papua New Guinea, *Amir* et al. (1990) in Peninsular Malaysia and *Poels* (1987) in Suriname. The information provided by these studies is somewhat more difficult to interpret, however, in view of the non-stratified sampling procedures followed. In the Papuan study, the rather fertile volcanic soil showed a return to near former levels of organic matter after 18 months but concentrations of calcium, potassium, and, to a lesser extent, magnesium were thought to take much longer to recover to pre-logging levels (*Enright*, 1978). No details were given as to where the samples were taken with respect to location of skidder tracks and mounds, etc. On the basis of a random sampling design *Amir* et al. (1990) found rather serious losses of nutrients down to a depth of 30 cm immediately after logging, particularly of calcium, magnesium and nitrogen in the case of fertile (but rather sandy) soils, and of phosphorus in all three soil associations studied. Potassium on the other hand increased, which was ascribed to decompostion of slash and 'parent material' even though the rise in concentration was observed regardless of geological substrate. Recovery after one year was good for nitrogen, reasonable for magnesium, and poor for calcium, whereas potassium remained high. No data were given for phosphorus. The data on changes in soil nutrient concentrations in a selectively logged forest subjected to 'refinement' (poison girdling) in Suriname by *Poels* (1987) were re-calculated by *Bruijnzeel* (1992a) who derived substantial decreases in amounts of calcium and potassium stored in the upper 120 cm of the soil profile. Most of the decreases could be ascribed to vigorous uptake by regenerating vegetation rather than to leaching (*Poels*, 1987; see also below).

Thus far, three studies have attempted to quantify nutrient losses via deep leaching after selectively logging lowland tropical rain forests, viz. *Poels* (1987) in Suriname; *Zulkifli* (1989) in Peninsular Malaysia; and *Brouwer* (1996) in Guyana. The results of the first two studies were discussed in detail by *Bruijnzeel* (1992a). On the bais of the overall nutrient budgets presented by the latter, tentative conclusions were drawn with respect to the time span required for atmospheric contributions of nutrients to compensate the losses associated with timber removal and enhanced leaching (Table 1).

In view of various uncertainties associated with the two studies cited above, the recent work by *Brouwer* (1996) in an area underlain by infertile sandy soils in Guyana, constitutes a welcome addition to the literature. Two gaps of 730 and 3440 $m^2$, respectively, were created using heavy machinery after which soil water content and composition within and below the main root zone were monitored for almost three years using tensiometry and vacuum tube lysimetry. In addition, soil samples were taken at several points in time according to a stratified sampling schedule. Finally, the chemical compositions of the shallow groundwater and streamflow in a nearby small (6.2 ha) catchment area before and after low-intensity logging (21 $m^3$ $ha^{-1}$; *Jetten*, 1994) were monitored as well.

Differences in groundwater levels and stream discharge before and after logging were within the accuracy of the respective measurements and as such the effect of logging, if any, could not be demonstrated at the catchment scale (*Jetten*, 1994). Similar results were previously obtained for lightly logged forests in Queensland (*Gilmour*, 1977) and India (*Subba Rao* et al., 1985). *Bruijnzeel* (1990) concluded from a survey of the literature that at least 20% of the vegetation would need to be removed before any effects on water yield would become noticeable. Changes in the chemistry of both groundwater and stream baseflow in the Guyanese study were negligible as well, with the exception of sodium concentrations which became more variable after logging. *Brouwer* (1996) ascribed this to a combination of the low intensity of the logging (which left a large part of the catchment undisturbed) and a nutrient 'filtering' effect of the vegetation in the riparian zone (cf. *Hsia* and *Horng*, 1990). Apparently, sodium was not in short supply and thus 'permitted' to bypass the riparian zone without being taken up. Some support for this contention may be derived from the results obtained by the Malaysian study referred to already. Here, solute losses in streamflow following a 'commercial' logging operation (removing 40% of the stocking) were two to three times those associated with the 'supervised' harvesting of 33% of the stocking (*Zulkifli*, 1989). The observed difference is hardly explainable in terms of the difference in volume of harvested timber but it remains possible that the retaining of a 40 m wide buffer strip in the 'supervised' case masked the true extent of leaching on the sideslopes (*Bruijnzeel*, 1992a).

As in the Costa Rican example referred to earlier (*Parker*, 1985), the soils in the gaps of the Guyanese study were consistently wetter than those under the adjacent undisturbed forest (leading to approximately 555 mm $yr^{-1}$ of extra percolation during the first 1.8 years; *Brouwer*, 1996). However, rises in soil water nutrient concentrations at 120 cm depth were both more pronounced (up to 20 times higher than observed under undisturbed forest in the case of nitrate and 4–10 times for calcium, magnesium and potassium) and commenced earlier (within a month after disturbance) than in the Costa Rican experiment. Such differences probably reflect the lower water and nutrient retention capacity of the sandy soils in Guyana. Although concentrations started to decrease again 12–15 months after gap creation they generally remained stable at levels that exceeded those observed in the undisturbed plots until at least 34 months after disturbance (*Brouwer*, this volume; cf. *Uhl* et al., 1982). Total amounts of nutrients leached from the two gaps during the entire period were substantial, with losses of nitrogen, calcium, magnesium and potassium from the larger gap being *ca.* 17, 5, 10 and 4 times those observed under undisturbed forest (*Brouwer*, this volume). The corresponding leaching losses from the smaller gap were roughly 50% of those associated with the bigger gap, except for calcium losses which were similar in both cases (*Brouwer*, 1996; Table 1). These results suggest that the contention of *Parker* (1985) that the amounts of nutrients leached from his two large gaps were very similar, may well be in need of modification (cf. *Brouwer*, this volume). Further work is necessary.

In view of the virtual absence of weatherable minerals in the substratum of the gaps in the Guyanese study, replenishment of the nutrients lost via increased leaching and timber extraction will have to come again primarily from the atmosphere (except for nitrogen). Approximate estimates for the corresponding periods required to compensate the respective losses have also been included in Table 1. On the basis of the results presented in Table 1 it remains to be seen whether the very high intensities of timber extraction that are common in parts of Borneo (up to 120 $m^3$ $ha^{-1}$; *Whitmore*, 1990), may not lead to an unacceptable degree of ecosystem nutrient depletion. Serious decreases in nutrient reserves would certainly occur after a few rotations in areas where contributions of rock weathering are negligible. One could argue that soils in steep terrain tend to be shallower and the weathering front closer to the surface and therefore whithin easier reach of the roots (*Baillie* and *Mamit*, 1983; *Burnham*, 1989). Under such conditions, nutrients released by weathering could theoretically constitute an important component of the nutrient budget. Naturally, actual amounts so contributed will depend on rainfall totals and geological substrate (*Bruijnzeel*, 1991). The observations of *Grip* et al. (1994) at Sipitang, Sabah, indicate that amounts of calcium, magnesium, and potassium released by the weathering of sandstones amount to only a few kg $ha^{-1}$ $yr^{-1}$ but that the corresponding amounts produced by shales (10–20 kg $ha^{-1}$ $yr^{-1}$) would be sufficient.

The information compiled in Table 1 suggests that the prime loss of nutrients upon logging tropical forest occurs in the form of harvested timber rather than through enhanced leaching. Nevertheless, there is a need to expand the data set on the various gains and losses of nutrients associated with logging operations for different harvesting regimes and geological substrates (cf. *Bruijnzeel,* 1992a; *Nykvist*, 1992). Similarly, published information on the changes in streamflow totals after logging is scarce and contradictory (see *Bruijnzeel* (1996) for a detailed discussion).

In contrast to the gaps created by (selective) logging, which tend to be of limited areal extent, some truly massive gap creation c.q. forest disturbance in the form of canopy defoliation, the snapping or uprooting of trees as well as landsliding, may occur during the passing of tropical hurricanes (*Scatena* and *Larsen*, 1991; *Walker*, 1991; *Scatena* et al., 1993; *Waterloo*, 1994). The sudden addition of large masses of fresh organic material to the forest floor, coupled with dramatic increases in light levels, soil temperatures and soil water content, bring about an array of changes in the processes governing release, cycling and loss of nutrients, particularly nitrogen (*Lodge* et al., 1991; *Blood* et al., 1991; *Steudler* et al., 1991). For example, ammonium availability, net nitrogen mineralization and nitrification rates were all observed to be elevated four months after the passage of Hurricane Hugo in Puerto Rico (*Steudler* et al., 1991). Similarly, concentrations of nitrate and base cations in streamwater also increased, although overall solute losses remained low due to the rather small amounts of streamflow during this period (F. N. Scatena, personal communication). *Waterloo* (1994) also reported significant changes in streamflow chemistry during the first month after extensive damage occurred to a pine forest estate in Viti Levu, Fiji during the passage of a cyclone (Sina). Interestingly, ammonium and nitrate concentrations were not affected here, while concentrations of base cations increased and levels of sulphate were reduced. *Waterloo* (1994) explained the latter finding in terms of the chemical composition of the rainfall and sea spray associated with the cyclone which happened to be low in sulphate. No reduction in sulphate concentration was observed in the flow from an adjacent grassland catchment which may reflect the lower trapping efficiency of the grass for cyclone sea spray (*Waterloo*, 1994). The lack of response of the nitrogen constituents may be related to the stimulation of nitrogen immobilization by the microbial community through the large addition of fresh organic matter via needle fall during the cyclone (cf. *Vitousek* and *Matson*, 1985).

Despite the potentially more widespread areal impact of hurricanes, *Steudler* et al. (1991) considered the effects of *forest clearcutting* on the soil nitrogen cycle and trace gas fluxes to be more intense and of longer duration. In the hurricane case, the root system of much of the (natural) vegetation generally remains in place and many trees and seedlings respond rapidly (*Brokaw* and *Walker*, 1991). As a result, the nitrogen cycle can be expected to recover quickly, whereas the vegetation is slower to re-establish itself after clear cutting, particularly so after the use of heavy machinery (*Uhl* et al., 1982, 1988a; see also the next section). *Malmer* and *Grip* (1994) carried out an interesting experiment in this respect in Sabah, Malaysia. They monitored the flow and solute losses from an entire catchment area ($W_4$, 3.4 ha) before, during and after it was subjected to clearcutting and timber extraction. Some 146 $m^3$ of timber was removed from the area using manual labour to minimize surface disturbance (*Malmer* and *Grip*, 1990; *Sim* and *Nykvist*, 1991). The remaining slash was left on the slopes which were then planted to *Accacia mangium*. During the first 1.5 and 3.7 years after planting the *Acacia* trees acquired above-ground biomass figures of 10.5 and 44.6 t $ha^{-1}$, respectively (*Malmer*, 1993). The nutrient losses associated with timber extraction and leaching have been included in Table 1 for comparison. With the exception of potassium, which was leached severely, overall losses for the other macro-nutrients were modest and could theoretically be recovered within 35–55 years by atmospheric inputs alone and within 5–20 years once contributions of rock weathering are taken into account as well (Table 1).

### Forest fires and shifting cultivation

The occurrence of fire may increase overall site nutrient losses in various ways. Firstly, there are the losses associated with volatilization and ash particle transport during the burn itself (*Raison* et al., 1985; *Ewel* et al., 1981;

*Mackensen* et al., 1996). Secondly, nutrients contained in the ash residue will be more prone to leaching and erosion (by both water and wind) than those in gradually decomposing slash (*Toky* and *Ramakrishnan*, 1981; *Uhl* and *Jordan*, 1984; *Waterloo*, 1994). Finally, surface erosion may increase dramatically after burning (*Wiersum*, 1984). Changes in soil physical characteristics due to fire range from negligible to dramatic (*Van Lear* et al., 1985; *Leitch* et al., 1983). Much depends on the intensity of the fire which is governed mainly by amounts and moisture content of the material available as fuel (undergrowth, logging debris and litter) and ambient weather conditions (*Scott*, 1993; *Mackensen* et al., 1996). Different soils may exhibit a different response to excess heat, with some soil types even becoming temporarily water repellent. Naturally, water repellence of soils impedes infiltration, and this may lead to the generation of overland flow (*Burch* et al., 1989; *Scott*, 1993). When the overland flow occurs on sites where the protective litter layer was just destroyed by the fire, some truly dramatic surface and gully erosion may follow (*Brown*, 1972; *Leitch* et al., 1983). However, such effects are hard to distinguish from those exerted by soil compaction.

Nutrient losses to the atmosphere during the burning of slash in tropical regions have been quantified by only a few studies (Table 2), some of which (e.g. *Toky* and *Ramakrishnan*, 1981) seem to have produced anomalously high values. Contrary to earlier assumptions (e.g. *Ewel* et al., 1981), losses due to volatilization do not merely concern those elements having a gaseous phase (nitrogen, sulphur) but also the base cations calcium, magnesium, potassium as well as phosphorus (Table 2). Whilst some of the higher figures quoted for the respective cations in Table 2 may be due to wind effects (*Waterloo*, 1994), the ones pertaining to the Amazonian study were obtained under conditions of still air and must therefore represent upward molecular transport related to thermally induced kinetic effects (*Mackensen* et al., 1996). In contrast to losses of nitrogen and sulphur, those for calcium etc. appear influenced by the amount of fuel and the intensity of the burn (*Mackensen* et al., 1996; Table 2).

There have been numerous studies of the changes in soil chemical contents at the respective stages of the *shifting cultivation* cycle (see reviews by *Sanchez*, 1976; *Bruijnzeel*, 1990; *Ramakrishnan*, 1992). Such changes are the net result of an array of nutrient inputs and outputs, of which volatilization and erosion, uptake by regenerating vegetation, and leaching losses are generally considered to be the most important (*Uhl* and *Jordan*, 1984, *Wiesum*, 1984; *Ramakrishnan*, 1992). The lack of reliable estimates of the amounts lost via enhanced leaching after tropical forest cutting and burning has been commented upon already. As such, the results of an ongoing study of the water and nutrient dynamics of freshly cleared and burned as well as regenerating sites in eastern Amazonia using a combination of micro-meteorological, plant physiological and soil hydrological techniques (*Hölscher*, 1995; *Mackensen* et al., 1996; *Roberts* et al., 1996; *Klinge*, 1997) are a welcome addition to the literature. More work is needed, however, covering different combinations of rainfall regime and soil type.

There are virtually no data on the hydrological impacts (including enhanced solute losses) of *natural forest fires* at the catchment scale in the humid tropics for the simple reason that river gauging stations (if present at all) are usually destroyed by such fires, and their re-establishment tends to take longer than that of a vegetative cover (*Brown*, 1972). Arguably, the best way of collecting such information is through experimental studies (*O'Loughlin* et al., 1982). A particularly interesting case has been reported by *Malmer* (1993) for experimentally burned secondary scrub (biomass $< 5$ t ha$^{-1}$) in Sabah that had been hit by a forest fire 5–6 years previously. After the experimental burn, nutrient concentrations in streamwater increased more vigorously in catchments that had experienced the great fires of 1982–83 than for a mechanically logged catchment where far greater amounts of biomass (logging debris) were burned (see Table 3 below). The effect was much more short-lived, however, and within 6 months concentrations in the basins that had been burned twice started to fall below predicted levels. *Malmer* (1993) interpreted

**Table 2:** Relative nutrient losses (per cent) due to volatilization during the burning of residual biomass in selected tropical forest areas as a function of fuel mass (t ha$^{-1}$) and mass reduction (per cent).

| Location | Fuel mass (t ha$^{-1}$) | Mass reduction (per cent) | N | P | K | Ca | Mg | S |
|---|---|---|---|---|---|---|---|---|
| | | | | | (per cent) | | | |
| Turrialba, Costa Rica[1] | 38 | 83 | 23 | | negligible? | | | 44 |
| Belém, Brazil[2] site 2 | 33 | 90 | 95 | 26 | 16 | 9 | 17 | 67 |
| Belém, Brazil[2] site 3 | 95 | 96 | 98 | 33 | 31 | 25 | 43 | 68 |
| Viti Levu, Fiji[3] | 40 | 86 | 84 | 52 | 79 | 78 | 60 | – |

[1]Ewel et al. (1981);
[2]Mackensen et al. (1996);
[3]Waterloo (1994).

**Table 3:** (a) Nutrient losses (kg ha$^{-1}$) in harvested timber and (b) deep leaching after clearcutting and/or burning tropical forest; (c) approximate time span (yr) required to compensate the losses via nutrient inputs from atmospheric sources only, and (d) *idem* plus rock weathering; values of estimated time spans above 10 yr rounded off to the nearest 5.

| Location | Treatment | Harvested timber (t ha$^{-1}$) | Burned slash (t ha$^{-1}$) | | N | P | K | Ca | Mg |
|---|---|---|---|---|---|---|---|---|---|
| | | | | | | | (kg ha$^{-1}$) | | |
| Sipitang, Malaysia[1–3@] | | | | | | | | | |
| Basin $W_{1+2}$* | Burning only | – | 5 | (a) | – | – | – | – | – |
| | | | | (b) | 17 | 2 | 84 | 28 | –5 |
| | | | | (c) | | 6 | 15 | 10 | – |
| | | | | (d) | | – | 9 | 2 | – |
| Basin $W_5$[#] | Mechanized logging + burning | 75 | 170 | (a) | 80 | 2 | 57 | 112 | 32 |
| | | | | (b) | 40 | 1 | 189 | 27 | 16 |
| | | | | (c) | | 10 | 50 | 50 | 50 |
| | | | | (d) | | – | 35 | 15 | 6 |
| Basin $W_4$[o] | Manual logging + no burn | 90 | 185 | (a) | 100 | 2 | 71 | 139 | 39 |
| | | | | (b) | 27 | 1 | 106 | 25 | 8 |
| | | | | (c) | | 10 | 35 | 55 | 45 |
| | | | | (d) | | – | 20 | 10 | 4 |
| Viti Levu, Fiji[4+] | Mechanized logging + burning | 55 | 40 | (a) | 45 | 5 | 29 | 23 | 11 |
| | | | | (b) | 6 | < 1 | 3 | 2 | 4 |
| | | | | (c) | | 20 | 10 | 15 | 6 |
| | | | | (d) | | 9 | 6 | 1 | < 1 |

[1]Sim & Nykvist (1991);
[2]Grip et al. (1994);
[3]Malmer & Grip (1994);
[4]Waterloo (1994);
[@]high rainfall environment (3350 mm yr$^{-1}$); paired catchment technique; leaching losses computed for a period of 34 months; weathering contributions estimated as indicated in Table 1;
[*]6.45 ha catchment; 38 per cent podzols, 62 per cent ultisols; vegetation burned five years prior to the experiment and since then consisting of scrub, grasses and ferns;
[#]9,67 ha catchment; 62 per cent spodosols, 38 per cent ultisols; selectively logged about 10 years before experimental cutting and burning;
[o]3.4 ha catchment; 34 per cent spodosols, 66 per cent ultisols; selectively logged about 10 years before experimental logging;
[+]single catchment experiment; low rainfall environment (1800 mm yr$^{-1}$; leaching losses for the first 16 months only; basin size 62,9 ha; moderately fertile ultisols; basin 18 per cent covered by native riparian vegetation and remainder by 15-year-old *Pinus caribaea*; 42 per cent of plantation damaged by hurricane in November 1990; intermittent tractor logging between December 1990 and July 1991; burned in August 1991; weathering contributions calculated from catchment nutrient input-output budget, taking into account amounts immobilized in biomass and assuming no net changes in soil nutrient stores during a rotation.

this as an exhaustion effect. However, the phenomenon was observed most strongly for silica and it would seem that the reported changes in streamwater solute concentrations reflected a shift in water pathways (lowered baseflows and increased peakflows) following burning rather than just nutrient uptake processes alone. Such findings illustrate the need for integrated process studies to illuminate the traditional paired catchment approach which, although statistically sound, remains essentially a black box approach (*Grip*, 1994; *Bruijnzeel*, 1996). More work is needed with respect to the fate of nutrients released by the burning of residual biomass (see also below).

### 3.2 Response to disturbances of high intensity (forest conversion)

It is well-established that manual methods of forest clearing are superior to most wheeled or tracked equipment in terms of damage to the forest floor (*Lal*, 1987; *Fritsch*, 1992; *Bruijnzeel*, 1996). Indeed, there are strong arguments for the use of alternative mechanized timber extraction methods that tread lightly on the soil surface, such as skyline yarding (*Ludwig*, 1992), elephants where the size of the logs allow this (*Whitmore*, 1990), or even helicopters in the case of well-stocked forests in steep terrain (*Blakeney*, 1992). As will be shown below, there is also increasing evidence that leaving slash on site to rot rather than 'windrowing' and burning it has various beneficial effects on the hydrological and chemical characteristics of the soils. However, when converting rain forest land to other land use types, the use of heavy mechanized equipment is often unavoidable. *Martin* (1970) and *Couper* et al. (1981) have offered various suggestions as to how damage to the soil can be minimized.

Naturally, nutrient losses associated with large-scale mechanized forest clearfelling and burning large amounts of residual biomass are potentially much greater than those for the less intensive types of forest disturbance discussed in the previous sections, particularly under conditions of high rainfall. A review of changes in water yield following tropical forest clearance by *Bruijnzeel* (1990) revealed that reported first-year increases in streamflow ranged between 125 and 820 mm. There can be little doubt that such increases reflect the different evaporative characteristics of mature tropical forest and young secondary or planted vegetation, and, therefore, differences in soil water status (cf. *Parker*, 1985; *Brouwer*, this volume). Interestingly, the large variation in initial streamflow response to clearing can be explained only partially by differences in rainfall between locations (Figure 3). Other factors include differences in elevation (and thus evaporation), catchment steepness and soil depth (govering the residence time of the water in the catchment, baseflow recession and stormflow patterns; *Ward*, 1984), and above all the degree of disturbance of the soil by machinery or fire (influencing runoff patterns; *Malmer* 1992; *Swindel* et al., 1982, 1983), and the fertility status of the soil (affecting, like the intensity of soil disturbance, post-clearing plant productivity; *Uhl* et al., 1982, 1988a; *Brown* and *Lugo*, 1990). The relative importance of the respective factors will vary between locations and the resulting high variability of the data (Figure 3) does not allow any detailed quantitative predictions. Again, additional process studies (particularly on runoff generation mechanisms) are required if the results of black box catchment experiments (Figure 3) are to be more fully understood (*Bonell* and *Balek*, 1993; *Bruijnzeel*, 1996).

Most of the factors listed above also affect the amounts of nutrients lost via erosion and deep leaching, illustrating the close links between water and nutrient pathways (*Likens* et al., 1977). As noticed earlier for studies of small-scale forest cutting and burning (*Toky* and *Ramakrishnan*, 1981; *Uhl* et al., 1982; *Uhl* and *Jordan*, 1984), most early investigations of the effects of tropical forest conversion on soil and ecosystem nutrient reserves (*Hase* and *Fölster*, 1983; *Russell*, 1983; *Buschbacher*, 1984) used a 'false time series' approach. Leaching losses were generally estimated on the basis of rather flimsy hydrological budgets or simply as average values taken from the literature while nutrient losses via surface erosion were usually not taken into account either (*Bruijnzeel*, 1990). Despite the shortcomings of such studies, their overall conclusions as to which macro-nutrients were likely to become limiting first, are probably justified, although some authors (e.g. *Hase* and *Fölster*, 1983) were quick to point out that their nutrient budgets should be regarded as preliminary and semi-quantitative.

Table 3 summarizes the losses of nutrients in harvested timber and enhanced leaching as determined by two recent studies of mechanized forest clearing and burning of slash under contrasting environmental conditions. The Malaysian site experiences high annual rainfall totals (> 3300 mm; *Malmer*, 1992) while the Fijian site is characterized by relatively low rainfall (c. 1800 mm $yr^{-1}$) and a long dry season (*Waterloo*, 1994). Also illustrated in Table 3 are the contrasting effects of (i) burning different amounts of slash (catchments $W_{1+2}$ vs. $W_5$) and (ii) mechanized clearing plus burning (catchment $W_5$) vs. manual clearing and leaving slash to rot (catchment $W_4$).

Despite the severity of the treatment, amounts of nutrients lost from the Fijian catchment during the first 16 months after (pine) forest clearcutting and burning some 40 t $ha^{-1}$ of slash were small. Various explanations may be offered for this finding, including below-average rainfall totals during the first rainy season after the burn; the filtering effect of the riparian vegetation which remained undisturbed (cf. *Zulkifli*, 1989; *Brouwer*, 1996); the vigorous regrowth of the understorey vegetation (cf. *Uhl* and *Jordan*, 1984); and the high nutrient retention capacity of the clayey subsoils in the area (*Waterloo*, 1994).

As could be expected on the basis of the higher rainfall and/or larger amounts of burned residual biomass,

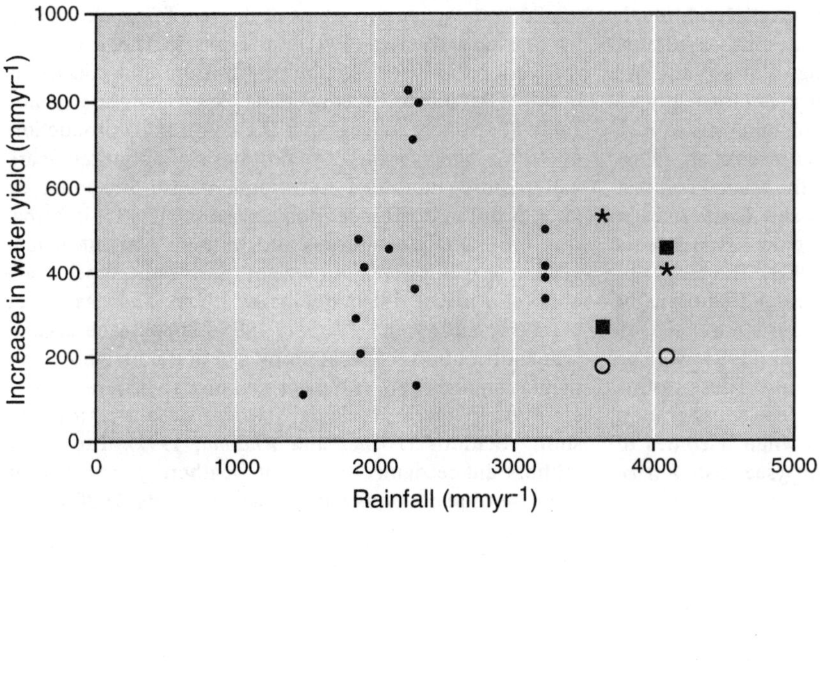


**Figure 3:** Increases in water yield (mm $yr^{-1}$) during the first three years after clearing tropical forest vs. corresponding amounts of annual rainfall (after Bruijnzeel, 1997). ○ catchment $W_4$ (manual clearing), ■ catchment $W_5$ (mechanized clearing plus burning of slash), and * catchment $W_{1+2}$ (burning only) of Malmer (1992).

cationic losses in solution were generally much larger in the Malaysian experiment (even when allowing for the longer observation perid, 34 vs. 16 months; Table 3). This was the case regardless of the treatment, with the exception of magnesium losses after logging only, which were similar to those observed in the Fijian study (Table 3). As indicated earlier (section 3.1), the negative result obtained for magnesium after burning only (catchment $W_{1+2}$) probably reflects a shift in runoff patterns. Unfortunately, it is not known to what extent this phenomenon affected leached amounts of calcium and potassium as well (*Malmer* and *Grip*, 1994). Leaching of calcium was similar for all three treatments (Table 3), despite large differences between catchments in water yield after conversion (more than twice as high for $W_{1+2}$ and $W_4$ compared to $W_5$; *Malmer*, 1992) and a different soils association in catchment $W_5$. Solute losses of potassium were about twice as high after mechanized harvesting and burning of slash ($W_5$) in comparison with either burning ($W_{1+2}$) or manual logging only ($W_4$). Similarly, magnesium losses in streamflow after the most severe treatment ($W_5$) were also double those associated with logging only ($W_4$) (Table 3).

Strictly speaking it is difficult to draw any firm conclusions on the basis of these results with respect to solute losses as a function of treatment, other than that additional process work is still required. For example, a direct comparison of the leaching losses obtained for catchments $W_5$ (tractor logging + burning) and $W_4$ (manual logging only) is hampered because of the different proportions of sandy and clayey soils found in the respective basins. The overall water and nutrient retention capacities of the predominantly sandy soils in $W_5$ could be expected to be lower than those of the predominantly clayey soils in $W_4$ (*Grip* et al., 1994; cf. footnotes to Table 3). Also, the similarity in losses of calcium, but not in magnesium or potassium, for the two catchments is difficult to explain without additional information on soil nutrient reserves and nutrient uptake by the vegetation. Similarly, although the areal distribution of sandy and clayey soils was almost identical for catchments $W_{1+2}$ (burning of slash) and $W_4$ (rotting of slash), the effect of the burn cannot really be predicted because of the vastly differing amounts of residual biomass involved (Table 3; cf. *Mackensen* et al., 1996). A further complicating factor concerns the related contrast in growth exhibited by the understorey and *Acacia* trees in the two areas (22.2 vs. 50.0 t $ha^{-1}$ during the first 3.5 years in $W_{1+2}$ and $W_4$, respectively; *Malmer*, 1993). The results of the inventories of soil chemical reserves and amounts of nutrients incorporated in forest biomass announced by *Sim* and *Nykvist* (1991) and *Nykvist* et al. (1994) will be needed to obtain a more complete picture. These results illustrate once more the importance of close collaboration between pedologists and hydrologists when it comes to experimental design and site location (see also *Fritsch* et al., 1987; *Fritsch*, 1993).

Nutrient losses associated with the harvesting of stems were generally greater than those in streamflow in both studies, with the notable exception of potassium which was leached in large quantities from the Malaysian catchments (Table 3). All in all, cationic losses in the latter environment were such that at least 35 to 55 years of nutrient additions via bulk precipitation would be required to compensate them (Table 3). The corresponding recovery periods for the respective elements in the Fijian case, where both leaching losses and volume of harvested timber were lower, ranged from about five years (magnesium) to 15 years (calcium).

The information presented in Table 3 may be used to help assess the eventual effects of plantation forestry on soil nutrient reserves. Overall nutrient inputs from atmospheric sources throughout a typical pine plantation rotation period of 15–20 years in the Fijian example were such that they would cover the losses associated with stemwood (plus bark) removal and temporarily increased leaching (*Waterloo*, 1994). Because amounts of nutrients (particularly calcium and magnesium) released by rock weathering in this area were substantial and the weathering front usually within reach of the pine root network (*Waterloo*, 1994) one may conclude that the cultivation of *Pinus caribaea* in south-western Viti Levu will not lead to unacceptable site degradation. However, *Waterloo* (1994) added the precaution that both whole-tree harvesting and surface erosion should be avoided.

It is well-established that nutrient concentrations in the stemwood and bark of non-coniferous species (particularly teak and *Gmelina*) are often much higher than those contained in most pine species (*Halenda*, 1993; *Fölster* and *Khanna*, 1997). It follows that amounts of nutrients lost from the site in harvested hardwood timber are potentially much greater than for pines or, indeed, natural forest. For example, amounts of nutrients accumulated in the stems of a 6.6-year-old stand of *Gmelina arborea* on a very acid ultisol in Sarawak (*Halenda*, 1993) exceeded those contained in the harvested stems of the rain forest clearfelled by *Sim* and *Nykvist* (1991) in Table 3. The estimated recovery periods for calcium, magnesium and potassium in the Malaysian forest conversion experiment (c. 50 years, Table 3) strongly suggest that the continuous production of *Acacia mangium* on a rotation basis of about ten years will require additional nutritional measures sooner or later. A full evaluation will only be possible after information on plantation biomass and nutrient content at harvesting as well as on soil nutrient reserves becomes available (cf. *Ruhiyat*, 1989; *Bruijnzeel*, 1992b; *Spangenberg* et al., 1996; *Fölster* and *Khanna*, 1997). It could be argued that contributions by weathering had better be ignored by plantation managers as long as we know so little about the amounts involved, particularly where trees are grown in short rotations (*Fölster* and *Khanna*, 1997). The latter authors did recognize a role for weathering, however, in the case of plantations with longer rotations ($>$ 20 years) and certain soil groups that still contained weatherable minerals. Nevertheless, the provisional estimates of the amounts of nutrients released by rock weathering in the Malaysian example (notably of the shales in catchment $W_3$) suggest that recovery periods could be substantially reduced by the inclusion of nutrient inputs from weathering, provided the trees would have access to them (Table 3). The quantification of the amounts of nutrients released by mineral weathering is notoriously difficult (*Clayton*, 1979; *Verstraten*, 1980) and its estimation has only rarely been attempted within the framework of tropical forestry (*Bruijnzeel*, 1983; *Hase* and *Fölster*, 1983; *Tandy*, 1987; *Grimaldi*, 1988). Yet this aspect may be expected to become increasingly important in view of the rapidly expanding area of fast-growing forest plantations in the humid tropics and the high costs of fertilizer (*Evans*, 1992).

## 4 Research perspectives

Several important gaps in our knowledge have become apparent in the previous discussion. Arguably, the most striking of these is the *almost complete lack of reliable information on the hydrological behaviour (rainfall interception, transpiration, percolation) of young secondary vegetation and tree plantations* (*Hölscher*, 1995; *Bruijnzeel*, 1996, 1997). Needless to say, such information is a *sine qua non* for the estimation of amounts of nutrients lost via leaching. As shown in Tables 1 and 3, leaching losses upon forest disturbance and conversion can be substantial, particularly under conditions of high rainfall and soils with low nutrient retention capacity. The methodology for the estimation and modelling of rainfall interception and water use is well-established (*Lloyd* and *Marques*, 1988; *Lloyd* et al., 1988; *Roberts* et al., 1993, 1996), although the scattered nature of many secondary stands in the tropics (*Uhl* et al., 1998a; *Denich*, 1989) may present methodological problems (*Hölscher*, 1995). A useful role may be played in this respect by such approaches as the isotope tracer injection technique (*Calder*, 1991) and the heat pulse velocity (or heat balance) method (*Hatton* and *Vertessy*, 1990) which may (also) be used to study water uptake rates at different topographic positions (e.g. dry ridge tops vs. wet valley bottoms) or from soils of different degrees of compaction.

Although amounts of slowly percolating soil water (unsaturated matrix flow) can be quantifed adequately using the various techniques listed above, additional observations will be required of the amounts and chemical composition of the rapid flow through macropores and pipes during periods of intense rainfall if total amounts of leached nutrients are to be determined accurately. Indeed, the *generation of runoff during storms (both at and below the surface) constitutes one of the more poorly documented aspects of tropical forest hydrology* (*Bonell* and *Balek*, 1993; *Elsenbeer* and *Lack*, 1996; *Klinge*, 1997). Naturally, the study of such 'bypass flow' or 'quickflow' (depending on one's standpoint: the soil profile or the stream/catchment) assumes particular importance in relation to the proper understanding of the effects of forest disturbance or conversion. *Bishop* (1991) and *Bonell* (1993) have provided recent in-depth discussions of the direction that future runoff generation investigations could take. Both stressed the need for an integrated approach that would combine classical hydrological (volumetric) techniques and detailed event sampling (for solutes and stable isotopes in soil- and streamwater) with physically-based distributed hydrological models. Because of the great depth to bedrock in many tropical catchments (*Burnham*, 1989), however, the application of such models may not be as straightforward as it looks (*Quinn* et al., 1991). *McDonnell* et al. (1996), working on deeply weathered soils in the south-eastern U.S., recently demonstrated the close correspondence between subsurface flow paths and bedrock topography. Similar studies under humid tropical conditions could be revealing.

There is also a *need for more studies of nutrient losses via eroded topsoil material after forest clearance and burning as a function of rainfall regime, soil type and degree of disturbance* (*Hudson* et al., 1983; *Bruijnzeel* and *Wiersum*, 1985; *Malmer*, 1996). Ideally, such work should be conducted at various levels of scale (hillslope plots vs. small catchments) so as to enable the assessment of on-site (erosion) and off-site (basin sediment yield) effects. Such experiments could be usefully *combined with research into the filtering effects of streamside buffer strips* whose role with respect to nutrient interception is still poorly defined (*Zulkifli*, 1989; *Hsia* and *Horng*, 1990; *Brouwer*, 1996). These aspects assume particular importance in the context of fast-growing forest plantations with short harvesting cycles (< 5–8 years) and, consequently, frequent surface disturbance (*Wiersum*, 1984; *Fölster* and *Khanna*, 1997).

Finally, the *potential relevance of nutrient inputs to tropical forest ecosystems by mineral weathering* has already been commented upon in previous sections. Research to this end is probably conducted best via the small catchment nutrient budget approach (*Clayton*, 1979), supplemented by water-rock interaction studies (*Bruijnzeel*, 1983; *Tandy*, 1987; *Grimaldi*, 1988). Such work could well be combined with the study of the subsurface topography of the fresh bedrock in relation to hillslope hydrological pathways and nutrient export routes (*McDonnell* et al., 1996).

Evidently, there are distinct opportunities for closer collaboration between tropical pedologists, hydrologists and plant ecologists. Thus far, no single study has addressed the soils and water aspects of a tropical forest ecosystem in a fully integrated manner, either in undisturbed forest or after disturbance or conversion (*Bruijnzeel*, 1983; *Fritsch*, 1992; *Malmer*, 1993). In view of the very considerable investments in manpower and time required by such studies (*Likens* et al. 1977; *Swank* and *Crossley*, 1988) it would seem advisable to focus future integrated research efforts on a limited number of sites for which detailed soils and vegetation inventories are already available and where the basic fluxes of water and nutrients have been measured for a sufficiently long time (cf. *Bruijnzeel* and *Abdul Rahim*, 1992). Similarly, the renewed assessment of amounts of nutrients currently stored in the soil and vegetation of sites that were the subject of investigation in the past could prove more effective than the establishment of entirely new locations. Examples include the teak forests in southwestern Venezuela studied by *Hase* and *Fölster* (1983) or the pine plantations in Fiji sampled by *Waterloo* (1994). Since there are a number of existing research sites scattered all over the humid tropics (*Bruijnzeel*, 1990), there may be scope for a screening exercise on the basis of which decisions could be taken as to where additional sites would be established best (*Bruijnzeel* 1993, 1996). For instance, there are virtually no studies evaluating the hydrological and fertility aspects of reforesting degraded areas at the catchment scale (cf. *Waterloo*, 1994), or the impacts of clearing montane cloud forests (*Bruijnzeel* and *Proctor*, 1995). Arguably, the coordination of such a (possibly rather informal) pantropical 'network' of research sites might be handled best by the Humid Tropics Programme within UNESCO's International Hydrological Programme.

## 5 Acknowledgement

This paper was completed during prolonged sickness of the author. The assistance rendered by Rai 'Leave it to me' Hafkenscheid, Jaap, 'No sweat' Schellekens, Maarten 'Dr M.' Waterloo and Mieke Zadel is gratefully acknowledged. Special thanks are due to Andreas Schulte for his patience and above all to Irene Sieverding for her unstinting care. Mr Henri Sion provided the diagrams.

## 6 References

*Amir, H. M. S., Zakaria, M., Ghazali Hasan, M.* and *Ahmad, R.* (1990). Nutrient dynamics of Tekam Forest Reserve, Peninsular Malaysia, under different logging phases. J. Trop. For. Sci. 2, 71–80.

*Baillie, I. C.* (1989). Soil characteristics and mineral nutrition of tropical wooded ecosystems. In: Mineral Nutrition in Tropical Forest and Savanna Ecosystems (J. Proctor, Ed.), Brit. Ecol. Soc. Spec. Publ. no. 9, 15–26.

*Baillie, I. C.* and *Mamit, J. D.* (1983). Observations on rooting in mixed dipterocarp forest, Central Sarawak. Malay. For. 46, 369–374.

*Bishop, K. H.* (1991). Episodic Increases in Stream Acidity, Catchment Flow Pathways and Hydrograph Separation. PhD Thesis, Univ. Cambridge, 246 pp.

*Blakeney, K. J.* (1992). Cable and helicopter logging for reduced damage. In: Proceedings of the International Symposium on Harvesting and Silviculture for Sustainable Forestry in the Tropics, Kuala Lumpur, October 1992, IUFRO, Vienna, 6 pp.

*Blood, E. R., Anderson, P., Smith, P. A., Nybro, G.* and *Ginsberg, K. A.* (1991). Effects of hurricane Hugo on coastal soil solution chemistry in South Carolina. Biotropica 23, 348–355.

*Bonell, M.* (1993). Progress in the understanding of runoff generation dynamics in forests. J. Hydrol. 150, 217–275.

*Bonell, M.* and *Balek, J.* (1993). Recent scientific developments and research needs in hydrological processes of the humid tropics. In. Hydrology and Water Management in the Humid Tropics. Hydrological Research Issues and Strategies for Water Management (M. Bonell, M. M. Hufschmidt and J. S. Gladwell, Eds.), Cambridge University Press, Cambridge, 167–260.

*Brasell, H. M.* and *Gilmour, D. A.* (1980). The cation composition of precipitation at four sites in far North Queensland. Austr. J. Ecol. 5, 397–405.

*Brinkmann, W. L. F.* (1985). Studies on hydrobiogeochemistry of a tropical lowland forest system. GeoJournal 11, 89–101.

*Brokaw, N. V. L.* and *Walker, R. L.* (1991). Summary of the effects of Caribbean hurricanes on vegetation. Biotropica 23, 442–447.

*Brouwer, L. C.* (1996). Nutrient Cycling in Pristine and Logged Tropical Rain Forest. A Study in Guyana. Tropenbos – Guyana Series no. 1. Tropenbos Foundation, Wageningen, The Netherlands, 224 pp.

*Brown, J. A. H.* (1972). Hydrologic effects of a bushfire in a catchment in south-eastern New South Wales, J. Hydrol. 15, 77–96.

*Brown, S.* and *Lugo, A. E.* (1990). Tropical secondary forests. J. Trop. Ecol. 6, 1–32.

*Brünig, E. F., Herrera, R., Heuveldop, J., Jordan, C. F., Klinge, H.* and *Medina, E.* (1978). The international Amazon project coordinated by Centro de Ecologia, IVIC: organization and recent advances. Spec. Report no. 1, Chair of World Forestry, Hamburg-Reinbeck, pp. 104–131.

*Bruijnzeel, L. A.* (1983). Hydrological and Biogeochemical Aspects of Man-made Forests in South-central Java, Indonesia. PdD Thesis, Vrije Universiteit, Amsterdam, 256 pp.

*Bruijnzeel, L. A.* (1990). Hydrology of Moist Tropical Forests and Effects of Conversion, a State-of-Knowledge Review. UNESCO, Paris, and Free Univ., Amsterdam, 224 pp.

*Bruijnzeel, L. A.* (1991). Nutrient input-output budgets of tropical forest ecosystems, a review, J. Trop. Ecol. 7, 1–24.

*Bruijnzeel, L. A.* (1992a). Managing tropical forest watersheds for production, where contradictory theory and practice co-exist. In: Wise Management of Tropical Forests 1992 (F. R. Miller and K. L. Adam, Eds.), Oxford Forestry Institute, Oxford, 37–75.

*Bruijnzeel, L. A.* (1992b). Sustainability of fast-growing plantation forests in the humid tropics, with special reference to nutrition. In: Taungya: Forest Plantations with Agriculture in South-east Asia (C. F. Jordan, J. Gajaseni and H. Watanabe, Eds.), C.A.B. International, Wallingford, 51–67.

*Bruijnzeel, L. A.* (1993). Land-use and hydrology in warm humid regions, where do we stand? In: Hydrology of Warm Humid Regions (J. S. Gladwell, Ed.), Int. Assoc. Hydrol. Sci. Publ. 216, 1–34.

*Bruijnzeel, L. A.* (1996). Predicting the hydrological impacts of tropical forest conversion: the need for an integrated approach. In: Amazonian Deforestation and Climate (J. H. C. Gash, C. A. Nobre, J. Roberts and R. Victoria, Eds.), J. Wiley, New York, 15–55.

*Bruijnzeel, L. A.* (1997). Hydrology of forest plantations in the humid tropics. In: Bekker Management of Soil, Nutrients, and Water in Tropical Plantation Forests (E. K. S. Nambiar and A. G. Brown, Eds.), ACIAR, Canberra 125–167.

*Bruijnzeel, L. A.* and *Abdul Rahim, N.* (1992). Ecological and environmental services provided by tropical trees and forests. In: Center for International Forestry Research (CIFOR). Strategic Planning Thematic Papers, ACIAR, Canberra, 23–26.

*Bruijnzeel, L. A.* and *Proctor, J.* (1995). Hydrology and biogeochemistry of tropical montane cloud forests: what do we really know? In: Tropical Montane Cloud Forests (L. S. Hamilton, J. O. Juvik and F. N. Scatena, Eds.). Springer Ecological Studies no. 110, 38–78.

*Bruijnzeel, L. A.* and *Wiersum, K. F.* (1985). A nutrient budget sheet for *Agathis dammara* Warb. plantation forest under various management conditions. For Ecol. Manage. 10, 195–208.

*Burch, G. J., Moore, I. D.* and *Burns, J.* (1989). Soil hydrophobic effects on infiltration and catchment runoff. Hydrol. Proc. 3, 211–222.

*Burghouts, Th. B. A.* (1993). Spatial Variability of Nutrient Cycling in a Bornean Rain Forest. PhD Thesis, Vrije Universiteit, Amsterdam, 156 pp.

*Burnham, C. P.* (1989). Pedological processes in temperate and tropical soils. In: Mineral Nutrients in Tropical Forest and Savanna Ecosystems (J. Proctor, Ed.), Brit. Ecol. Soc. Spec. Publ. no. 9, 27–41.

*Buschbacher, R. J.* (1984). Changes in Productivity and Nutrient Cycling Following Conversion of Amazonian Rainforest to Pasture. PhD Thesis, Univ. Georgia, Athens, U.S., 193 pp.

*Calder, I. R.* (1991). Deuterium tracing for the estimation of transpiration from trees. Part 2. Estimation of transpiration rates and transpiration parameters using a time-averaged deuterium tracing method. J. Hydrol. 130, 27–35.

*Clayton, J. J.* (1979). Nutrient supply to soil by rock weathering. In: Impacts of Intensive Harvesting on Forest Nutrient Cycling (A. L. Leaf, Ed.), New York State Univ., Ithaca, New York, 75–96.

*Cooper, J. D.* (1979). Water use of a tea estate from soil moisture measurements. East Afr. Agric. For. J. 43, 102–121.

*Couper, D. C., Lal, R.* and *Claassen, S. L.* (1981). Land clearing and development for agricultural purposes. In: Tropical Agricultural Hydrology (R. Lal and E. W. Russell, Eds.), J. Wiley, New York, 119–130.

*Denich, M.* (1989). Untersuchungen zur Bedeutung junger Sekundärvegetation für die Nutzungssystemproduktivität im östlichen Amazonasgebiet, Brasilien. Gött. Beitr. zur Land- und Forstwirtschaft in den Tropen und Subtropen 46, 1–265.

*Eernisse, N.* (1993). Relative Depths of the Weathering Front and the Fine Root Network, a Pilot Study, Mabura Hill, Guyana. Tropenbos Programme, Guyana, 21 pp.

*Elsenbeer, H.* and *Lack, A.* (1996). Hydrological pathways and water chemistry in amazonian rain forests. In: Advances in Hillslope Processes (M. G. Anderson and S. M. Brooks, Eds.). J. Wiley, Chichester, 939–959.

*Elsenbeer, H., West, A.* and *Bonell, M.* (1994). Hydrologic pathways and stormflow chemistry at South Creek, northeast Queensland. J. Hydrol. 162, 1–21.

*Enright, N. J.* (1978). The effects of logging on the regeneration and nutrient budget of *Araucaria cunninghamii* dominated tropical rainforest in Papua New Guinea. Malay. For. 41, 303–318.

*Evans, J.* (1992). Plantation Forestry in the Tropics. Clarendon Press, Oxford, 403 pp.

*Ewel, J. J.* (1976). Litterfall and leaf decomposition in a tropical forest succession in Eastern Guatemala. J. Ecol. 64, 297–307.

*Ewel, J. J., Berish, C., Brown, B., Price, N.* and *Raich, J.* (1981). Slash and burn impacts on a Costa Rican wet forest site. Ecology 62, 816–829.

*Fölster, H.* and *Khanna, P. K.* (1997). Dynamics of nutrient supply in plantation soils. In: Bekker Management of Soil, Nutrients, and Water in Tropical Plantation Forests (E. K. S. Nambiar and A. G. Brown, Eds.), ACIAR, Canberra (in press).

*Fritsch, J. M.* (1992). Les Effets du Défrichement de la Forêt Amazonienne et de la Mise en Culture sur l'Hydrologie des Petits Bassins Versants. ORSTOM Editions, Paris, 392 pp.

*Fritsch, J. M.* (1993). The hydrological effects of clearing tropical rainforest and of the implementation of alternative land uses. In: Hydrology of Warm Humid Regions (J. S. Gladwell, Ed.), Int. Assoc. Hydrol. Sci. Publ. 216, 53–66.

*Fritsch, J. M., Dubreuil, P. L.* and *Sarrailh* (1987). De la parcelle au petit bassin-versant: effêt d'échelle dans l'écosystème forestier amazonien. Int. Assoc. Hydrol. Sci. Publ. 167, 131–142.

*Fritsch, J. M.* and *Sarrailh J. M.* (1986). Les transports solides dans l'écosystème forestier tropical humide en Guyane: les effets du défrichement et de l'installation de paturages. Cahiers ORSTOM, Série Pédologie 22, 93–106.

*Gholz, H. L., Fisher, R. F.* and *Pritchett, W. L.* (1985). Nutrient dynamics in slash pine plantation ecosystems. Ecology 68, 647–659.

*Gillman, G. P., Sinclair, D. F., Knowlton, R.* and *Keys, M. G.* (1985). The effect on some soil chemical properties of the selective logging of a north Queensland rainforest. For. Ecol. Manage. 12, 195–214.

*Gilmour, D. A.* (1977). Effect of rainforest logging and clearing on water yield and quality in a high rainfall zone of north-east Queensland. In: Proceedings of the Brisbane Hydrology Symposium 1977, Institution of Engineers Australia, Canberra, 156–160.

*Grimaldi, C.* (1988). Origine de la composition chimique des eaux superficielles en milieu tropicale humide: example de deux petits bassins versants sous forêt en Guyane Française. Sci. Géol. Bull. 41, 247–262.

*Grip, H.* (1994). Forest hydrochemistry, from budget calculation to process understanding. In: International Symposium on Forest Hydrology, Tokyo, October 1994, IUFRO, Vienna, 1–7.

*Grip, H., Malmer, A.* and *Wong, F. K.* (1994). Converting tropical rain forest to forest plantation in Sabah, Malaysia. I. Dynamics and net losses of nutrients in control catchment streams. Hydrol. Proc. 8, 179–194.

*Halenda, C. J.* (1993). Above-ground biomass production and nutrient accumulation of a *Gmelina arborea* plantation in Sarawak, Malaysia. J. Trop. For. Sci. 5, 429–439.

*Hamilton, L. S.* and *King, P. N.* (1983). Tropical Forested Watersheds. Hydrologic and Soils Response to Major Uses or Conversions. Westview Press, Boulder, 168 pp.

*Hase, H. and Fölster, H.* (1983). Impact of plantation forestry with teak *(Tectona grandis)* on the nutrient status of young alluvial soil in West Venezuela. For Ecol. Manage. 6, 33–57.

*Hatton, T. J.* and *Vertessy, R. A.* (1990). Transpiration of plantation *Pinus radiata* by the heat pulse method and the Bowen ratio. Hydrol. Proc. 4, 289–298.

*Herrera, R., Jordan, C. F., Klinge, H.* and *Medina, E.* (1978). Amazon ecosystems, their structure and functioning, with particular emphasis on nutrients. Interciencia 3, 223–232.

*Hölscher, D.* (1995). Wasser- und Stoffhaushalt eines Agrarökosystems mit Waldbrache im östlichen Amazonasgebiet. Göttinger Beiträge zur Land- und Forstwirtschaft in den Tropen und Subtropen no. 106, 134 pp.

*Hsia, Y. J.* and *Horng, F. W.* (1990). The route of nutrient outflow through a riparian buffer zone. In: Research Needs and Applications to Reduce Erosion and Sedimentation in Tropical Steeplands (R. R. Ziemer, C. L. O'Loughlin and L. S. Hamilton, Eds.), Int. Assoc. Hydrol. Sci. Publ. 192, 387.

*Hudson, J., Kellman, M., Sanmugadas, K.* and *Alvarado, C.* (1983). Prescribed burning of *Pinus oocarpa* in Honduras. II. Effects on nutrient cycling. For. Ecol. Manage. 5, 283–300.

*Jaffré, T.* (1985). Composition minérale et stocks de bioéléments dans la biomasse épigée de recrûs forestiers en Côte d'Ivoire. Acta Oecol. 6/20, 233–246.

*Jetten, V. G.* (1994). Modelling the Effects of Logging on the Water Balance of a Tropical Rain Forest. A Study in Guyana. Tropenbos Series no. 6, Tropenbos Foundation, Wageningen, The Netherlands, 196 pp.

*Johnston, M. H.* (1992). Soil-vegetation relationships in a tabonuco forest community in the Luquillo Mountains of Puerto Rico. J. Trop. Ecol. 8, 253–263.

*Jordan, C. F.* (1985). Nutrient Cycling in Tropical Forest Ecosystems. J. Wiley, Chichester, 190 pp.

*Jordan, C. F.* (1989). An Amazonian Rain Forest. Man and the Biosphere Series, Volume 2, Parthenon Publ., Carnforth U. K., 176 pp.

*Kang, B. T.* and *Lal, R.* (1981). Nutrient losses in water runoff from agricultural catchments. In: Tropical Agricultural Hydrology (R. Lal and E. W. Russell, Eds.), J. Wiley, New York, 153–161.

*Klinge, R.* (1997). Wasser- und Nährstoffdynamik im Boden und Bestand beim Aufbau einer Holzplantage im östlichen Amazonasgebiet. Göttinger Beiträge zur Land- und Forstwirtschaft in den Tropen und Subtropen (in press).

*Lal, R.* (1987). Tropical Ecology and Physical Edaphology. J. Wiley, Chichester, 732 pp.

*Lal, R.* (Ed.) (1988). Soil Erosion Research Methods. Soil and Water Conserv. Soc., Ankeny, Iowa, USA.

*Lee, R.* (1970). Theoretical estimates versus forest water yield. Water Resour. Res. 6, 1327–1334.

*Leitch, C. J., Flinn, D. W.* and *Van de Graaff, R. H. H.* (1983). Erosion and nutrient loss resulting from Ash Wednesday (February 1983) wildfires, a case study. Austr. For. 46, 173–180.

*Lesack, L. F. W.* (1993a). Export of nutrients and major ionic solutes from a rain forest catchment in the Central Amazon basin. Water Resour. Res. 29, 743–758.

*Lesack, L. F. W.* (1993b). Water balance and hydrologic characteristics of a rain forest catchment in the Central Amazon basin. Water Resour. Res. 29, 759–773.

*Lieberman, D., Lieberman, M., Peralta, R.* and *Hartshorn, G. S.* (1985). Mortality patterns and stand turnover rates in a wet tropical forest in Costa Rica. J. Ecol. 73, 915–924.

*Likens, G. E., Bormann, F. H., Pierce, R. S., Eaton, J. S.* and *Johnson, N. M.* (1977). Biogeochemistry of a Forested Ecosystem. Springer, New York, 146 pp.

*Lima, W. de P.* (1988). Escoamenta superficial, perdas de solo e de nutrientes em microparcelas reflorestadas com eucalipto em solos arenosos no municipio de São Simão, SP. IPEF, Piracicaba 38, 5–16.

*Lloyd, C. R., Gash, J. H. C., Shuttleworth, W. J.* and *Marques, A. de O.* (1988). The measurement and modelling of rainfall interception by Amazonian rainforest. Agric. For. Meteo. 43, 277–294.

*Lloyd, C. R.* and *Marques, A. de O.* (1988). Spatial variability of throughfall and stemflow measurements in Amazonian rain forest. Agric. For. Meteo. 42, 63–73.

*Lodge, D. J., Scatena, F. N., Asbury, C. E.* and *Sanchez, M. J.* (1991). Fine litterfall and related nutrient inputs resulting from hurricane Hugo in subtropical wet and lower montane rain forests of Puerto Rico. Biotropica 23, 336–342.

*Ludwig, R.* (1992). Cable crane yarding: an economical and ecologically suitable system for commercial timber harvesting in logged-over rain forests of the Philippines. In: Proceedings of the International Symposium on Harvesting and Silviculture for Sustainable Forestry in the Tropics, Kuala Lumpur, October 1992, 12 pp.

*Mackensen, J., Hölscher, D., Klinge, R.* and *Fölster, H.* (1996). Nutrient transfer to the atmosphere by burning of debris in eastern Amazonia. For. Ecol. Manage. 86, 121–128.

*Malmer, A.* (1992). Water-yield changes after clear-felling tropical rain forest and establishment of forest plantation in Sabah, Malaysia. J. Hydrol. 134, 77–94.

*Malmer, A.* (1993). Dynamics of Hydrology and Nutrient Losses as Response to Establishment of Forest Plantation. A Case Study on Tropical Rain forest Land in Sabah, Malaysia. PhD Thesis, Swedish Univ. Agric. Sci., Umeå, Sweden, 181 pp.

*Malmer, A.* (1996). Phosphorus loading to tropical rain forest streams after clear-felling and burning in Sabah, Malaysia. Water Resour. Res. 32, 2213–2220.

*Malmer, A.* and *Grip, H.* (1990). Soil disturbance and loss of infiltrability caused by mechanized and manual extraction of tropical rain forest in Sabah, Malaysia. For. Ecol. Manage. 38, 1–12.

*Malmer, A.* and *Grip, H.* (1994). Converting tropical rain forest to forest plantation in Sabah, Malaysia. II. Effects on nutrient dynamics and net losses in streamwater. Hydrol. Proc. 8, 195–209.

*Martin, G.* (1970). Le défrichement mécanique pour la création de palmeraies industrielles. Oléagineux 25, 575–580.

*McDonnell, J. J., Freer, J., Hooper, R. P., Burns, D., Beven, K.* and *Peters, J.* (1996). A new method developed for studying flow on hillslopes. EOS. 77, 465–472.

*Nepstad, D. C., Carvalho, C. R. de, Davidson, E. A., Jipp, P. H., Lefebvre, P. A., Negreiros, Silva, E. D., Stone, T. A., Trumbore, S. E.* and *Vieira, S.* (1994). The role of deep roots in the hydrological and carbon cycles of Amazonian forests and pastures. Nature 372, 666–669.

*Nortcliff, S., Ross, S. M.* and *Thornes, J. B.* (1990). Soil moisture, runoff and sediment yield from differentially cleared tropical rain forest plots. In: Vegetation and Erosion (J. B. Thornes, Ed.), J. Wiley, Chichester, 419–436.

*Nortcliff, S.* and *Thornes, J. B.* (1978). Water and cation movement in a tropical rain forest environment. 1. Objectives, experimental design and preliminary results. Acta Amaz. 8, 245–258.

*Nykvist, N.* (1992). The amounts of plant nutrients in the stems of tropical rain forests. In: Harvesting and Silviculture for Sustainable Forestry in the Tropics, IUFRO, Vienna, 20 pp.

*Nykvist, N., Grip, H., Sim, B. L., Malmer, A.* and *Wong, F. K.* (1994). Nutrient losses in forest plantations in Sabah, Malaysia. Ambio, 23, 210–215.

*O'Loughlin, E. M., Cheney, N. P.* and *Burns, J.* (1982). The Bushrangers experiment: hydrological response of a eucalypt experiment. In: The First National Symposium on Forest Hydrology (E. M. O'Loughlin and L. J. Bren, Eds.), Nat. Conf. Publ. no. 82/6, Instit. Eng. Austr., 132–138.

*Parker, G. G.* (1985). The Effect of Disturbance on Water and Solute Budgets of Hillslope Tropical Rain forest in Northeastern Costa Rica. PhD Thesis, Univ. Georgia, Athens, 161 pp.

*Poels, R. L. H.* (1987). Soils, water and Nutrients in a Forest Ecosystem in Suriname. PhD Thesis, Agric. Univ., Wageningen, 253 pp.

*Proctor, J.* (1987). Nutrient cycling in primary and old secondary rain forests. Appl. Geogr. 7, 135–152.

*Proctor, J., Anderson, J. M., Chai, P.* and *Vallack, H. W.* (1983). Ecological studies in four contrasting lowland rain forests in Gunung Mulu National Park, Sarawak. I. Forest environment, structure and floristics. J. Ecol. 71, 237–260.

*Quinn, P. P. F., Beven, K. J., Chevallier, P.* and *Planchon, O.* (1991). The prediction of hillslope flow paths for distributed hydrological modelling using digital terrain models. Hydrol. Proc. 5, 59–79.

*Raison, R. J., Khanna, P. K.* and *Woods, P. V.* (1985). Mechanisms of element transfer to the atmosphere during vegetation fires. Can. J. For. Res. 15, 132–140.

*Ramakrishnan, P. S.* (1992). Shifting Agriculture and Sustainable Development. Man and the Biosphere Series, Vol. 10, Parthenon Publ., Carnforth U. K., 424 pp.

*Richter, D. D.* and *Babbar, L. I.* (1991). Soil diversity in the tropics. Adv. Ecol. Res. 21, 315–389.

*Roberts, J. M., Cabral, O. M. R., Fisch, G., Molion, L. C. B., Moore, C. J.* and *Shuttleworth, W. J.* (1993). Transpiration from an Amazonian rain forest calculated from stomatal conductance measurements. Agric. For. Meteo. 65, 175–196.

*Roberts, J., Cabral, O. M. R., da Costa, J. P., McWilliam, A. L. C.* and *Sá, T. D. de A.* (1996). An overview of the leaf area index and physiological measurements during ABRACOS. In: Amazonian Deforestation and Climate (J. H. C. Gash, C. A. Nobre, J. Roberts and R. Victoria, Eds.), J. Wiley, New York, 287–306.

*Robertson, G. P.* (1984). Nitrification and nitrogen mineralization in a lowland rain forest succession in Costa Rica, Central America. Oecol. 61, 99–104.

*Roose, E. J.* (1981). Dynamique actuelle des sols ferrallitiques et ferrugineaux tropicaux d'Afrique Occidentale. Trav. Doc. ORSTOM no. 130, ORSTOM, Paris, 571 pp.

*Rose, C. W.* (1993). Erosion and sedimentation. In: Hydrology and Water Management in the Humid Tropics. Hydrological Research Issues and Strategies for Water Management (M. Bonell, M. M. Hufschmidt and J. S. Gladwell, Eds.), Cambridge University Press, Cambridge, 301–343.

*Ross, S. M.* and *Dykes, A.* (1996). Soil conditions, erosion and nutrient loss on steep slopes under mixed dipterocarp forest in Brunei Darussalam. In: Tropical Rain Forest Research – Current Issues, D. S. Edwards et al., Eds.), Kluwer Academic Publ., The Netherlands, 259–270.

*Ross, S. M., Thornes, J. B.* and *Nortcliff, S.* (1990). Soil hydrology, nutrient and erosional response to the clearance of terra firme forest, Maraca Island, Roraima, northern Brazil. Geogr. J. 156, 267–282.

*Ruhiyat, D.* (1989). Die Entwicklung der standörtlichen Nährstoffvorräte bei naturnäher Waldbewirtschaftung und im Plantagenbetrieb, Ostkalimantan. Gött. Beitr. zur Land- und Forstwirtschaft in den Tropen und Subtropen 35, 1–206.

*Russell, A. E.* and *Ewel, J. J.* (1985). Leaching form a tropical andept during big storms, a comparison of three methods. Soil Sci. 139, 181–189.

*Russell, C. E.* 1983). Nutrient Cycling and Productivity in Native and Plantation Forests at Jari Florestal, Para, Brazil. PhD Thesis, Univ. Georgia, Athens, USA. 130 pp.

*Sanchez, P. A.* (1976). Properties and Management of Soils in the Tropics. J. Wiley, New York, 618 pp.

*Scatena, F. N.* and *Larsen, M. C.* (1991). Physical aspects of Hurricane Hugo in Puerto Rico. Biotropica 23, 317–323.

*Scatena, F. N., Silver, W. L., Siccama, T., Johnson, A.* and *Sanchez, M. J.* (1993). Biomass and nutrient content of the Bisley experimental watersheds, Luquillo Experimental Forest, Puerto Rico, before and after Hurricane Hugo, 1989. Biotropica 25, 15–27.

*Scott, D. F.* (1993). The hydrological effects of fire in South African mountain catchments. J. Hydrol. 150, 409–432.

*Sim, B. L.* and *Nykvist, N.* (1991). Impact of forest harvesting and replanting. J. Trop. For. Sci. 3, 251–284.

*Sollins, P.* and *Radulovich, R.* (1988). Effects of soil structure on solute transport in a weathered tropical soil. Soil Sci. Soc. Amer. J. 52, 1168–1173.

*Spangenberg, A., Grimm, U., da Silva, J. R. S.* and *Fölster, H.* (1996). Nutrient store and export rates of *Eucalyptus urograndis* plantations in eastern Amazonia (Jari). For. Ecol. Manage. 80, 225–234.

*Steudler, P. A., Melillo, J. M., Bowden, R. D.* and *Castro, M. S.* (1991). The effects of natural and human disturbances on soil nitrogen dynamics and trace gas fluxes in a Puerto Rican wet forest. Biotropica 23, 356–363.

*Stocking, M. A.* (1984). Erosion and Soil Productivity: a Review. FAO Consultant's Working Paper no. 1, FAO, Rome, 102 pp.

*Stoorvogel, J. J.* (1993). Gross Inputs and Outputs of Nutrients in Undisturbed Forest, Taï, Côte d'Ivoire. Tropenbos Series no. 5, Tropenbos Foundation, Wageningen, 148 pp.

*Subba Rao, B. K., Ramola, B. C.* and Sharda, V. N. (1985). Hydrologic response of a forested mountain watershed to thinning. Indian For. 111, 681–690.

*Swank, W. T.* and *Crossley, D. A.* (Eds.) (1988). Forest Hydrology and Ecology at Coweeta. Springer Ecological Studies no. 66, 466 pp.

*Swindel, B. F., Lassiter, C. J.* and *Riekerk, H.* (1982). Effects of clearcutting and site preparation on water yield from slash pine forests. For. Ecol. Manage. 4, 101–113.

*Swindel, B. F., Lassiter, C. J.* and *Riekerk, H.* (1983). Effects of clearcutting and site preparation on stormflow volumes of streams in *Pinus elliottii* flatwood forests. For. Ecol. Manage. 5, 245–253.

*Tandy, J. C.* (1987). Contribution á l'Etude des Couvertures Pédologiques sur Socle en Guyane Française: Rélations entre la Nature Minéralogique, l'Organisation des Particules et le Comportement Hydrologique. DEA Pédologie, INRA, Versailles, 50 pp.

*Toky, O. P.* and *Ramakrishnan, P. S.* (1981). Runoff and infiltration losses related to shifting agriculture (Jhum) in northeastern India. Env. Conserv. 8, 313–321.

*Uhl, C.* (1982). Tree dynamics in a species rich Tierra Firme forest in Amazonia, Venezuela. Acta Cient. Venez. 33, 72–77.

*Uhl, C., Buschbacher, R.* and *Serrao, E. A. S.* (1988a). Abandoned pastures in eastern Amazonia. I. Patterns of plant succession. J. Ecol. 76, 663–681.

*Uhl, C., Clark, K., Dezeo, O. N.* and *Maquirino, P.* (1988b). Vegetation dynamics in Amazonan treefall gaps. Ecology 69, 751–763.

*Uhr, C.* and *Jordan, C. F.* (1984). Succession and nutrient dynamics following forest cutting and burning in Amazonia. Ecology 65, 1476–1490.

*Uhl, C., Jordan, C. F., Clark, K., Clark, H.* and *Herrera, R.* (1982). Ecosystem recovery in Amazon caatinga forest after cutting, cutting and burning and bulldozer clearing treatments. Oikos 38, 313–320.

*Van Lear, D. H., Douglass, J. E., Cox, S. K.* and *Augspurger, M. K.* (1985). Sediment and nutrient export in runoff from burned and harvested pine watersheds in the South Carolina Piedmont. J. Env. Qual. 14, 169–174.

*Verstraten, J. M.* (1980). Water-rock Interactions. A Case Study in a Very Low-grade Metamorphic Shale Catchment in the Ardennes, NW Luxembourg. Brit. Geomorph. Res. Group Res. Monogr. no. 2, Geobooks, Norwich, U.K., 243 pp.

*Vitousek, P. M.* and *Denslow, J. S.* (1986). Nitrogen and phosphorus availability in treefall gaps of a lowland tropical rain forest. J. Ecol. 74, 1167–1178.

*Vitousek, P. M.* and *Matson, P. A.* (1985). Disturbance, nitrogen availability, and nitrogen losses in an intensively managed loblolly pine plantation. Ecology 66, 1360–1376.

*Vitousek, P. M.* and *Sanford, R. L.* (1986). Nutrient cycling in moist tropical forest. Ann. Rev. Ecol. Syst. 17, 137–167.

*Walker, L. R.* (1991). Tree damage and recovery from hurricane Hugo in Luquillo Experimental Forest, Puerto Rico. Biotropica 23, 379–385.

*Walling, D. E.* (1983). The sediment delivery problem. J. Hydrol. 65, 209–237.

*Ward, R. C.* (1984). On the response to precipitation of headwater streams in humid areas. J. Hydrol. 74, 171–189.

*Waterloo, M. J.* (1994). Water and Nutrient Dynamics of *Pinus Caribaea* Plantation Forests on Former Grassland Soils in Southwest Viti Levu, Fiji. PhD Thesis, Vrije Universiteit, Amsterdam, 478 pp.

*Whitmore, T. C.* (1990). Introduction to Tropical Rain Forests. Clarendon Press, Oxford, UK, 226 pp.

*Wiersum, K. F.* (1984). Surface erosion under various tropical agroforestry systems. In: Effects of Forest Land Use on Erosion and Slope Stability (C. L. O'Loughlin and A. J. Pearce, Eds.), IUFRO, Vienna, 231–239.

*Young, A.* (1989). Agroforestry for Soil Conservation. CAB International, Wallingford, U.K., 276 pp.

*Zulkifli, Y.* (1989). Effects of selective logging methods on dissolved nutrients exports in Berembun watershed, Peninsular Malaysia. Paper presented at the FRIM-IHP-UNESCO Regional Seminar on Tropical Forest Hydrology, Kuala Lumpur, September 1989, 26 pp.

*Zulkifli, Y.* and *Abdul Rahim, N.* (1994). Nutrient exports from secondary forest catchments of Bukit Tarek experimental forest. Paper presented at the Soil Science Conference of Malaysia, Kuala Lumpur, April 1994, 20 pp.

# Hydrology of Acrisols beneath Dipterocarp forests and plantations in East Kalimantan, Indonesia

**Walter W. Wenzel*, Hansjörg Unterfrauner*, Andreas Schulte**, Daddy Ruhiyat***, Dicky Simorangkir***, Václav Kuráz****, A. Brandstetter*** and **Winfried E. H. Blum***

*Institute of Soil Science, University of Agriculture, Gregor-Mendel-Straße 33, A-1180 Wien, Austria
**University of Paderborn, Forest Ecology, An der Wilhelmshöhe 44, D-37671 Höxter, Germany
***Faculty of Forestry, Mulawarman University, Samarinda 75123, Kaltim, Indonesia
****Department of Irrigation and Drainage, Czech Technical University in Prague

## Abstract

The hydrology of Haplic Acrisols on sandstone beneath old-growth and slightly logged Dipterocarp forests, and beneath Paraserianthes and Eucalyptus plantations in the lowlands of East Kalimantan was investigated to determine possible ecological constraints for forest and plantation management. The soils are characterized by 40–60 cm thick E horizons, texture silt clay, above clayey Bt horizons. Soil moisture ($\theta$), measured during a period of few weeks at the end of the rain season, was typically close to field capacity. Field-saturated hydraulic conductivities ($K_{GP}$) indicate that the E horizons are moderately permeable (medians of $K_{GP}$ 2.5–42.4 cm $d^{-1}$), while the Bt horizons are virtually impermeable (medians of $K_{GP} < 0.17$ cm $d^{-1}$). Macropores generated by biota are likely the cause of the observed large within-plot variability of $K_{GP}$ in the E horizons and may provide pathways for preferential water flow. Soil water budgets constructed from measurements of precipitation (245 cm $y^{-1}$), throughfall (210–230 cm $y^{-1}$), using soil hydrological constraints obtained from $K_{GP}$ values measured in the Bt horizons, and assumed transpiration rates (90–110 cm $y^{-1}$) suggest that 40–140 cm $y^{-1}$ of the soil water is drained laterally in the E horizons. Accordingly, $<$ 5–40 cm $y^{-1}$ are available for deep seepage below 40–60 cm depth. Since lateral water flow is expected to increase with distance from the top of slopes, lower slopes receive large amounts of lateral water, depending on the slope length. It is concluded that the thickness, pore size distribution and permeability of the E horizons along with slope aspect are master variables for soil hydrology in the study area and determine site quality for forestry. Degradation of the E horizons due to compaction and erosion will increase surface flow, the frequency of flooding events and the impact of droughts on Dipterocarp forests and plantations of fast growing tree species in the lowlands of East Kalimantan. The conservation of topsoil physical and hydrological characteristics is therefore a prerequisite for sustainable forestry and landuse in this region.

## 1 Introduction

During the last few decades Dipterocarp forest ecosystems have being exploited at unprecedented rates and concern about the ecological and economical impacts are growing. In many regions of Southeast Asia, man are today confronted with the situation that the natural primary forests have either already disappeared or will do so in the very near future. Left over are residual forests, which are considerably different to the primary forests in respect of species composition, structure, dynamics and stability (*Schulte*, 1996).

During the last two decades, many logged-over forests have been converted to forest plantations of fast growing tree species, primarily planted with monocultures of Paraserianthes sp., Eucalyptus sp. and Acacia sp. Many of these plantations will soon be ready to harvest for the emerging pulp industry. The Indonesian government has established guidelines known as the Indonesian Selective Cutting System that require the remaining indigenous Dipterocarp forests be utilized in more environmentally sound methods (*Ruslim*, 1994, *Schulte*, 1996).

Facing these evolving landuse policies, essential questions remain. Little is known about the ecology of indigenous Dipterocarp forests and even less on the effect of forest management and landuse practices on the ecology of the whole landscape. In turn, ecological constraints for forest and landuse have hardly been investigated yet. Water and nutrient cycles as well as physical and chemical degradation of the forest soils may constrain sustainable utilization of Dipterocarp forests and the establishment of forest plantations (*Schulte*, 1996).

Therefore, a comprehensive research program on the ecology, management, and harvest of Dipterocarp forest ecosystems, and plantations of Paraserianthes and Eucalyptus was designed under the Indonesian-German Forestry Project at Mulawarman University in Samarinda, East Kalimantan. Four experimental plots in the vicinity of Kiani Lestari (Figure 1) have been installed under this program. Studies at these sites include: sustainability of various forest and plantation management systems, nutrient fluxes, and root ecology (*Murach*, 1995). There are few reports on soil physical characteristics in the region of East Kalimantan (*Ohta* et al., 1992), but none for the experimental sites.

This contribution reports on soil hydrology beneath Dipterocarp oldgrowths and Eucalyptus deglupta and Paraserianthes falcataria plantations grown on similar sites after clearcutting of Dipterocarp stands. Soil water budgets constructed from input measurements (precipitation, throughfall) and soil hydrological characteristics are presented and discussed in scope of landscape ecology and forest management.

## 2 Materials and methods

### 2.1 Study area

Investigations were carried out at four experimental plots in the forest concession Kiani Lestari, in the Batu Ampar region, East Kalimantan. The climate is warm perhumid tropics, with average annual temperatures of 26°C and average annual rainfall of 2450 mm. The geomorphology is characterized by low mountains and rugged hills with moderately steep slopes (*Ohta* et al., 1992).

### 2.2 Sites and soils

Some characteristics of the experimental plots are summarized in Table 1. The soils, formed on weakly lithified silt and sandstone of Pliocene/Miocene origin, are strongly acid, Haplic Acrisols with diagnostic argic B horizons. Stagnic features, e.g. mottles in the Eg horizon of the soil in plot D are not sufficiently strong to meet the criteria of stagnic properties according to FAO (1988).

### 2.3 Field studies and analytical procedures

Precipitation (P) was measured near plots A and D in three replicates between December 1994 and November 1995. During the same period, throughfall (TF) was collected at four plots described in Table 1. TF measurements were obtained in 20 replicates in the Dipterocarp plots A and B, and in 15 replicates in the less heterogenous and smaller plantation plots D and E.

Soil sampling and field measurements of soil hydrological characteristics were obtained in March and April 1995. A soil pit was described at each plot according to FAO (1988). Composite samples were collected from each horizon, air-dried, and passed through a 2-mm sieve. Organic carbon (OC) was measured on dried (105°C, 24 h) and ground samples by dry combustion (Carlo Erba CNS Analyzer NA 1500 Series 2); pH ($pH_{cacl_2}$) was measured in a soil – 0.01 M $CaCl_2$-suspension (1:2.5 g $mL^{-1}$) after 2 h using an ORION glass electrode (*Blum* et al., 1996).

Three soil texture fractions (sand, 2000–63 µm; silt, 63–2 µm; clay, < 0.2 µm) were determined using a combined sieve and pipette method (*Blum* et al., 1996). Bulk density was measured on 200 $cm^3$ soil cores.

Soil cores from E, Bt and Cw/CwBt horizons were used to measure the soil-water retention curves (SWRC) and the saturated hydraulic conductivity ($k_s$). The matric potentials (log h) were fitted to experimental data using van Genuchten's model (*van Genuchten*, 1980)

$$S_e = (1/[1 + (-\alpha h)^n]^m \quad (m = 1-1/n) \qquad \text{(Equation 1)}$$

with

$$S_e = (\theta - \theta_r)/(\theta_s - \theta_r) \qquad \text{(Equation 2)}$$

where $S_e$ is the normalized moisture content, $\theta$ the volumetric moisture content ($m^3\ m^{-3}$), with the subscripts r and s denoting residual and saturated, and $\alpha$ ($cm^{-1}$),

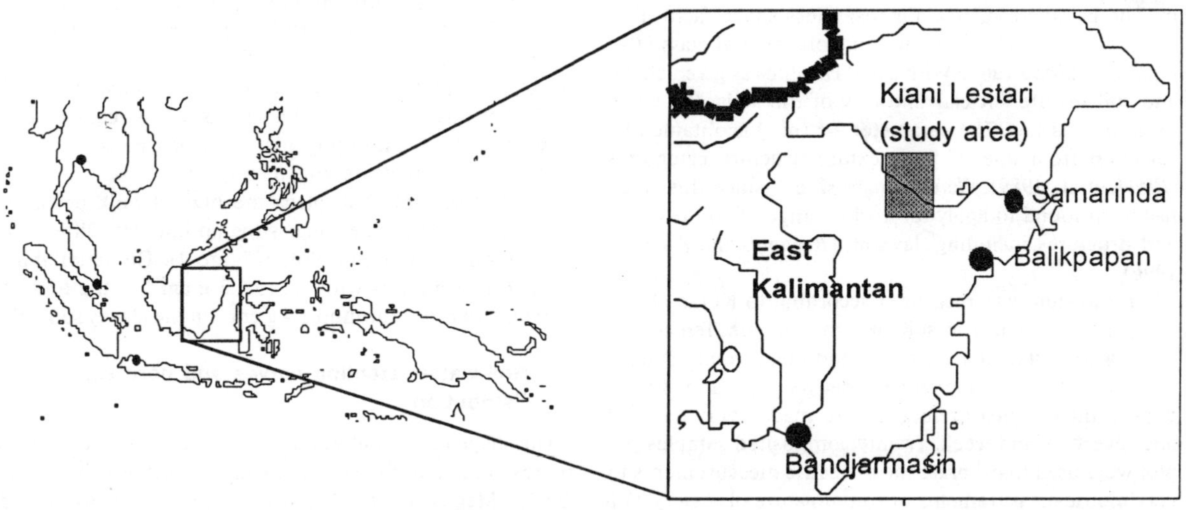


**Figure 1:** Location of the experimental sites in East Kalimantan, Indonesia.

**Table 1:** Characteristics of the experimental plots.

| | Plot A | Plot B | Plot D | Plot E |
|---|---|---|---|---|
| Plot size | 1 ha | 1ha | 0.25 ha | 0.25 h |
| Latitude | 0°30'N-1°15' | 0°30'N-1°15'N | 0°30'41"N-<br>0°52'18"N | 0°30'41"N-<br>0°52°18"N |
| Longitude | 116°40'E-<br>117°20'E | 116°40'E-<br>117°20'E | 116°49'43"E-<br>117°05'08"E | 116°49'43"E-<br>117°05'08"E |
| Altitude [m] | 250 | 250 | 250 | 250 |
| Position of site | crest | upper slope | middle slope | middle slope |
| Exposure | SE | WNW | ENE | E |
| Slope | 5–10% | 15–20% | 15–20% | 20% |
| Vegetation | Dipterocarp<br>forest | Dipterocarp<br>forest | Plantation<br>Eucalyptus<br>deglupta | Plantation<br>Paraserianthes falcataria |
| Age (y) | old growth | old growth | 9 | 9 |
| Utilization type | indigenous | selective cutting | clearcut | clearcut |

n and m are empirical parameters that affect the shape of the curve. The parameters estimated with the retention curve code RETC (*van Genuchten* et al., 1991; *Yates* et al., 1992) are presented in Table 2. Unsaturated hydraulic conductivities ($K_u$) were modeled simultanously using measured saturated hydraulic conductivities obtained on soil cores in the laboratory.

The Guelph permeameter method (*Reynolds* and *Elrick*, 1985) was used to measure field-saturated hydraulic conductivity ($K_{GP}$) in 9–15 replicates at 30, 50, and 80 cm depth. The single-head Richards analysis (*Elrick* et al., 1989) was employed to calculate $K_{GP}$ [m s$^{-1}$] from the rate of fall of the liquid level in the permeameter, $Q_{GP}$ [m$^3$ s$^{-1}$], applying the relationship

$$K_{GP} = C\ Q_{GP}/[2\pi H^2 + C\pi\ a^2 + (2\pi\ H/\alpha_E^*)] \qquad \text{(Equation 3)}$$

where a denotes the radius of the well (0.0375 m), and H the constant depth of liquid in the well maintained at 0.18 m. The parameter C is a dimensionless shape factor that depends on the H/a ratio and on soil textural class (*Reynolds* and *Elrick*, 1985). With a and H values as given above and soil texture silt clay and clay of the soils, the value used for C is 1.54. The parameter $\alpha_E^*$ (m$^{-1}$) is obtained by selection from one of four texture-structure categories (*Elrick* et al., 1989). We used $\alpha_E^*$=12 m$^{-1}$ since this value has been found to apply for a wide range of soil textures and structures, including clay soils (*Reynolds* and *Zebchuk*, 1996).

Soil moisture was measured according to Kuráz (1982) using a field dielectric soil moisture meter. Ten access tubes were installed at each plot and measurements were obtained in 10-cm increments between 10 and 80 cm depth; initially measurements were made every day and later every second week. Twenty composited samples per plot were used to calibrate the dielectric measurements to gravimetric measurements of soil moisture ($\theta_{G1}$ gg$^{-1}$). Volumetric water contents ($\theta$, m$^3$ m$^{-3}$) were calculated using measurements of bulk density (BD) and the known specific mass of water ($\gamma_w$, g cm$^{-3}$):

$$\theta = BD\ \theta_G/\gamma_w \qquad \text{(Equation 4)}$$

A linear calibration curve obtained in laboratory experiments (*Kuráz* et al., 1995) was used to calculate changes in volumetric soil water content ($\Delta\theta$, m$^3$ m$^{-3}$) from differences of the readings ($\Delta$d) obtained with the soil moisture meter:

$$\Delta\theta = \Delta d/270 \qquad \text{(Equation 5)}$$

## 3 Results and Discussion

### 3.1 Soil texture and bulk density

The soils, low in sand fractions, contain considerable amounts of silt and clay (Table 3). The increase in clay content from the lower E to the upper Bt horizon by about 45 to 120 g kg$^{-1}$ suggests clay accumulation in the Bt horizons due to in-situ formation or translocation. In the soils of the plots A and B beneath Dipterocarp forests, the clay content continually increases from the Ah to the Bt horizon; in plots E and D beneath plantations, the distribution of clay in the topsoil horizons is less regular, possibly indicating anthropogenic disturbance as a consequence of clearcutting and plantation.

The soil of plot D displays the highest bulk densities throughout the profile, suggesting an inherent difference in relation to the other plots (Figure 2). The latter show similar distributions with 1.2 to 1.3 g cm$^{-3}$ in the topsoil, and bulk densities around 1.4 g cm$^{-3}$ in the clayey subsoils.

### 3.2 Soil-water retention curves and pore size distribution

The shape of the soil water retention curves (Figure 3) is representative of finetextured soils (Schachtschabel et al., 1992). Means of total porosity, calculated from the SWRC's, were 0.48 m$^3$ m$^{-3}$ in the E and Bt horizons and

0.46 $m^3$ $m^{-3}$ in the BtCw and Cw horizons (Table 4). Consistent with the high clay contents, fine pores with Ø < 0.2 μm were most abundant, increasing with depth from 0.28 to 0.34 $m^3$ $m^{-3}$. In turn, the amount of mesopores decreased with depth from 0.13 to 0.09, that of macropores from 0.04 to 0.02 $m^3$ $m^{-3}$. Large macropores decreased on average from 0.03 in E horizons to 0.02 $m^3$ $m^{-3}$ in the subsoils (Table 4).

Although E and Bt horizons display the same average total porosity, the E horizons contain more meso- and macropores; this indicates that the E horizons are more permeable, better aerated and display a higher capacity for plant-available water than the subsoils. In spite of average precipitation rates as high as 245 cm $y^{-1}$ the capacity for plant-available water may become limiting to site quality during dry periods. This is consistent with reports on severe droughts with an annual precipitation as low as 70 cm in the region of Balikpapan (*Goldhammer* et al., 1996; *Richards*, 1996).

Differences in pore-size distribution are related to higher clay and lower OC contents in the Bt horizons. The higher amount of pores > 10μm in the topsoils seems to

**Table 2:** Values for van Genuchten parameters as obtained from Equ. 1 using the RETC code of van Genuchten et al. (1991). N denotes the number of observations per horizon using data from all four plots.

| Horizon | N | | Depth | $\theta_r$ [a] | $\theta_s$ [b] | $\alpha$ [c] | n [c] | $R^2$ [d] |
|---|---|---|---|---|---|---|---|---|
| | | | cm | $m^3$ $m^{-3}$ | | $cm^{-1}$ | | |
| E | 15 | x̄ | 24 | 0.1028 | 0.4752 | 0.0145 | 1.2191 | 0.9977 |
| | | SD | 13 | 0.1095 | 0.0184 | 0.0234 | 0.2019 | 0.0067 |
| | | Min | 10 | 0.0000 | 0.4518 | 0.0025 | 1.0675 | 0.9747 |
| | | Max | 45 | 0.2699 | 0.5079 | 0.0936 | 1.6749 | 1.0000 |
| Bt | 10 | x̄ | 74 | 0.2528 | 0.4774 | 0.0076 | 1.3083 | 0.9952 |
| | | SD | 18 | 0.1224 | 0.0297 | 0.0119 | 0.5446 | 0.0112 |
| | | Min | 55 | 0.0181 | 0.4467 | 0.0012 | 0.0072 | 0.9639 |
| | | Max | 100 | 0.3470 | 0.5428 | 0.0402 | 1.8447 | 1.0000 |
| Bt/Cw/Cw | 7 | x̄ | 112 | 0.2074 | 0.4603 | 0.0075 | 1.2367 | 0.9969 |
| | | SD | 28 | 0.1080 | 0.0271 | 0.0055 | 0.2631 | 0.0073 |
| | | Min | 80 | 0.0592 | 0.4303 | 0.0016 | 1.0550 | 0.9804 |
| | | Max | 130 | 0.3521 | 0.5022 | 0.0173 | 1.8067 | 1.0000 |

[a]residual moisture content
[b]saturated moisture content
[c]van Genuchten parameters affecting the shape of the retention curce
[d]goodness of fit of the van Genuchten model to the experimental data

**Table 3:** Soil chemical and textural characteristics.

| Plot | Horizon | Depth | $pH_{CaCl_2}$ | OC | Clay | Silt | Sand |
|---|---|---|---|---|---|---|---|
| | | cm | | g $kg^{-1}$ | | | |
| A | Ah | 0–1 | 3.40 | 60.2 | 346 | 613 | 41 |
| | E1 | 1–20 | 3.60 | 12.2 | 351 | 627 | 23 |
| | E2 | 20–40 | 3.66 | 9.0 | 397 | 580 | 23 |
| | Bt | 40–70 | 3.75 | 6.1 | 441 | 535 | 24 |
| | BtCw | 70–95 | 3.75 | 4.5 | 475 | 475 | 50 |
| | Cw | 95–125+ | ND | 3.9 | 349 | 589 | 62 |
| B | Ah | 0–4 | 3.54 | 19.4 | 268 | 570 | 162 |
| | E1 | 4–30 | 3.67 | 8.8 | 326 | 529 | 146 |
| | E2 | 30–60 | 3.67 | 5.5 | 357 | 504 | 139 |
| | Bt1 | 60–90 | 3.73 | 4.1 | 463 | 428 | 109 |
| | Bt2 | 90–150+ | 3.78 | 4.1 | 489 | 409 | 101 |
| D | Ah | 0–1 | 3.76 | 25.5 | 309 | 652 | 39 |
| | E | 1–18 | 3.70 | 9.8 | 371 | 601 | 28 |
| | Eg | 18–40 | 3.75 | 7.3 | 334 | 637 | 30 |
| | Bt1 | 40–70 | 3.68 | 6.5 | 380 | 567 | 54 |
| | Bt2 | 70–110 | 3.73 | 4.4 | 448 | 498 | 54 |
| | Cw | 110–150+ | 3.73 | 3.9 | 372 | 541 | 88 |
| E | Ah | 0–5 | 3.90 | 46.7 | 336 | 495 | 169 |
| | E1 | 5–20 | 3.72 | 12.8 | 325 | 520 | 154 |
| | E2 | 20–40 | 3.70 | 7.5 | 309 | 549 | 142 |
| | Bt | 40–100 | 3.75 | 3.8 | 430 | 465 | 106 |
| | BtCw | 100–150+ | 3.71 | 3.4 | 424 | 434 | 142 |

be associated with the observed maximum of biological activity. The contribution of macroscopic biopores is probably underestimated since the size of soil cores (200 $cm^3$) used to estimate pore volumes is below the scale of pore systems created by larger biota. Holes and caves with diameters up to 50 cm were observed during field work, e.g. in the E horizon of plot B.

The variability of the pore volumes expressed by SD and ranges (Table 4) is typically higher in the E horizons. This is again indicative for the abundance of biopores. This secondary pore system is confounded with a primary system associated with soil texture. At depth, the development of biopores seems to be constrained by an increase in clay content (Table 3) and bulk density (Figure 2). Lower amounts of mesopores in the subsoils are related to a decrease in silt fractions with depth.

### 3.3 Field measurements of soil moisture

During the monitoring period, the volumetric soil moisture content ($\theta$) varied less than 0.1 $m^3\ m^{-3}$ (Figure 4). Depending on the experimental plot and soil depth, the maximal moisture contents ($\theta_{max}$) ranged between 0.42 and 0.55 $m^3\ m^{-3}$. Values for $\theta_{max}$ and total porosity estimated with van Genuchten's model are in general consistent, however, in the topsoil of plot A the total porosity was clearly exceeded by $\theta_{max}$ (Figure 4). The apparent oversaturation in the E horizon of plot A may be associated with water-filled macroscopic biopores during and imme-

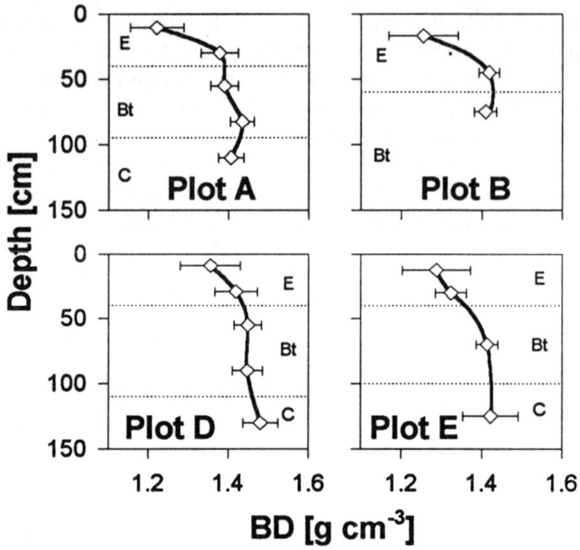


**Figure 2:** Distribution with depth of bulk density (BD) in the soils of the experimental plots beneath Dipterocarp forests (plots A, B) and plantations (plots D, E). Dots denote means, error bars standard deviations ($6 \leq n \leq 8$ per horizon and plot).

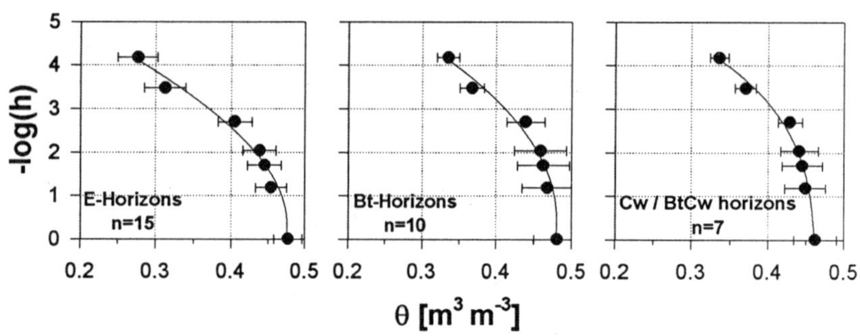


**Figure 3:** Soil-water retention curves for E, Bt, and BtCw/Cw horizons. Dots denote medians lumped for all four plots, error bars denote standard deviations.

**Table 4:** Pore size distribution in E, Bt, and BtCw/Cw horizons lumped for all four plots calculated from the soil-water retention curve after fitting the data according to Van Genuchten (1980).

| Horizon | Parameter | Depth | Volume of pores with Ø | | | | Total |
|---|---|---|---|---|---|---|---|
| | | | < 0.2 µm | 0.2–10 µm | 10–50 µm | > 50 µm | porosity |
| | | cm | $m^3\ m^{-3}$ | | | | |
| E | Mean | 24 | 0.28 | 0.13 | 0.04 | 0.03 | 0.48 |
| | SD | 13 | 0.03 | 0.02[a] | 0.03 | 0.02 | 0.02 |
| | Min | 10 | 0.24 | 0.10 | 0.01 | 0.00 | 0.45 |
| | Max | 45 | 0.32 | 0.16 | 0.10 | 0.08 | 0.51 |
| Bt | Mean | 74 | 0.33 | 0.10 | 0.02 | 0.02 | 0.48 |
| | SD | 18 | 0.02 | 0.01 | 0.02 | 0.01 | 0.03 |
| | Min | 55 | 0.32 | 0.07 | 0.00 | 0.01 | 0.45 |
| | Max | 100 | 0.37 | 0.13 | 0.05 | 0.04 | 0.55 |
| BtCw/Cw | Mean | 112 | 0.34 | 0.09 | 0.02 | 0.02 | 0.46 |
| | SD | 28 | 0.01 | 0.02 | 0.01 | 0.01 | 0.03 |
| | Min | 80 | 0.31 | 0.07 | 0.00 | 0.01 | 0.43 |
| | Max | 130 | 0.35 | 0.12 | 0.05 | 0.03 | 0.50 |

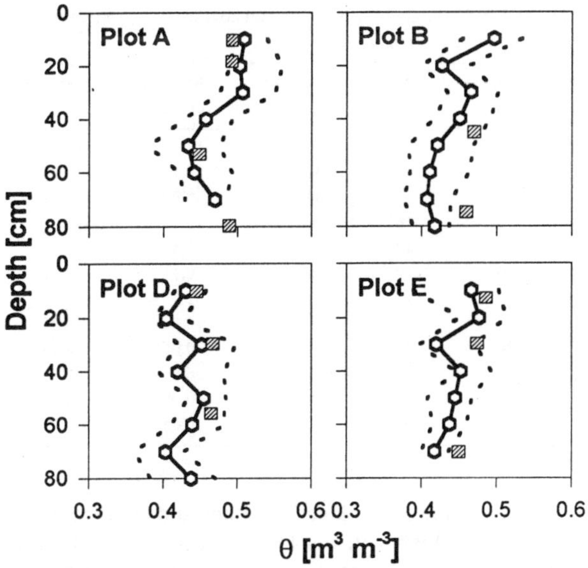


**Figure 4:** Soil moisture distribution with depth in the experimental plots. Medians (open circles), minima and maxima (dotted lines) for the period March 26 to April 17, 1995. The filled rectangulars represent total porosity obtained from SWRC's (mean of 3 soil cores).

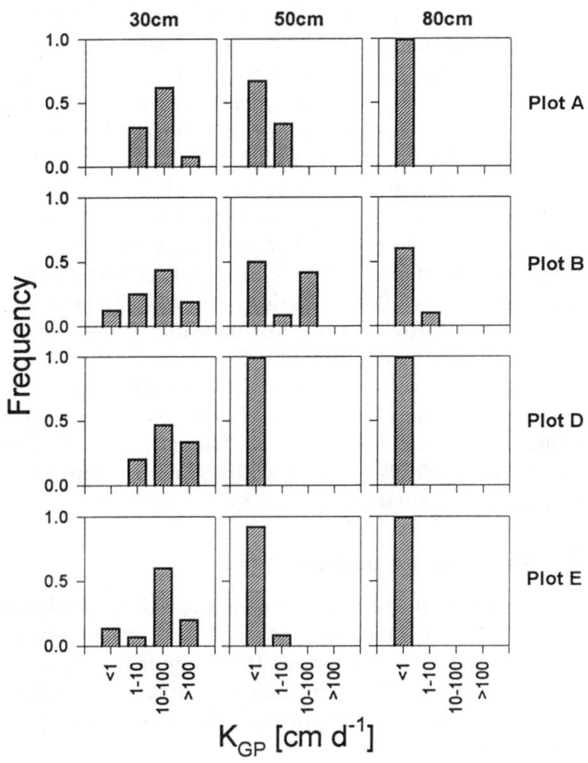


**Figure 5:** Frequency distribution of saturated hydraulic conductivity ($K_{GP}$).

diately after precipitation events. As discussed in section 3.2, these macropores apparently occur on a scale larger than the size of the soil cores used for the determination of the SWRC's and pore-size distributions, respectively. The medians of soil moisture (Figure 4) show that the soils were typically close to saturation during the monitoring period. Therefore, aggregate dynamics in the clayey soil is considered to be of minor importance to the development of macropores. Field observations reveal that macropores are most abundant in topsoils. This observation is consistent with highly variable hydraulic conductivities in the topsoils (Figure 5).

These findings indicate that during the rain season even fast-draining macropores may temporarely be saturated with water. Provided that the soil is at field capacity prior to a precipitation event, 1.5 cm precipitation were sufficient to increase the soil moisture content of the top 30 cm by 0.05 $m^3$ $m^{-3}$. This is consistent with the range of oversaturation observed in the topsoils of plot A (Figure 4), and with the finding that vertical seepage is limited due to the low permeability of the Bt horizons. During precipitation events the soil-water regime is likely characterized by (almost) saturated lateral flow along the slopes. Between rain events, matric potentials (–log h) between 2.5 and 3 may be attained, indicating that the macropores and large mesopores are quickly drained, while the bulk of micropores maintains saturated.

### 3.4 Hydraulic conductivity

Frequency distributions of field-saturated hydraulic conductivities ($K_{GP}$) are depicted in Figure 5. The within-plot variation of $K_{GP}$ is large, covering > 5 orders of magnitude. This variation compares to $10^3$ to $10^{-2}$ m $s^{-1}$ found by *Reynolds* and *Zebchuk* (1996) in cultivated clay soils near Ottawa, Canada, using the same method of measurement. The high variability was particularly associated with the topsoils and is consistent with the large variation of macropore volumes at these depths (Figure 3).

The majority of measurements of $K_{GP}$ are >10 cm $d^{-1}$ at 30 cm depth, and are < 10 cm $d^{-1}$ at 50 and 80 cm depth. The low permeability in the subsoils is linked to the Bt horizons. This is revealed by a close correlation between clay contents and $K_{GP}$ (Figure 6). The sharp decrease in $K_{GP}$ with depth is also associated with a decrease in OC (Figure 6; Table 3) indicating the effect of biopores on $K_{GP}$ in the topsoils. In contrast, $K_{GP}$ measurements obtained in a Canadian clay soil were rather correlated to the antecedent volumetric moisture content than to OC or clay content (*Reynolds* and *Zebchuk*, 1996). As evidenced by Figure 4, the soils of the present study were close to saturation during field measurements, indicating that the variation in antecedent water content was less important (*Peige* and *Hillel*, 1993; *Selverda* and *Done*, 1993). Like in the study of *Reynolds* and *Zebchuk* (1996), saturated hydraulic conductivity was not related to OC or clay content in a number of soils under Dipterocarp forests from the Bukit Soeharto Conservation Forést region, East Kalimantan

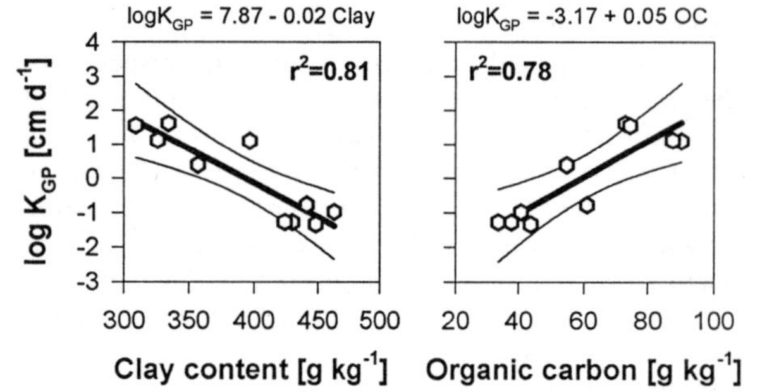


**Figure 6:** Scatter plots, regression lines and 99%-confidence intervals for field-saturated hydraulic conductivity ($K_{GP}$) versus clay content, and organic carbon.

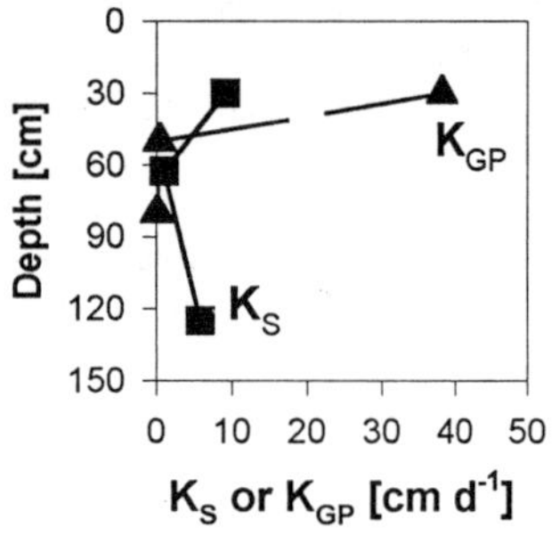


**Figure 7:** Medians of saturated hydraulic conductivities versus soil depth. Comparison of field measurements obtained with Guelph permeameters ($K_{GP}$) and laboratory assessment on soil cores ($K_s$).

(*Ohta* et al., 1992). This may be due to different methodology, since *Ohta* et al. (1992) measured saturated hydraulic conductivity ($K_s$) using soil cores in the laboratory. We compared $K_{GP}$ with $K_s$ values obtained on soil cores (Figure 7). Field measurements for all plots were in topsoils (E horizons/30 cm depth) about four times higher than corresponding $K_s$ values, but, in turn, $K_s$ values were about two times higher than $K_{GP}$ in Bt horizons. In spite of different measurement techniques, the higher $K_{GP}$ values in topsoils may be associated with abundant macroscopic biopores in the E horizons that are not addressed employing laboratory measurement on small soil cores. Lower values for field measurements in the subsoils may be explained by air-entrapment (*Paige* and *Hillel*, 1993). Although the walls of the well were worked with a wire brunch to minimize smearing, compaction and siltation due to augering of the well, these effects may contribute to the lower hydraulic conductivities obtained with the Guelph permeameters at depths below 40 cm (*Reynolds* and *Zebchuk*, 1996).

The $K_{GP}$ values at 30 cm depth in plots D and E beneath plantations are higher than those measured in the plots A and B beneath Dipterocarp forests (Figure 5, Table 5). The reverse trend was found at 50 cm depth (Figure 5). These differences between Dipterocarp forests and forest plantations are not simply related to differential bulk densities, clay or OC contents. Differences in the total living root biomass provide no explanation either, and living root biomass is even higher in the Dipterocarp forests (*Murach*, 1995). Only the substantially larger biomass of living roots of the understorey vegetation (i.e. grass and ferne species; *Murach*, 1995) may account for the higher permeability of the topsoils under plantations.

**Table 5:** Field-saturated hydraulic conductivities ($K_{GP}$) measured at three depths in the experimental plots.

| Plot | Depth cm | n | Minimum | Median cm $d^{-1}$ | Maximum |
|---|---|---|---|---|---|
| A | 30 | 13 | 2.12 | 12.7 | 123 |
| | 50 | 9 | 0.00 | 0.17 | 2.83 |
| | 80 | 9 | 0.00 | 0.00 | 0.19 |
| B | 30 | 16 | 0.00 | 13.3 | 551 |
| | 50 | 12 | 0.00 | 2.50 | 153 |
| | 80 | 10 | 0.00 | 0.11 | 2.61 |
| D | 30 | 15 | 1.41 | 42.4 | 878 |
| | 50 | 13 | 0.00 | 0.00 | 0.85 |
| | 80 | 10 | 0.00 | 0.05 | 0.33 |
| E | 30 | 15 | 0.17 | 36.1 | 505 |
| | 50 | 12 | 0.00 | 0.05 | 4.24 |
| | 80 | 9 | 0.00 | 0.05 | 0.21 |

In plot B the permeability at 50 and 80 cm is higher than in the other plots (Figure 5, Table 5). At both, 50 and 80 cm depth the frequency distribution shows two peaks, which is indicative for the development of a secondary pore system (*Schachtschabel* et al., 1992). The higher permeability at depth is consistent with a thicker E horizon (60 cm) evidenced by the profile description.

Unsaturated hydraulic conductivities ($K_u$) are depicted in Figure 8. $K_u$ strongly depends on the soil water content $\theta$ (Figure 8). The range of $\theta$ monitored in the field indicates that the moisture regime in the topsoils was dominated by saturated flow conditions, while at 50 and 80 cm depth it was partly controlled by unsaturated flow conditions. This may be due to trapped air in the fine textured, slowly diffusive soils. From the $K_u$ values obtained at the observed medians of soil moisture (Figure 4), the hydraulic conductivities at 50 and 80 cm depth are predicted to be typically 1–2 orders of magnitude below those measured at saturation (Figure 8).

### 3.5 Soil moisture budgets

In 1995, the study area received about 245 cm precipitation (Figure 9), which is consistent with the long-term annual mean of precipitation. The pronounced seasonal variation of precipitation indicates dryer and wetter periods. In some months, precipitation was as low as or even below the rate of evapotranspiration (about 11 cm per month) estimated from data compiled by *Bruijnzeel* (1990).

Evaporation due to interception in the canopy (E), calculated as the difference between P and T, was 25 cm $y^{-1}$ in the Dipterocarp plots A and B (Figure 11). In the planta-

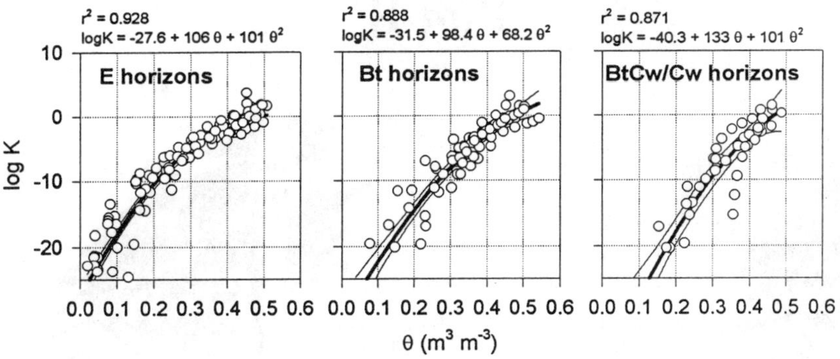


**Figure 8:** Hydraulic conductivity (K) versus soil moisture content (θ). K values were estimated using the model of Mualem (1976) after fitting the soil-water retention curve according to Van Genuchten (1980).

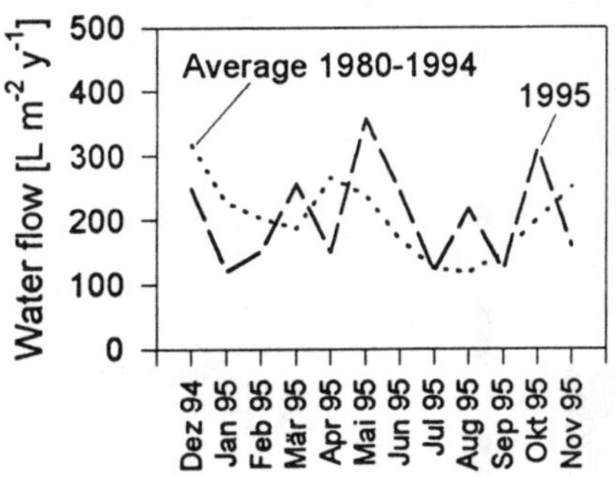


**Figure 9:** Seasonal variation of precipitation in the study area measured in 1994/95 and at the climate station Batu Ampar (average of 1980–1994).

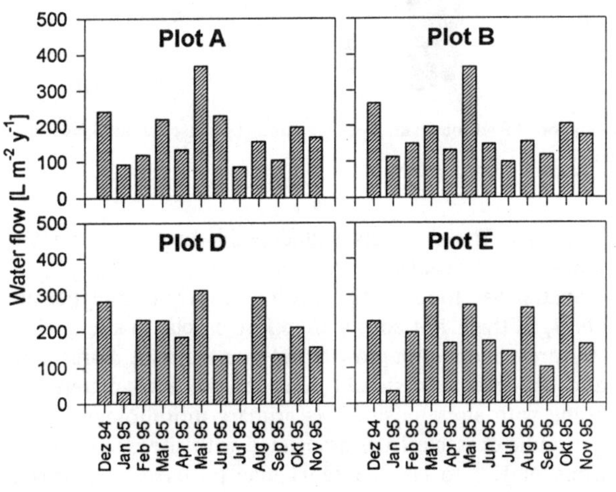


**Figure 10:** Seasonal variation of throughfall in the experimental plots during the monitoring period.

tions plots D and E, the interception was 15 cm $y^{-1}$, about 43% of that measured in the Dipterocarp forests (Figure 12). This is in accord with the lower stand density and leaf area index of the nine years old plantations.

Throughfall was 210 and 230 cm $y^{-1}$ in the Dipterocarp forests and plantations, respectively. Accordingly, the forest floor received about 86% (Dipterocarps) and 94% (plantations) of precipitation. The most reliable data for lowland tropical forests suggest a similar range of throughfall (TF) between 77 and 93% of precipitation (*Bruijnzeel*, 1990).

According to *Bruijnzeel* (1990), the rate of evapotranspiration (ET) in moist tropical forest ecosystems worldwide almost invariably amounts to 130–150 cm $y^{-1}$. Since ET was not measured in this study, this range was used to construct soil water budgets for the experimental plots (Figures 11 and 12). Assuming that evaporation primarily occurs in the canopy, the difference between Bruijnzeel's range for ET and measured interception E is labeled as transpiration (T) of the forest stands. For simplicity, it is assumed that 100% of T occurs in the top 30 cm of the soils. The error caused by this simplification is probably small, since 67 to 78% of the living roots were counted in this depth increment (*Murach*, 1995). In consequence, 95–115 cm $y^{-1}$ (Dipterocarps) and 115–135 cm $y^{-1}$ (plantations) are available for deep seepage, lateral surface and subsurface flow. Based on the measured field saturated hydraulic conductivities (Table 5), a maximum of 60 cm water is expected to permeate below 50 cm soil depth at plot A. Note that $K_{GP}$ at 50 cm depth was probably underestimated by a factor 2; however, its use as a hydraulic constraint is considered safe since unsaturated conditions with $K_u$ values 1–2 orders of magnitude below $K_s$ dominated the moisture regime at this depth even during the rain season (Figure 4). In consequence, 35–55 cm $y^{-1}$ are available for lateral subsurface and surface flow (Figure 11). Although virtually no signs of surface flow were observed during field work, it is recognized that it may contribute to the water fluxes labeled as SL in Figures 11 and 12. $K_{GP}$ values even lower at 80 cm depth suggest a lateral flow of up to 60 cm $y^{-1}$ at this depth. Less than 5 cm $y^{-1}$ (4.3–5.3% of total seepage) may permeate to depths below 80 cm.

Subsurface flow contributes also significantly to soil water fluxes in plots B, D, and E. Differences among the plot are basically caused by variations in the lower boundary of the E and in the permeability of the Bt horizons (Figures 11 and 12).

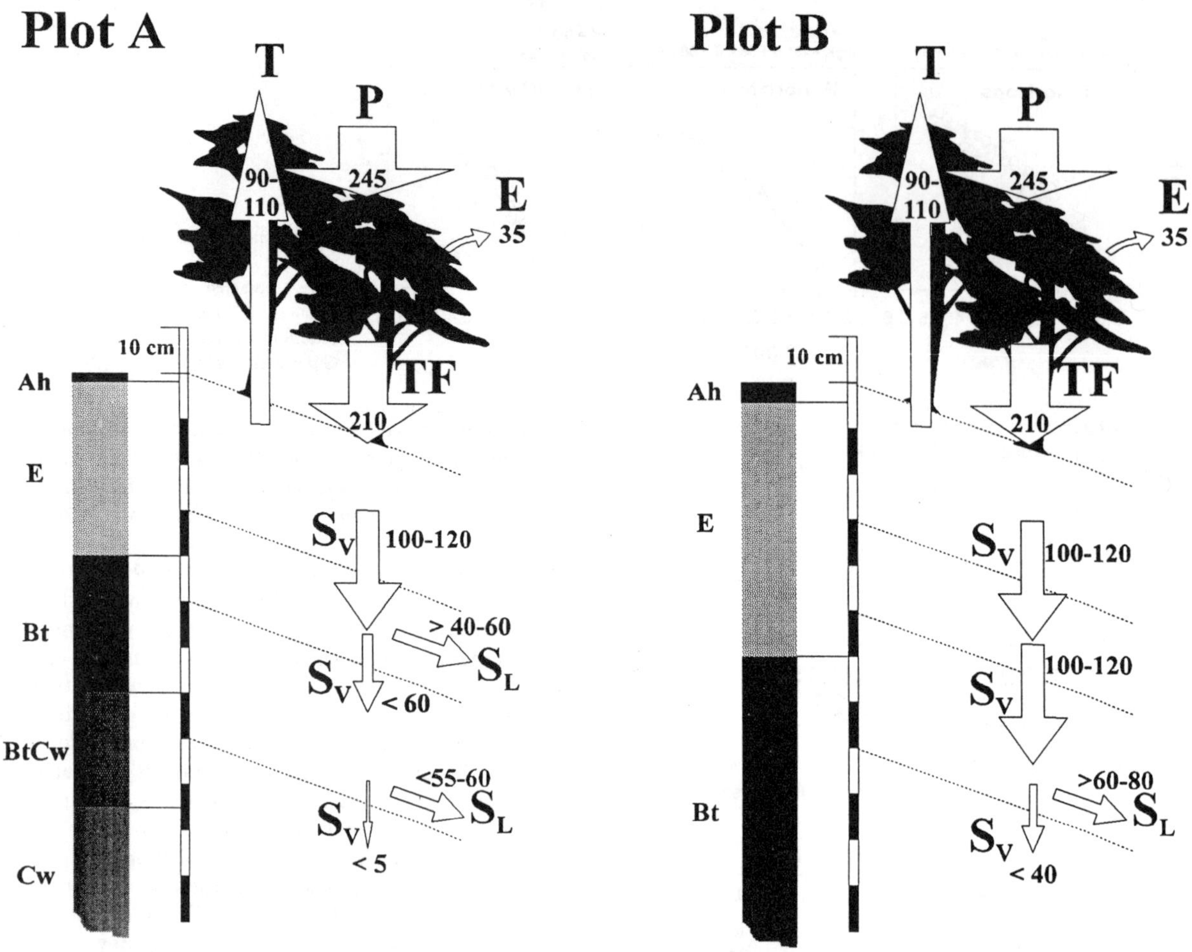


**Figure 11:** Soil water budgets (cm water flow) for plots A and B (P=precipitation, TF=throughfall, E=evaporation, T=transpiration, $S_V$=vertical seepage, $S_L$=lateral water flow (without consideration of inputs due to lateral flow).

## 4 Conclusions

Soil hydrology at the investigated sites is dominated by lateral subsurface flow in the E horizons. Vertical seepage below 40–60 cm depth is almost neglectible because of the low amount of macropores in the clayey subsuface (Bt) horizons.

The rate of lateral flow is expected to increase with distance from the top of slopes. If lateral flow in E horizons is assumed to increase linearly along the slopes, the water flow would increase from 400–1400 L $m^{-2}$ $y^{-1}$ (top of slope) to 40 000–140 000 L $m^{-2}$ $y^{-1}$ at 100 m distance from the top. Using the medians of hydraulic conductivity (12.7–42.4 cm $y^{-1}$) measured in the E horizons (Table 5) and a horizon thickness of 40 to 60 cm (Table 3), lateral water flow through the permeable cross section is constrained to 18500–92900 L $m^2$ $y^{-1}$. Therefore, stagnic properties associated with reductive conditions and low aeration may be characteristic for the topsoils at lower slope positions. These effects will depend on the permeable cross section available for lateral flow.

Spatial variation in the natural thickness and permeability of the E horizons is therefore crucial to site quality for forest/plantation growth and management. Forest site and soil mapping should consider these characteristics along with slope position as primary morphological and hydraulic characteristics associated with site quality. Relations of $K_{GP}$ to clay contents may be used to estimate hydraulic conductivities from information on soil texture.

Soil erosion due to increasing forest exploitation and land use is considered a severe risk. Partial or complete degradation of the permeable E horizons would have dramatic impacts on the ecology and on sustainable landuse of the lowlands of East Kalimantan since the amount of surface flow and the frequency of flooding events would increase. Moreover, a decrease in the capacity of plant available moisture content (middle pores) would cause a loss of resiliance of Dipterocarp forest ecosystems and

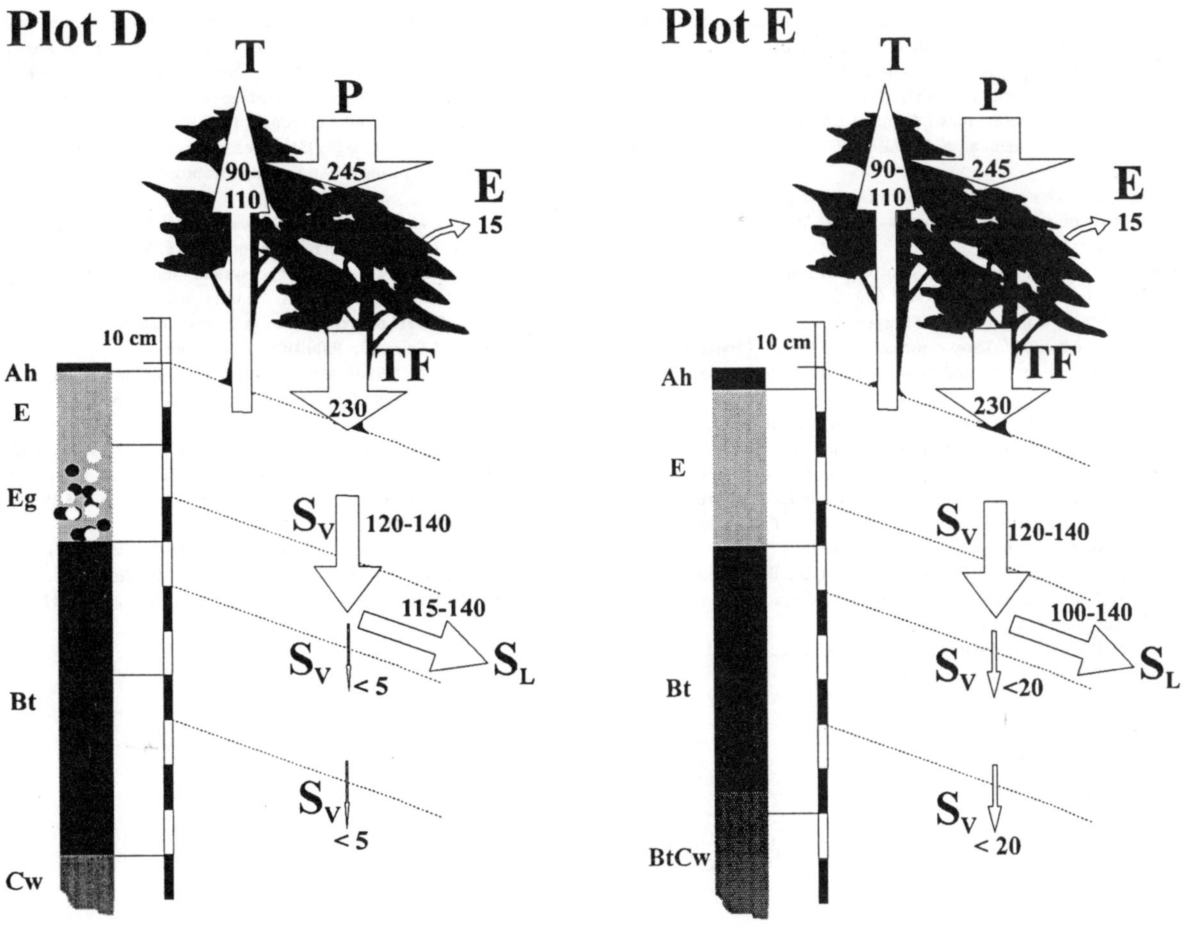


**Figure 12:** Soil water budgets (cm water flow) for plots D and E (P=precipitation, TF=throughfall, E=evaporation, T=transpiration, $S_V$=vertical seepage, $S_L$=lateral water flow (without consideration of inputs due to lateral flow).

plantations against occasionally occuring draughts. This is particularly important at sites located at the top and upper parts of slopes, since these sites receive less additional water through lateral flow. Conversion of indigenous Dipterocarp forests to plantations increases the amount of throughfall received by the soil surface and may thus increase the risk of surface flow and erosion.

## 5 Acknowledgements

This study was funded by the Indonesian-German Forestry Project at the Mulawarman University in Samarinda/Indonesia, Deutsche Gesellschaft für Technische Zusammenarbeit GmbH (GTZ).

## 6 References

*Blum, W. E. H., H. Spiegel* and *W. W. Wenzel* (1996): Bodenzustandsinventur. Konzeption, Durchführung und Bewertung. Federal Ministry for Agriculture and Forestry, Vienna, p. 105.

*Bruijnzeel, L. A.* (1990): Hydrology of moist tropical forests and effects of conversion: A state of knowledge review. UNESCO International Hydrological Programm, 224 p.

*Durner, W.* (1991): Vorhersage der hydraulischen Leitfähigkeit strukturierter Böden. Bayreuther Bodenkundliche Berichte 20.

*Elrick, D. E.* and *W. D. Reynolds* (1992): Infiltration from constant-head well permeameters and infiltrometers. In G.C. Topp et al. (eds.), Advances in measurement of soil physical properties: Bringing theory into practice. SSSA Spec. Publ. 30, p. 1–24, SSSA, Madison, WI.

*Elrick, D. E., W. D. Reynolds* and *K. A. Tan* (1989): Hydraulic conductivity measurements in the unsaturated zone using improved well analysis. Ground Water Monit. Rev. 9, 184–193.

*FAO* (1988): Soil map of the world. Revided legend. FAO, Rome, 1988.

*Goldammer, J. G., B. Seibert* and *W. Schindele* (1996): Fire in Dipterocarp forests. In A. Schulte and D. Schöne (eds.), Dipterocarp forest ecosystems toward sustainable management, p. 155–186, World Scientific Publishing, Singapore.

*Kuráz, V., A. Brandstetter, H. Unterfrauner* and *W. W. Wenzel* (1995): Field and laboratory calibration of a dielectric soil moisture meter. Poster presentation at the Third Conference on Tropical Forest Soils, Balikpapan, East Kalimantan, 1995.

*Kuráz, V.* (1982): Testing a field dielectric soil moisture meter. Geotechnical Testing Journal, GTJODJ 4, 111–116.

*Malmer, A.* (1996): Nutrient losses from Dipterocarp forests – a case study of forest plantation establishment in Sabah, Malaysia. In A. Schulte and D. Schöne (eds.), Dipterocarp forest ecosystems toward sustainable management. p. 52–73, World Scientific Publishing, Singapore.

*Mualem, Y.* (1976): A new model for predicting the hydraulic conductivity of unsaturated porous media. Water Resour. Res. 12, 513–522.

*Murach, D.* (1995): Bericht zu den Wurzeluntersuchungen auf den Demonstrationsflächen des IGFP. Unpublished research report, Inst. für Waldbau, University of Göttingen.

*Ohta, S., S. Effendi, N. Tanaka* and *S. Miura* (1992): Characteristics of major soils under lowland Dipterocarp forest in East Kalimantan, Indonesia.

*Peige, G. B.* and *D. Hillel* (1993): Comparison of three methods for assessing soil hydraulic properties. Soil Science 155, 175–189.

*Reynolds, W. D.* and *D. E. Elrick* (1985): Measurement of field-saturated hydraulic conductivity, sorptivity and the conductivity-pressure head relationship using the "Guelph permeameter". Proc. National Water Well Ass. Conf., Denver, Colorado, November 1985.

*Reynolds, W. D.* and *N. D. Zebchuk* (1996): Hydraulic conductivity in a clay soil: Two measurement techniques and spatial characterization. Soil Sci. Soc. Am. J. 60, 1679–1685.

*Richards, P. W.* (1996): The tropical rain forest. An ecological study. Cambridge University Press, 2nd ed., 575 p.

*Ruslim, Y.* (1994): Der Beitrag eines planmäßigen Erschließungs- und Nutzungskonzeptes zur pfleglichen Holzernte im tropischen Regenwald, untersucht am Beispiel eines Dipterocarpwaldes in Ostkalimantan, Indonesien. Göttinger Beiträge zur Land- und Forstwirtschaft in den Tropen und Subtropen, Heft 95.

*Salverda, A. P.* and *J. H. Done* (1993): An examination of the Guelph permeameter for measuring the soil's hydraulic properties. Geoderma 57, 405–421.

*Schachtschabel, P., H.-P. Blume, G. Brümmer, K.-H. Hartge* and *U. Schwertmann* (1992): Lehrbuch der Bodenkunde. Enke, Stuttgart.

*Schulte, A.* (1996): Dipterocarp forest ecosystem theory based on matter balances and biodiversity. In A. Schulte and D. Schöne (eds.), Dipterocarp forest ecosystems toward sustainable management. p. 3–28, World Scientific Publishing, Singapore.

*Van Genuchten, M. Th.* (1980): A closed-dorm equation for predicting the hydraulic conductivity of unsaturated soils. Soil Sci. Soc. Am. J. 44, 892–898.

*Van Genuchten, M. Th., F. J. Leij* and *S. R. Yates* (1991): The RETC code for quantifying the hydraulic functions of unsaturated soils. U. S. Environ. Protection Agency Rep. 600/2–91/065. U.S. Environ. Protection Agency, Ada, OK.

*Yates, S. R., M. Th. van Genuchten, A. W. Warrick* and *F. J. Leij* (1992): Analysis of measured, predicted, and estimated hydraulic conductivity using the RETC computer code. Soil Sci. Soc. Am. J. 56, 347–354.

# Nutrient dynamics in intact and logged tropical rain forest in Guyana

**Leo C. Brouwer** and **Hans Th. Riezebos**

Department of Physical Geography, Utrecht University, P.O. Box 80.115, 3508 TC Utrecht, The Netherlands, e-mail: l.brouwer@frw.ruu.nl

## Abstract

In tropical forests growing on nutrient poor and acid soils nutrient cycling is tight and ecosystem nutrient losses are low. In this chapter the effects of selective logging on leaf decomposition rates and soil nutrient leaching are evaluated. Two logging gaps of different size (730 $m^2$ and 3440 $m^2$) were each divided into three zones: skidder (open, disturbed soil), crown (debris, no soil disturbance) and undisturbed (small trees still present, no soil disturbance).

Decomposition and nutrient release from brown (recently shed) *Chlorocardium rodiei* leaves was slower than from green (fresh) leaves. In the skidder zone of the large gap, decomposition was slower than in closed forest. No difference in decomposition rate was found between leaves in the crown zone of the gap and in closed forest.

The soils in the gaps were wetter than in closed forest, especially during dry periods. This led to about 50 % more water percolation through the soil in the skidder zone of the gaps during the first two years. The concentrations of $NO_3$-N, Al, Ca, K, Mg, Mn, Na and Cl in percolating water in the gaps were significantly higher, and the pH lower than in closed forest, particularly during the first 15 months after logging. In the large gap, nutrient losses through leaching were greater and extended for a longer period than in the medium gap. The logging induced nutrient losses decreased in the order: skidder zone, crown zone, undisturbed zone.

The absence of living roots through soil disturbance and by removing a group of trees is a more important cause for increased leaching than the sudden addition of large amounts of fresh, rapidly decomposable organic matter to the ground surface. After logging, nitrification is a key process since it enhances cation leaching and induces soil acidification and increased Al solubility.

Leaching losses may contribute up to 50 % of the total nutrient losses associated with logging in case of a large gap, but significantly less in smaller gaps. Therefore, sustainable forest management systems should aim to reduce leaching as much as possible. Limitation of skidder activity to prevent excessive soil disturbance and avoiding the creation of large canopy gaps are crucial to minimise leaching losses.

## 1 Introduction

In large parts of the humid tropics currently under tropical forest, soils are poor and the low nutrient availability seriously hampers land use other than extensive forestry. In the interior of Guyana, selective logging of the forests is common practice, and increasingly large areas are given out as timber concessions (*Colchester*, 1994; ter *Steege* et al., 1995). Traditionally the forestry sector has concentrated on a few hardwood species, of which *Chlorocardium rodiei* (Greenheart) is by far the most important. This species has a clumped distribution and during heavy exploitation (locally more than 100 $m^3$ $ha^{-1}$) fairly large clearings (up to 5000 $m^2$) are created (ter *Steege* et al., 1995).

In relation to nutrient cycling, selective logging has been classified as a disturbance of moderate intensity (*Jordan*, 1985; *Bruijnzeel*, 1990). Nevertheless there is much variation in the intensity of selective logging, depending for example on the proportion of trees that can be extracted commercially. In Guyana the proportion of commercial trees is generally low and the average exploitation intensity for hardwoods rarely exceeds 50 $m^3$/ha. However, not only the average intensity but also the spatial distribution of the newly created gaps is important. At a given exploitation intensity a well-distributed cut may have completely different effects than a cut in one contiguous area (*Parker*, 1985).

Selective logging can cause a disruption of the nutrient cycle. Large amounts of organic debris reach the forest floor where it is decomposed and nutrients are released in mineral form. If the exchange capacity of the soil is low and rainfall is abundant, the risk of leaching is high and valuable nutrients will be lost.

In this paper two experiments are described that aimed 1) to estimate decomposition rates of leaf material present on the soil surface after a selective logging operation and 2) to quantify nutrient losses through leaching after a selective logging operation and to evaluate whether gap size and the degree of local disturbance influence leaching rates.

A decomposition study using litterbags was used to monitor dry weight and nutrient losses from *C. rodiei* leaves, the major timber tree in the area. Leaching of nutrients in two gaps of different size and in undistrubed closed forest was evaluated by monitoring soil water nutrient contents below the rooting zone, combined with soil moisture suction measurements and a vertical water balance model.

## 2 Study area

The study area is located in the forestry belt of north-central Guyana, South America (5°N, 59°W, elevation < 100 m). Average annual rainfall is 2500 mm, distributed over two wet and two relatively dry seasons. The driest month is usually October, with an average precipitation of about 100 mm (*Brouwer*, 1996). The wettest months are June and July. The average annual temperature is 26°C with little monthly variation (*Jetten*, 1994).

The topography consists of gently rolling hills which rise up to 30 m above the river level. The well-drained soils at the study site have formed in unconsolidated sandy sediments of Pliocene/Pleistocene age which cover the underlying rocks of the Precambrian Guyana shield. The brown sandy soils are classified as Ferralic Arenosols or Haplic Ferralsols (FAO classification) or as Paleudults (USDA Soil Taxonomy). These acid soils are characterised by a low content of weatherable minerals, low cation exchange capacity, low exchangeable bases, high content of Al and Fe and low P availability (*Brouwer*, 1996; *Van Kekem* et al., 1997). Some basic soil chemical characteristics are shown in table 1.

A mixed species tropical lowland rain forest is found on this soil, with sometimes local dominance of a few species (*Ter Steege*, 1993). The forest is about 30 m high and the average basal area is 30–35 $m^2$/ha (*Van Zanten*, 1991).

## 3 Methods

### 3.1 Plot preparation and description

An area containing a rich stock of commercial trees (mainly *C. rodiei*) was selected for the experiments. Within the area two, more or less circularly shaped, logging gaps of different size were created on 30 and 31 January 1992, by felling the commerical trees and extracting them using a heavy type skidder with rubber tyres. The largest gap (canopy opening 3440 $m^2$) resulted from the removal of an estimated 65 $m^3$ of timber in 17 trees. A medium sized gap (canpopy opening 730 $m^2$) resulted from the removal of 16 $m^3$ of timber in five trees. A third area with pristine forest, located about 60 m from the medium gap and 200 m from the large gap, was selected as control. The two gaps were subdivided into three zones depending on the amount of debris and the damage to vegetation and soil. The three zones, that occupied about equal surface areas in the gaps (Figure 1), were:

- the skidder zone; this is the area where skidder movement took place. It is characterised by more or less bare soil which is compacted. The vegetation did not survive the logging in this zone except for some shade adapted seedlings that died within a few weeks. Most of the skidder zone was located near the centre of the gaps.
- the crown zone; this is the area where the crowns of the felled trees were deposited. Except for a few large trees all vegetation in this zone was damaged and covered by a large amount of debris. The soil in this zone was generally not disturbed although locally there was some uprooting of smaller trees that had been smashed by falling larger trees. Most of the crown zone was located near the gap edges.
- the undisturbed zone; in this zone the soil was not disturbed and the "smaller" vegetation survived (seedlings, saplings, trees with a diameter < 15 cm). The major difference with the pre-logging situation was the absence of large canopy trees, with consequently a much higher irradiation at ground level.

**Table 1:** Chemical characteristics of the soil at various depths.

| depth | | 0–20 cm | 20–50 cm | 50–100 cm |
|---|---|---|---|---|
| texture | | loamy sand | loamy sand | sandy loam |
| pH | $H_2O$ | 4.7 | 5.1 | 5.2 |
| pH | KCl | 4.2 | 4.5 | 4.6 |
| Org. C | (%) | 0.9 | 0.5 | 0.3 |
| total N | (%) | 0.07 | 0.04 | 0.02 |
| total P | mg/kg | 69 | 67 | 53 |
| P-Bray | mg/kg | 3 | 0.8 | 0.3 |
| CEC | $cmol^+/kg$ | 4.1 | 3.8 | 3.5 |
| exchangeable Ca | $cmol^+/kg$ | 0.02 | 0.02 | 0.01 |
| exchangeable K | $cmol^+/kg$ | 0.01 | 0.01 | 0 |
| exchangeable Mg | $cmol^+/kg$ | 0.02 | 0.01 | 0.01 |
| exchangeable Na | $cmol^+/kg$ | 0.02 | 0.01 | 0.01 |
| exchangeable Al | $cmol^+/kg$ | 0.82 | 0.62 | 0.46 |

### 3.2 Decomposition

Leaf litter decomposition rates in the gaps and the closed forest were evaluated with a litterbag experiment. Both green (picked from felled trees) and recently shed (collected daily from well drained netting) brown leaves of

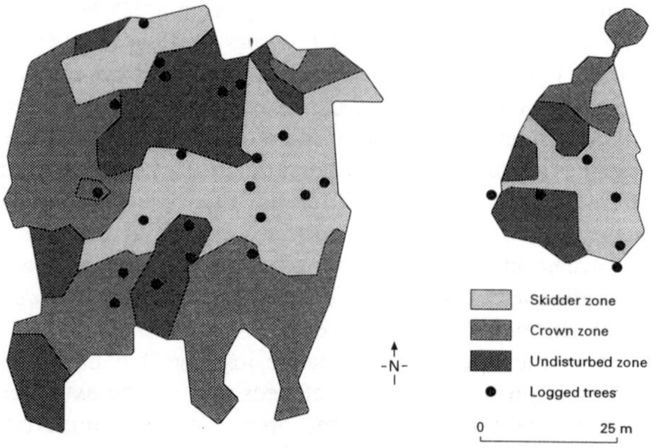


**Figure 1:** Layout and zonation of the experimental gaps.

*C. rodiei* were used. The leaves were dried at 40 °C to constant weight, cut into pieces of 5–10 $cm^2$ and mixed. Several pieces were put in one litterbag up to a weight of 2 g (green leaves) or 1.3 g (brown leaves). The 10 * 7 cm litterbags were prepared from 1 $mm^2$ polyethylene mesh, excluding soil macrofauna that could transport the leaf material but allowing access for soil mesofauna and microfauna. In total 140 litterbags were prepared, one half of them filled with green leaves and the other half of them filled with brown leaves. In the gap, the leaves in the litterbags were placed in the area where they "naturally" occurred, i.e. green leaves in the crown zone and brown leaves in the open (skidder) zone. In closed forest both litterbags filled with green and filled with brown leaves were placed. Wherever possible the litterbags were incorporated into the existing litter layer. Five replicate litter bags were recovered after 0, 2, 5, 9, 20, 34 and 49 weeks. After recovery the leaves were dried for 48 hours at 70 °C and weighed to 0.001 g. The samples were sealed in plastic and shipped to the laboratory of Physical Geography of Utrecht University in the Netherlands, where they were analysed for P, N, K, Ca, Mg and Na. Extracts were made by boiling the samples in a mixture of concentrated $H_2SO_4$ and $H_2O_2$ (after *Allen*, 1989) Cations and P were determined using an ICP-AES, N was determined using flow injection analysis.

### 3.3 Water flux estimation

The amount of water percolating through the soil at 120 cm depth was estimated with a one-dimensional water balance model called SOAP (SOil Atmosphere Plant model). Details on this model can be found in *Jetten* (1994 ab). The model calculates rainfall interception, transpiration, soil evaporation, infiltration into the soil and the redistribution of water in the soil. The atmospheric parameters required for the model were collected at the Tropenbos Meteorological station about 3 km north of the experimental plots.

In order to validate the performance of the water balance model, data were collected on the actual soil moisture suction. For this purpose, ten clusters of tensiometers were installed in closed forest (2), the large gap skidder (2), crown (2), and undisturbed (1) zone and the medium gap skidder (1), crown (1) and undisturbed (1) zone. Each cluster consisted of three tensiometers at 30, 75 and 120 cm depth. Readings of soil water tension were taken weekly during two periods: February–April 1992 (a dry period) and March–June 1993 (a wet period).

### 3.4 Soil water chemistry

Ceramic cup suction soil water samplers were used to collect soil water at 120 cm depth. The samplers were installed at an angle of 30 degrees from the vertical to avoid surface disturbance immediately above the porous cup. In total 27 samplers were used and about 700 samples were extracted and analysed over a period of almost three years. A depth of 120 cm was chosen because soil water at this depth can be considered essentially out of reach for (fine) roots. It is assumed that the effect of macropores (cf. *Sollins* and *Radulovich*, 1988; *Haines* et al., 1982; *Luxmoore* et al., 1990; *Nortcliff* and *Thornes*, 1989) is smoothened at that depth, and that the water sampled had a representative chemical composition for the percolating water.

A suction of 35 kPa was applied to the soil water samplers about 48 hours before sample collection. Immediately after sampling the electrical conductivity (EC) and pH of the samples were determined. Subsequently the samples were filtered through a 0.45 µm filter and two aliquots of the samples were taken. To one aliquot 25% $HNO_3$ was added (4% of the total sample volume), to the other 5% $H_2SO_4$ was added (4% of the total sample volume). The samples were stored at 4 °C and within three months shipped to the laboratory of Physical Geography of Utrecht University, The Netherlands for analysis. The $HNO_3$-acidified samples were analysed for Al, Ca, Fe, K, Mg, Mn, Na, Si and $SO_4$ on an ICP-AES. The $H_2SO_4$-acidified samples were analysed colorimetrically on a Skalar autoanalysor for $NH_4$, Cl and $NO_3$. The results of the chemical analysis were compared with the results of two independent high standard laboratories and the results corresponded well (*Brouwer*, 1996).

Leaching of nutrients was calculated by multiplying the aveage soil water nutrient concentration per zone with the water percolation in the corresponding period.

## 4 Results

### 4.1 Decomposition

Decomposition was significantly influenced by site and by leaf type, and the interaction term was not significant (two way ANOVA). Dry weight loss of shed leaves was significantly slower than that of fresh leaves. Brown leaves in the open area of the gap decomposed more slowly than leaves in closed forest (Scheffé test). Weight loss of fresh leaves was most rapid in the crown zone of the gap, although not significantly faster than that of fresh leaves in the forest (Table 2).

The two types of *C. rodiei* leaves (fresh green and shed brown) differed in initial nutrient concentrations, prob-

**Table 2:** Estimated value of k (and explained variance) by the single exponential model ($w_t/w_0 = e^{-k*t}$) for *C. rodiei* leaves (n = 35). Different letters indicate significant differences in decomposition rate (Scheffé test).

| treatment | k | $r^2$ | % weight loss after one year |
|---|---|---|---|
| brown leaves in closed forest | $0.62^b$ | 0.90 | 46 |
| green leaves in closed forest | $0.74^c$ | 0.87 | 52 |
| brown leaves in open zone of gap | $0.50^a$ | 0.91 | 39 |
| green leaves in crown zone of gap | $0.85^c$ | 0.92 | 58 |

ably as a result from leaf age and retranslocation processes before the leaf was shed (see also *Raaimakers*, 1994). In green leaves, the K, P and N concentrations were higher and the Ca concentrations were lower than in brown leaves, whereas the Mg concentrations were similar in both leaf types (Table 3).

In all treatments, N concentrations in the decomposing leaves were increased after 340 days, except for brown leaves in the skidder zone of the gap, where N concentrations were fairly constant. Between 55% (green leaves, crown zone of the gap) to 80% (brown leaves, closed forest) of the original N was still present after one year (Figure 2). P was immobilised in the brown leaves, whereas in the green leaves about 50% of the original P had been released. Mg concentrations decreased from about 3 mg $g^{-1}$ to about 1 mg $g^{-1}$ after 140 days of decomposition. K was rapidly released, particularly in the green leaves, which had a higher initial concentration. After 340 days less than 20% of the original K was remaining. Ca concentrations decreased with progressing decomposi-

**Table 3:** The initial concentration, the percentage of the original nutrient quantity still present in the leaf tissue after one year of decomposition and the amount of nutrients released per gram of original leaf material after a year. Different letters indicate significant differences in dry weights between groups ($p < 0.05$, Scheffé test).

| | type/ treatment | dry weight | Ca | K | Mg | Na | P | N |
|---|---|---|---|---|---|---|---|---|
| original | green | | 2.65 | 2.45 | 3.23 | 3.00 | 0.39 | 19.1 |
| concentration (mg $g^{-1}$) | brown | | 3.04 | 0.77 | 3.10 | 2.62 | 0.14 | 15.8 |
| % remaining after 340 | brown/forest | $56.6^{b}$ | 24.5 | 32.3 | 11.0 | 5.4 | 90.5 | 78.8 |
| days of decomposition | green/forest | $48.6^{a}$ | 13.1 | | 6.6 | 3.7 | 44.8 | 67.6 |
| | brown/gap | $63.0^{c}$ | 38.0 | | 7.1 | 5.8 | 97.3 | 67.4 |
| | green/gap | $42.2^{a}$ | 34.9 | 14.6 | 16.7 | 4.0 | 55.5 | 52.1 |
| quantity released | brown/forest | | 2.27 | 0.57 | 2.66 | 2.48 | 0.01 | 3.4 |
| after 1 year of | green/forest | | 2.37 | >1.84 | 3.14 | 2.76 | 0.21 | 6.2 |
| decomposition (mg $g^{-1}$) | brown/gap | | 1.93 | >0.45 | 3.02 | 2.49 | 0.00 | 5.1 |
| | green/gap | | 1.65 | 2.37 | 2.57 | 3.00 | 0.18 | 9.2 |

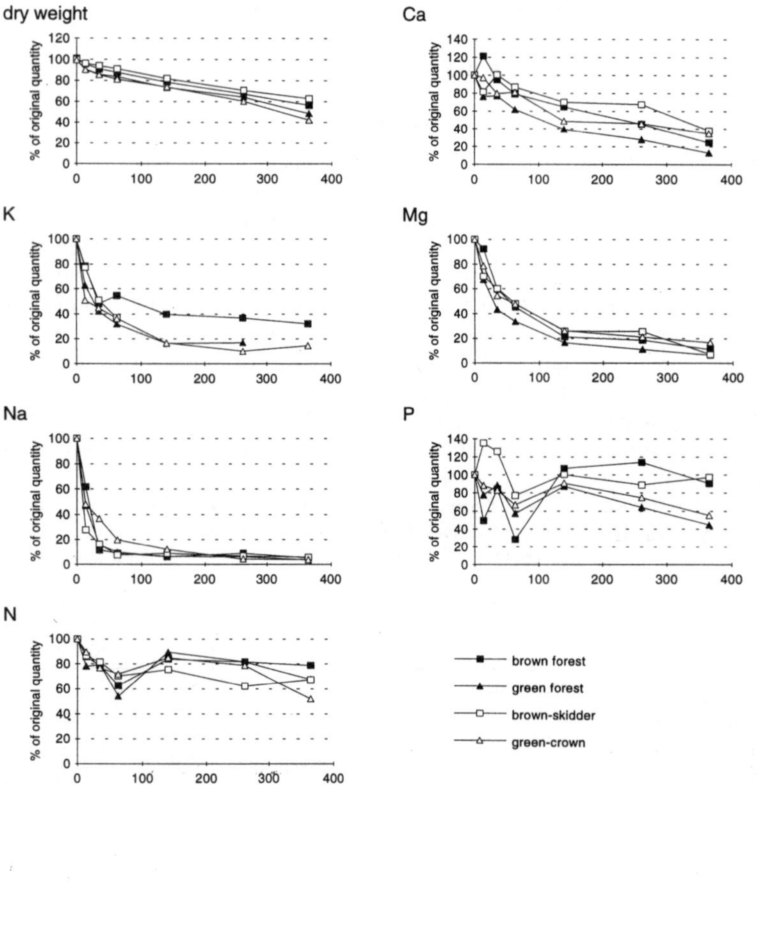


**Figure 2:** Dry weight and quantities of Ca, K, Mg, Na, P and N during a litterbag experiment over 1 year with green and brown leaves of *C. rodiei* in closed forest and in the skidder and crown zone in large logging gap (3440 $m^2$).

tion, except for green leaves in the crown zone of the gap, where the Ca concentration increased over the last 200 days. Between 13% (green leaves in closed forest) to 40% (green leaves in the gap) of the initial Ca was still present in the leaves after 340 days.

### 4.2 Water fluxes

The simulated soil water tension by the SOAP water balance model was compared with tensiometer readings over two periods: a relatively dry period in 1992 and a relatively wet period in 1993 (Figure 3). The correspondence between the simulated and the measured soil water tension was fairly good, so it was assumed that the performance of the model to calculate percolation at 120 cm depth was sufficiently accurate for the purpose of this study.

The soils in the gaps were consistently wetter than in closed forest, notably during dry periods (e.g. April 1992). Simulations by SOAP indicated that over the first 22 months after logging about 2800 mm water percolated to a depth greater than 120 cm in the skidder zones of the gaps, versus 1800 mm in the closed forest (out of 5000 mm rainfall). The percolation in the crown zone and the undisturbed zone of the gaps was intermediate. Because of the sandy nature of the soil, overland flow did not occur, despite soil compaction.

### 4.3 Soil moisture chemistry and nutrient leaching

#### *4.3.1 Effects of gap size*

Figure 4 shows the temporal courese of the average solute concentrations in soil water at 120 cm depth for the large gap, the medium gap and the closed forest. Solute concentrations in the soil moisture at 120 cm depth were generally greater in the gaps than in the closed forest, and greater in the large gap than in the medium gap (Figure 4). The EC, as a measure for total ionic content, started to rise within a month from a typical background value of 15–20 $\mu S\ cm^{-1}$ to more than 100 $\mu S\ cm^{-1}$ in the large gap within

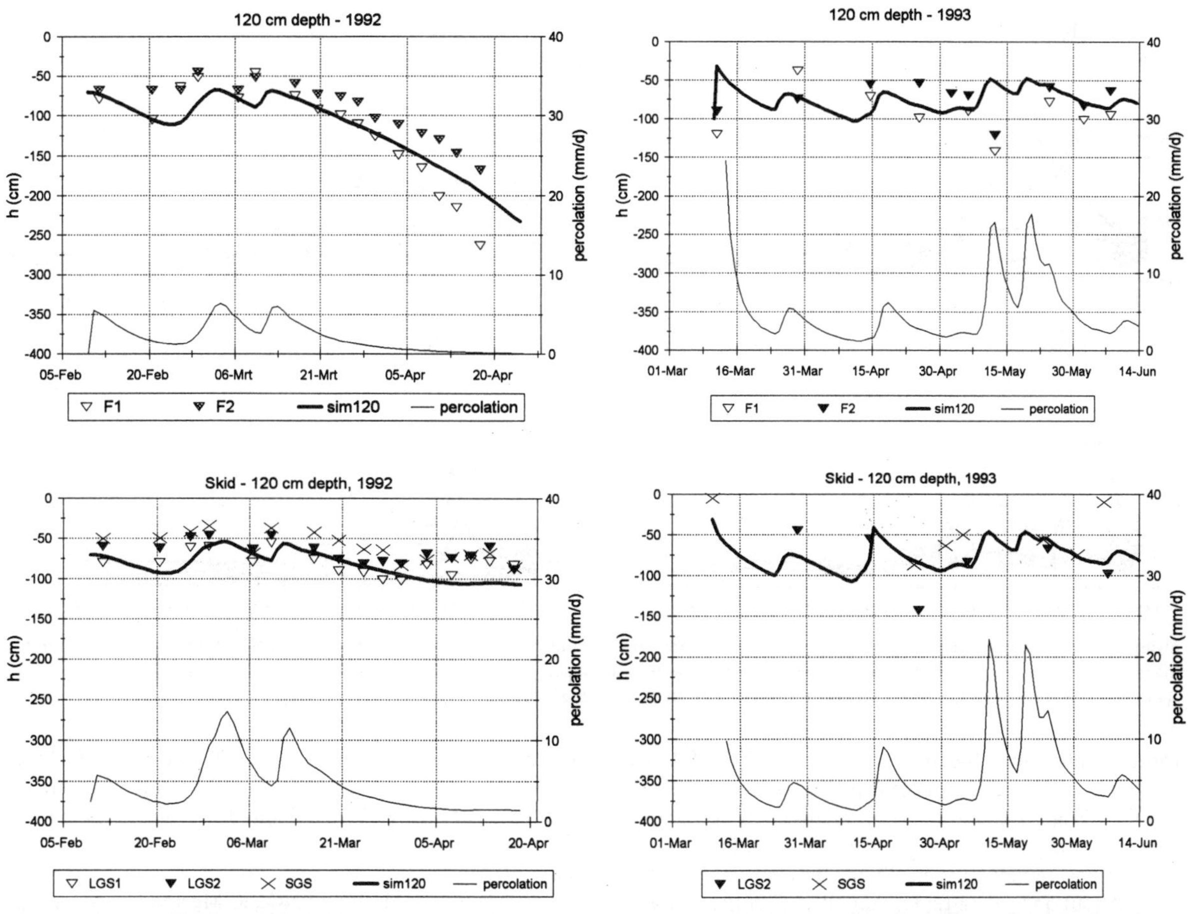


**Figure 3:** Measured and simulated soil water tension and the simulated percolation at 120 cm depth under closed forest (top) and in the skidder zone of logging gaps (bottom) for February–April 1992 (left) and for March–June 1993 (right). The symbols in the graphs indicate the tensiometer measurements, sim 120 shows the simulated soil water tension.

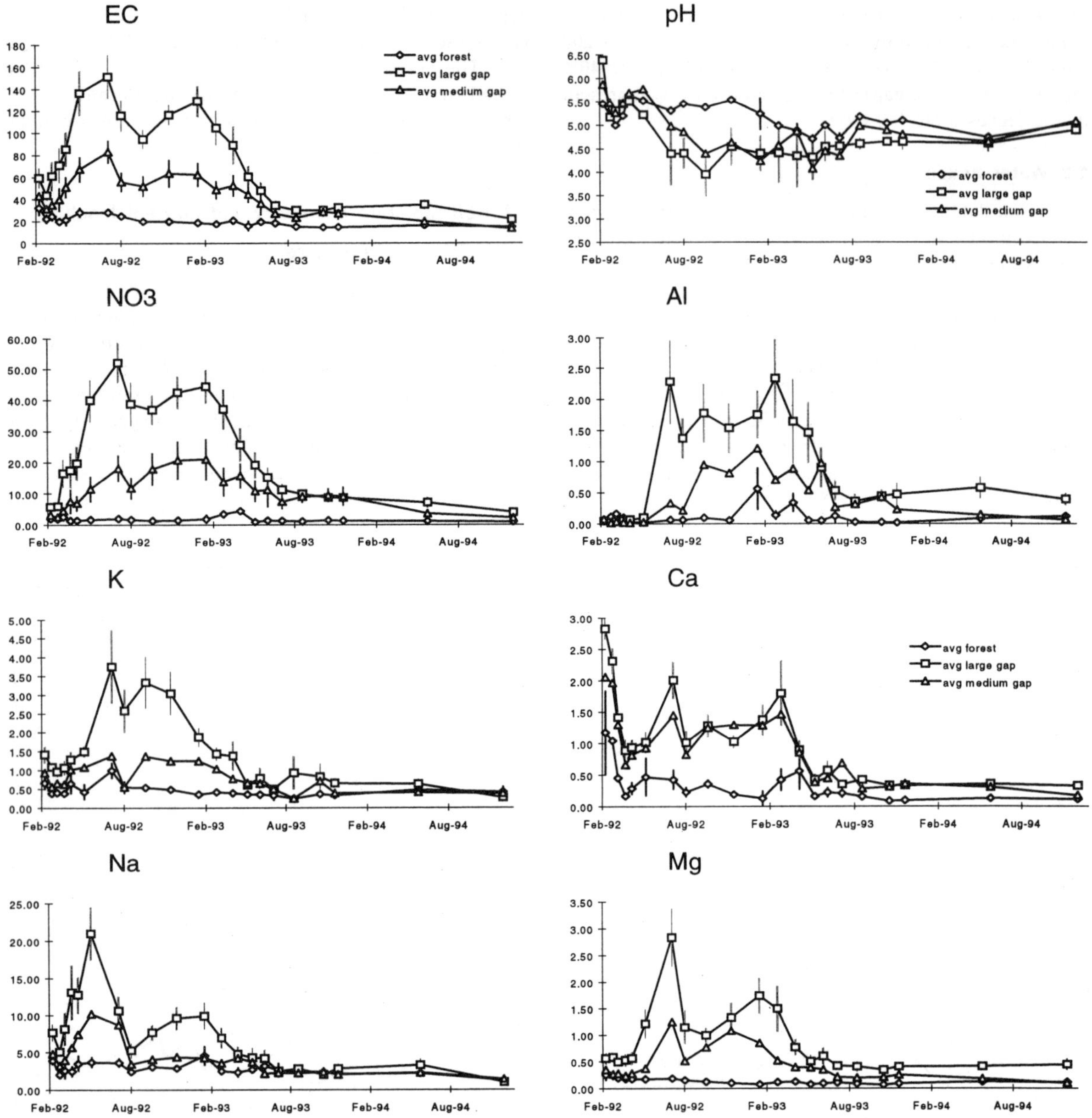


**Figure 4:** EC, pH and Al, Ca, K, Mg, Na and $NO_3$ concentrations (in mg $l^{-1}$, mean ± standard error) in soil water at 129 cm depth in the large and medium gap during the first 34 months after logging. For comparison the concentrations in closed forest are also shown. Where the bars indicating standard error are not visible, the standard error is smaller than the symbol size.

three months after logging. A maximum EC of 150 μS $cm^{-1}$ was reached in the large gap after approximately five months. High EC levels were maintained for another seven months. In the medium gap the rise in EC was more modest and remained typically in the range of 50–60 μS $cm^{-1}$. About one year after logging the EC of the soil water started to decrease both in the large and the medium sized gap. From 16 to 34 months after logging the EC of the soil water in the large gap remained relatively stable at approximately double the value of the soil water in the closed forest. In the meantime the EC of the soil water in the medium gap decreased gradually to values indistinguishable from background values in the closed forest (Figure 4).

The temporal variation of $NO_3$, K, Mg and Ca was similar to the temporal variation of the EC. The average concentration of $NO_3$ between 3 and 12 months after logging was typically 40–45 mg $l^{-1}$ (10 mg $l^{-1}$ $NO_3$-N) in the large gap, about 20 times more than the concentration in closed forest. In the medium gap the $NO_3$ concentrations were about half of the concentration in the large gap. $NH_4$ concentrations were low, both in the gaps and in closed forest. Ca, K and Mg concentrations showed a 3 to 10 times raise in the large gap compared to the closed forest and a 2 to 5 times raise in the medium gap. The difference between cation concentrations in the large gap and the medium gap were most obvious for K and almost non-existent for Ca. Na concentrations showed a somewhat different pattern than the EC. The concentrations of Na started to rise earlier than any other element and also peaked earlier. Maximum concentrations were reached after three months and then returned to background levels in the medium gap, but remained higher for another 9 months in the large gap.

The acidity of the soil water increased about five months after logging and the pH dropped from typical values between 5.0 and 5.5 to less than 4.5. Between 15 and 34 months the acidity of the soil water in the gaps decreased and approached the acidity of the soil water in the closed forest. As the acidity increased, the Al concentrations started to rise from less than 0.1 mg $l^{-1}$ to more than 1.5 mg $l^{-1}$ in the large gap. Even after 34 months the Al concentrations in the soil water remained clearly higher in the large gap than in the closed forest (Figure 4). Again, the changes in the medium gap were more modest than in the large gap.

#### *4.3.2 Effects of gap zonation*

Figure 5 shows the temporal course of solute concentrations in soil water per individual zone in the gaps and Table 4 gives the average concentrations in these zones for the first 15 months after logging and the period 16–34 months after logging.

In both gaps and for all solutes, except Na and Cl, the response to logging was strongest in the skidder zone (Figure 5). $NO_3$, Al, K and Mg concentrations in the skidder zone of the large gap were often about double the concentrations in the other zones. Even greater differences between skidder zone and the other zones occurred in the medium sized gap. Solute concentrations in the undisturbed zone and the crown zone of the medium gap responded similarly to logging (Figure 5, Table 4). In the crown zone of the large gap however, leaching of K, Na and Cl was greater than in the undisturbed zone. The increasing acidity of the soil water which was very clear in the skidder zone and undisturbed zone, was much more modest in the crown zone. Similarly, the raise in Al concentration was small in the crown zones of both gaps.

The highest K concentrations were observed in the skidder zone during the first 15 months after logging, but elevated K concentrations were most prolonged in the crown zones of the gaps (Figure 5, Table 4). The same pattern was observed for Mg in the large gap, but not in the medium gap. The skidder zone in the medium gap recovered more slowly than in the large gap, indicated for example by the longer duration of the $NO_3$ peak and the relatively low pH values during the last year of measurements (Figure 5). Re-growth of vegetation on the skidder zone of the medium gap was also slower than in the large gap, especially from May 1993 onwards.

#### *4.3.3 Nutrient leaching*

Table 5 presents the calculated nutrient leaching losses for the different zones. Leaching losses were calculated for the period 0–652 days after logging, covering the period with available meteorological measurements (February 1992–November 1993). Additionally, nutrient leaching was estimated for the period November 1993–December 1994, but now using the meteorological data for 1993 to estimate the water fluxes in 1994.

Most nutrient leaching occurred in the skidder zones of the gaps, then in the crown zones and the undisturbed zones. The large gap showed much larger nutrient losses through leaching than the medium gap. In general, the patterns observed in total nutrient losses were quite similar to the patterns described for nutrient concentrations, demonstrating that the variation in soil water nutrient concentrations dominates above differences in water percolation (see also *Parker*, 1985).

Most nutrient losses occurred within the first 652 days. However as the results for the period 652 to 1030 days after logging indicate, leaching losses continued to be larger in the gaps when compared with closed forest. Leaching losses in the undisturbed and the crown zone of the medium gap approached the leaching losses in unlogged closed forest during the last year (Table 5).

## 5 Discussion

### 5.1 Decomposition in a large logging gap

In closed forest, green leaves decomposed faster than brown leaves, and moreover, nutrient release patterns differed. After one year no P was released from brown leaves, contrasting to about 50% P release from green leaves (Table 2). Similar but smaller differences between release patterns in brown and green leaves were found for other nutrients. These results implicate that naturally shed leaves should be used and not green leaves picked from a tree if the purpose of a study is to describe the decomposition process representative for "undisturbed" conditions.

The comparison of decomposition rates between closed forest and two sites in a logging gap indicated that there was no uniform effect of logging gaps on decomposition rates. In the skidder zone of the gap decomposition rates were slower than in adjacent closed forest. The slow decomposition rate in the open area of the gap is in line with the results of *Ewel* (1977), who found 10 to 18% slower rates of decomposition in a cleared area, which is in

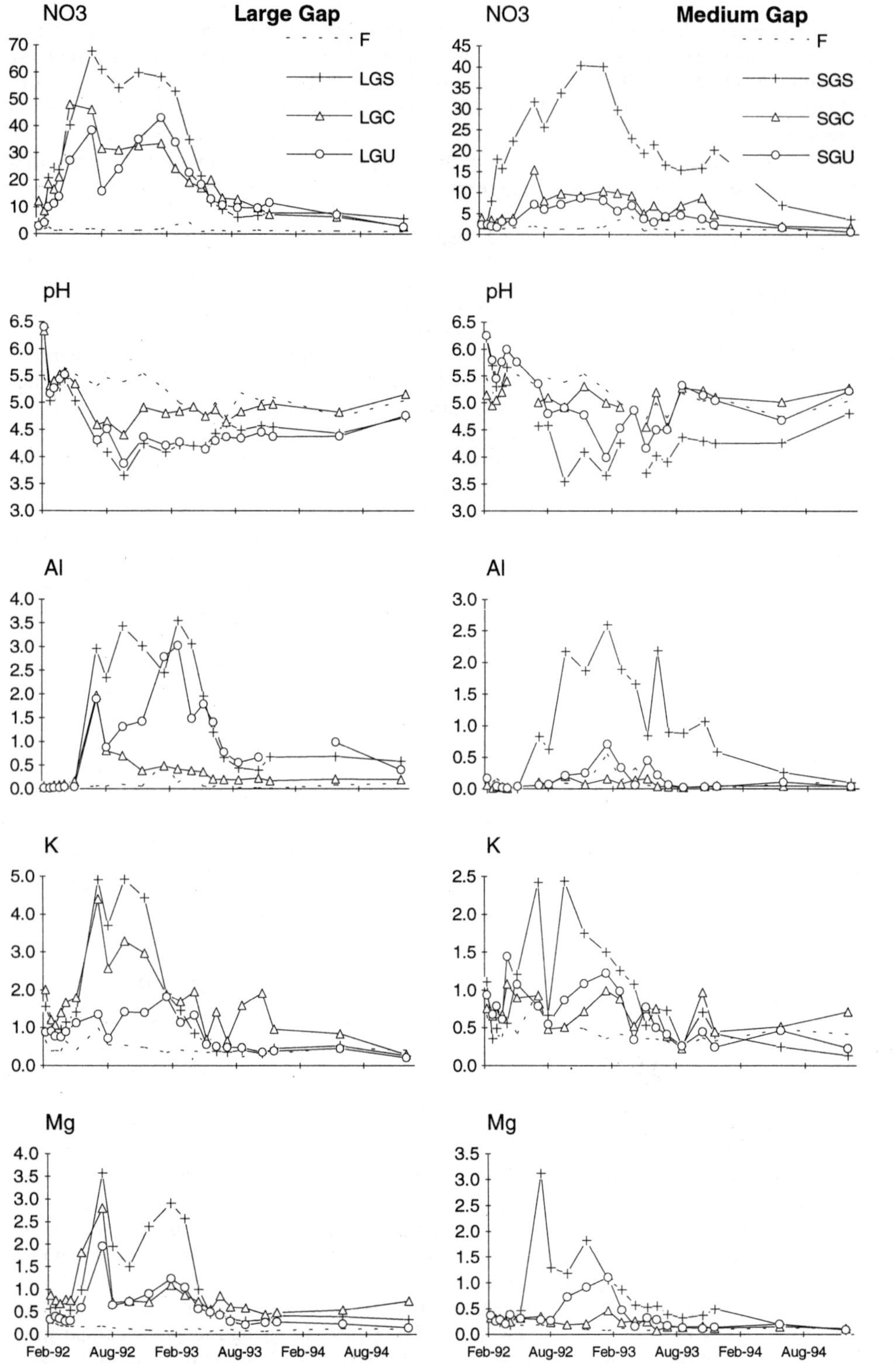


**Figure 5:** The pH and $NO_3$, Al, K and Mg concentrations (mg $l^{-1}$) in soil water during the first 34 months after logging for different zones in the large gap (left) and the medium gap (right). The concentrations in closed forest were added for comparison. Note that the scale of right hand and left hand graphs is not necessarily the same. F = closed Forest, LGS = Large Gap Skidder zone, LGC = Large Gap Crown zone, LGU = Large Gap Undisturbed zone, SGS = Medium Gap Skidder zone, SGC = Medium Gap Crown zone, SGU is Medium Gap Undisturbed zone.

**Table 4:** Average concentrations in soil moisture at 120 cm depth from February 1992–April 1993 and from May 1993 to December 1994. Abbreviations: LGS = Large Gap Skidder zone, LGC = Large Gap Crown zone, LGU = Large Gap Undisturbed zone, SGS = Medium Gap Skidder zone, SGC = Medium Gap Crown zone, SGU is Medium Gap Undisturbed zone.

| zone | EC µS/cm | pH | Al mg $l^{-1}$ | Ca mg $l^{-1}$ | K mg $l^{-1}$ | Mg mg $l^{-1}$ | Na mg $l^{-1}$ | Si mg $l^{-1}$ | $SO_4$ mg $l^{-1}$ | $NH_4$ mg $l^{-1}$ | Cl mg $l^{-1}$ | $NO_3$-N mg $l^{-1}$ |
|---|---|---|---|---|---|---|---|---|---|---|---|---|
| February 1992–April 1993 | | | | | | | | | | | | |
| forest | 22.9 | 5.3 | 0.14 | 0.43 | 0.51 | 0.15 | 3.07 | 3.76 | 1.23 | 0.19 | 3.71 | 0.45 |
| LGC | 111.2 | 5.1 | 0.54 | 1.32 | 2.46 | 1.18 | 13.90 | 4.99 | 4.40 | 0.24 | 15.21 | 7.68 |
| LGS | 120.7 | 4.7 | 1.81 | 1.66 | 2.46 | 1.60 | 8.97 | 4.26 | 1.58 | 0.27 | 7.36 | 9.64 |
| LGU | 75.0 | 4.7 | 1.13 | 1.22 | 1.15 | 0.76 | 6.64 | 4.03 | 2.46 | 0.42 | 8.83 | 6.90 |
| SGC | 40.1 | 5.1 | 0.10 | 0.91 | 0.76 | 0.29 | 5.27 | 3.85 | 1.72 | 0.30 | 9.47 | 2.22 |
| SGS | 76.5 | 4.7 | 1.20 | 1.57 | 1.27 | 0.96 | 5.51 | 3.66 | 1.38 | 0.28 | 5.07 | 7.18 |
| SGU | 45.7 | 5.3 | 0.18 | 1.19 | 0.92 | 0.46 | 5.31 | 4.24 | 1.57 | 0.33 | 10.83 | 1.49 |
| May 1993–December 1994 | | | | | | | | | | | | |
| forest | 15.9 | 4.9 | 0.07 | 0.14 | 0.37 | 0.09 | 2.31 | 3.23 | 0.71 | 0.12 | 1.64 | 0.27 |
| LGC | 43.0 | 4.9 | 0.21 | 0.43 | 1.12 | 0.60 | 5.18 | 4.69 | 4.15 | 0.13 | 2.54 | 2.47 |
| LGS | 30.5 | 4.5 | 0.80 | 0.47 | 0.41 | 0.38 | 1.19 | 3.65 | 0.82 | 0.11 | 1.24 | 2.13 |
| LGU | 35.3 | 4.4 | 0.94 | 0.23 | 0.44 | 0.30 | 1.92 | 3.73 | 0.96 | 0.21 | 1.85 | 2.43 |
| SGC | 20.7 | 5.1 | 0.06 | 0.23 | 0.57 | 0.14 | 2.73 | 3.50 | 1.02 | 0.18 | 1.31 | 1.11 |
| SGS | 39.3 | 4.2 | 0.88 | 0.56 | 0.45 | 0.36 | 1.85 | 4.01 | 0.34 | 0.21 | 1.14 | 3.30 |
| SGU | 19.7 | 4.9 | 0.13 | 0.29 | 0.42 | 0.17 | 2.31 | 3.34 | 0.76 | 0.12 | 2.13 | 0.75 |

**Table 5:** Calculated amounts of elements (kg $ha^{-1}$) leached to a depth greater than 120 cm. Quantities were calculated by multiplying the measured chemical composition of soil water at 120 cm depth with the percolated amount of water over the period in between two samplings as calculated with the SOAP water balance model. The 1994 figures (652–1030 days) were calculated using measured chemistry data, but using the 1993 meteorological data to estimate water percolation. LGS = Large Gap Skidder zone, LGC = Large Gap Crown zone, LGU = Large Gap Undisturbed zone, SGS = Medium Gap Skidder zone, SGC = Medium Gap Crown zone, SGU is Medium Gap Undisturbed zone.

| | period | Ca | K | Mg | Na | Cl | $NH_4$-N | $NO_3$-N | N-total |
|---|---|---|---|---|---|---|---|---|---|
| closed forest | 0–652 days | 5.1 | 8.5 | 2.2 | 50.5 | 54.2 | 1.7 | 7.2 | 8.8 |
| | 652–1030 days | 1.2 | 4.5 | 1.2 | 17.3 | 17.1 | 0.4 | 2.2 | 2.6 |
| | 0–1030 days | 6.3 | 13.0 | 3.4 | 67.8 | 71.3 | 2.1 | 9.3 | 11.4 |
| LGS | 0–652 days | 34.9 | 60.1 | 40.3 | 167.9 | 144.3 | 3.2 | 232.9 | 236.1 |
| | 652–1030 days | 5.4 | 5.5 | 4.9 | 15.9 | 14.9 | 0.3 | 21.2 | 21.6 |
| | 0–1030 days | 40.4 | 65.6 | 45.1 | 183.8 | 159.2 | 3.5 | 254.2 | 257.7 |
| LGC | 0–652 days | 24.6 | 50.7 | 27.8 | 248.8 | 248.3 | 4.3 | 154.8 | 159.1 |
| | 652–1030 days | 4.4 | 8.5 | 6.9 | 52.5 | 32.2 | 0.5 | 13.3 | 13.8 |
| | 0–1030 days | 29.0 | 59.2 | 34.8 | 301.3 | 280.5 | 4.8 | 168.1 | 172.9 |
| LGU | 0–652 days | 18.2 | 23.3 | 18.8 | 114.4 | 134.7 | 5.2 | 118.6 | 123.8 |
| | 652–1030 days | 2.2 | 4.1 | 2.3 | 13.8 | 20.3 | 0.4 | 14.0 | 14.4 |
| | 0–1030 days | 20.4 | 27.4 | 21.1 | 128.2 | 155.0 | 5.6 | 132.6 | 138.2 |
| SGS | 0–652 days | 37.7 | 31.6 | 29.5 | 101.5 | 112.0 | 4.4 | 152.9 | 157.3 |
| | 652–1030 days | 3.6 | 2.7 | 1.9 | 31.4 | 15.0 | 0.1 | 17.0 | 17.1 |
| | 0–1030 days | 41.3 | 34.3 | 31.4 | 132.9 | 127.0 | 4.6 | 169.9 | 174.5 |
| SGC | 0–652 days | 13.5 | 16.0 | 5.9 | 138.8 | 169.0 | 2.7 | 47.4 | 50.1 |
| | 652–1030 days | 2.6 | 7.1 | 1.6 | 21.2 | 11.7 | 0.4 | 5.1 | 5.4 |
| | 0–1030 days | 16.1 | 23.1 | 7.4 | 160.0 | 180.7 | 3.1 | 52.5 | 55.6 |
| SGU | 0–652 days | 17.9 | 16.8 | 8.3 | 117.3 | 165.8 | 4.6 | 29.0 | 33.6 |
| | 652–1030 days | 3.0 | 4.4 | 1.7 | 22.9 | 20.7 | 0.3 | 4.0 | 4.3 |
| | 0–1030 days | 20.9 | 21.1 | 10.0 | 140.2 | 186.5 | 4.9 | 32.9 | 37.8 |

the same order of magnitude as the difference found in this study. Similarly, *Maheswaran* and *Gunatilleke* (1988) found considerably slower decomposition in a fernland compared to an adjacent undisturbed rain forest.

The relative air humidity is relatively low in open areas of a logging gap (e.g. the skidder zone) and the temperature of the litter layer in such an area may reach up to 50 °C (*Brouwer*, 1996). These are harsh conditions for decomposer organisms and probably cause the slow decomposition rates here (cf. *Ewel* 1977, in Guatemala). An observation supporting the assumption of low activity of decomposer organisms is the only moderate increase of N concentration of decomposing leaves in the skidder zone compared to decomposing leaves in closed forest, because an increase in N concentration during decomposition is presumably due to microbial activity.

In the crown zone of the logging gap, large amounts of leaves are deposited with a relatively high nutrient content, which decompose relatively fast. So here, a large pulse of nutrients will be released through decomposition during the first year or so (Table 2). This pulse of nutrients can be: i) leached with infiltrating rain water, ii) taken up by surrounding or establishing vegetation or iii) stored temporarily in the soil.

The quantity of roots and fungi invading the leaves in the litterbags was assessed qualitatively for the bags recovered after 260 and 340 days. The amount of roots found in litterbags decreased in the order: green leaves in forest, brown leaves in forest, green leaves in gap, brown leaves in gap. The amount of fungi invading the litterbags decreased in the order: green leaves in gap, green leaves in forest, brown leaves in forest, brown leaves in gap. Litterbags heavily invaded by roots generally showed a relatively large weight loss and low Ca concentrations, suggesting that roots are actively involved in the decomposition process and perhaps cause preferential removal of Ca from decomposing material. This observation might be seen as supporting the hypothesis that tropical rain forests on infertile oxisols/ultisols are limited by Ca (cf. *Cuevas* and *Medina*, 1988). Also, the leaves in bags invaded by roots seemed to have relatively low Mg concentrations and relatively high N and P concentrations, but these observations were less conclusive.

### 5.2 Leaching

The creation of medium to large gaps during selective logging clearly leads to increased leaching. The increased leaching is due to increased percolation of water through the soil as well as increased solute levels in the percolating water. Over the first 22 months after logging a percolation of about 2800 mm was calculated in the skidder zone of the gaps, compared to 1800 mm in closed forest. The crown zone and undisturbed zone within the gaps experienced intermediate percolation rates (*Jetten*, 1994). Differences in soil moisture content between intact forest and the gap were most clear during dry periods. In closed forest the trees continuously extract water from the soil for their transpiration. In the bare zones of the gaps, soil moisture is only drained by gravity except for a rather thin layer at the topsoil which is rapidly desiccated by evaporation. Similar differences between gaps und forest have been observed in a few other tropical and temperate forests (*Parker*, 1985; *Bauhus* and *Bartsch*, 1995). *Parker* (1985), working in the high rainfall zone of Costa Rica on soils derived from volcanic material, reported a percolation of 2110 mm in a large clearing compared to 1252 mm in closed forest over a 404 day period. This finding corroborates the results of the present study. *Veenendaal* et al. (1995) also measured wetter soil conditions in a gap (diameter 26 m) in a forest in Ghana during the first dry season after gap creation and concluded that plants in the centre of the gap suffered less from drought stress than plants in the forest understorey, despite the much higher irradiation and vapour pressure deficit in the centre of the gap. Finally, *Uhl* et al. (1982) reported death of trees surrounding a clearing in poorly drained Amazon Caatinga forest, which they ascribed to the raise of groundwater levels after clearing. This indirectly supports the observation that where trees are removed, the soils become wetter.

The major increase in leaching rates after logging was caused by greater solute concentrations in the percolating soil water. Proportionally the largest increase in concentration was found for $NO_3$ which was 10–20 times greater in the gaps compared to closed forest during the first 15 months after logging. In the gaps Ca, K and Mg concentrations were 2–10 times greater than in closed forest, but rather large differences occurred between the different gap zones which are discussed in the following sections. A clear increase of $NO_3$ leaching after disturbance in forest ecosystems is common (cf. *Bauhus* and *Bartsch*, 1995; *Bormann* and *Likens*, 1967; *Herrera* et al., 1981; *Malmer* and *Grip*, 1993; *Parker*, 1985; *Uhl* and *Jordan*, 1984; *Uhl* et al., 1988; *Zulkifli*, 1989).

Increases in K leaching are usually reported to be proportionally greater than Ca and Mg (cf. *Herrera* et al., 1981; *Malmer*, 1993; *Poels*, 1987; *Uhl* et al., 1982; *Zulkifli*, 1989). In the present study the increase in Mg leaching in the gaps compared to closed forest was 5–10 times, versus 2–4 times for K and 5–6 times for Ca. Such a modest increase in K leaching was also reported by *Parker* (1985) for a Costa Rican forest. Another similarity between Parker's results and the present results is the rather large leaching of Na. The differences in leaching rates of individual solutes and possible explanations of these differences are discussed below in the section on the influence of zoning in the gaps.

#### 5.2.1 Gap size

In the large gap, leaching rates of $NO_3$, K and Mg were approximately double the rates in the medium gap, demonstrating that leaching indeed increased with a larger gap size, as expected. Other work supports these results. Neither *Uhl* et al. (1988) nor *Parker* (1985) found that single treefall gaps (47–164 $m^2$) had a significant effect on soil moisture composition, except for a slight increase in $NO_3$. Larger disturbances as multiple treefall gaps, clearings or clearing followed by burning caused considerably raised nutrient concentrations in percolating soil water (*Bauhus* and *Bartsch*, 1995; *Herrera* et al., 1981; *Parker*, 1985; *Uhl* and *Jordan*, 1984; *Uhl* et al., 1982). Few studies have focused on gaps in otherwise intact forest. In a temperate beech forest on acid soil, *Bauhus* and *Bartsch* (1995) evaluated soil water composition changes after gap creation (diameter 30 meter). They found levels of $NO_3$-N in soil water at 80 cm depth changing from 0 in an intact forest stand to about 15 mg $l^{-1}$ in the gaps. The main cation accompanying $NO_3$ during leaching appeared to be Al, although Ca and Mg also increased.

Arguably, the study that is most comparable to the present study is that of *Parker* (1985) in Costa Rica. Here two clearings, sized 50*10 and 50*50 meter, were made in a forest stand on a relatively fertile clayey ultisol, but no biomass was removed and soil disturbance was minimal. The calculated leaching losses in Parkers' study were of the same order of magnitude, although leaching of Ca, Mg and $NO_3$ under intact forest was substantially greater in Costa Rica, which is not surprising since the soils were more fertile. K and Na leaching was greater in Guyana, especially in gaps (Table 6, but note that the period used for calculation is much longer in Guyana).

The difference between leaching rates in the large and the medium gap is pronounced in this study, especially for the crown zone and the undisturbed zone contrary to the results in Costa Rica. Based on his calculations, *Parker* (1985) concludes that *"water and solute budgets appear to react to increasing disturbance intensity with a threshold for response and a maximal response at higher levels of disturbance, with a probable sharp transition in between"*, and, *"the threshold is likely to be less than 500 $m^2$ of forest cut and may be related to the extent that living roots from unaffected areas are present in the disturbed areas"*. The area (canopy opening) of the medium gap in this study was about 730 $m^2$ and round shaped, but in this area the soil water nutrient concentrations were distinctly lower than in the large gap. It is plausible that the extent of living roots from surrounding vegetation plays a decisive role in preventing high leaching rates (cf. *Silver* and *Vogt*, 1993). *Silver* and *Vogt* (1993) showed that after creation of a 900 $m^2$ gap in a premontane forest in Puerto Rico living fine root biomass initially remained constant or even increased slightly, but declined to 40% below control forest levels after two months. A larger scale disturbance caused by hurricane Hugo had much more impact and caused massive death of fine roots.

Given the importance of living roots in avoiding leaching after disturbance and the apparent negative relation with the intensity of disturbance, it is somewhat surprising that the calculated leaching losses in the rather narrow (10 m) clear-cut strip of *Parker* (1985) were comparable with leaching losses in the large clearing (50 x 50 m). In the narrow strip a much larger edge effect would have been expected since the distance to intact vegetation is not more than 5 m. It therefore remains to be seen if the calculations by *Parker* (1985) are representative. Individual examination of soil water content (p 76) and $NO_3$-N concentration (p 92) in Parkers' thesis suggests that there was a difference between the small and large clearing. Further work is necessary on hydrological and nutrient gradients from the gap centre to closed forest, preferably in combination with examination of root activity.

### 5.2.2 Effects of zones within gaps

The strongest pulse of solute leaching was observed in the skidder zone, except for Na and Cl. The skidder zone was also the zone with the most intensive disturbance, including locally complete destruction of the top soil roots, absence of surviving seedlings and higher maximum temperatures in the litter layer and top soil (*Brouwer*, 1996; *van Brunschot* and *de Lange*, 1992; *Schultz*, 1960). The amount of organic debris in the skidder zone was not more than already present before logging, and the decomposition rate of this debris was slow. In the crown zone, piles of fresh organic debris with relatively high nutrient concentrations were present after logging, and decomposition rates were equal, if not faster than in closed forest. The disturbance of the top soil in the crown zone was relatively minor. Given the fact that nutrient losses through leaching were larger in the skidder zone than in the crown zone, it is concluded that the disturbance of the top soil with roots therein, is a more important cause for increased leaching after logging then the release of large quantities of inorganic nutrients through decomposition of organic debris.

Na and Cl showed an alternative leaching behaviour compared to other solutes. Remarkably enough these elements were the only ones having larger concentrations in

**Table 6:** Leached nutrients (kg $ha^{-1}$ 1030 $days^{-1}$ for Guyana and kg $ha^{-1}$ 404 $days^{-1}$ for Costa Rica) under intact forest, single treefall gaps, and larger gaps resulting from the felling of trees. Data from Costa Rica derived from Parker (1985).

| | Ca<br>kg $ha^{-1}$ | K<br>kg $ha^{-1}$ | Mg<br>kg $ha^{-1}$ | Na<br>kg $ha^{-1}$ | N<br>kg $ha^{-1}$ |
|---|---|---|---|---|---|
| Guyana (1030 days after logging) | | | | | |
| closed forest | 6 | 13 | 3 | 68 | 11.4 |
| medium gap (730 $m^2$) | 26 | 26 | 16 | 144 | 90 |
| large gap (3440 $m^2$) | 30 | 51 | 34 | 204 | 190 |
| Costa Rica (404 days after logging) | | | | | |
| intact forest | 6 | 4 | 9 | 19 | 19 |
| single treefall gaps (± 100 $m^2$) | 7 | 3 | 8 | 15 | 12 |
| small clearing (500 $m^2$) | 23 | 5 | 36 | 53 | 108 |
| large clearing (2500 $m^2$) | 22 | 8 | 22 | 57 | 115 |

the crown zone (and generally also in the undisturbed zone) compared to the skidder zone. Two effects might play a role:

- rapid leaching of Na and Cl from the fresh organic debris in considerable quantities. Na is very mobile in the plant tissue of the rain forests under study as appeared from analysis of the throughfall chemistry and the rapid release of Na during decomposition of leaves (*Brouwer*, 1996). Less is known about Cl levels and mobility in plants, but concentrations of 2–10 mg $g^{-1}$ have been measured in leaf material of rain forest trees (*Fölster*, pers. comm.), which is comparable to normal K, Mg or Ca concentrations in tropical rain forest trees.
- an evapotranspiration-concentration effect. The skidder zones were wetter and experienced faster percolation rates, and thus less evapotranspiration concentration. The gap edges, where most of the crown zones were located, might have experienced relatively high evapotranspiration rates and therefore a relatively large evapotranspiration-concentration effect (*Camargo* and *Kapos*, 1995).

Increased Na and Cl leaching after disturbance is commonly observed (*Malmer*, 1993; *Poels*, 1987). It is most likely that these elements originate from organic debris and are very quickly released.

Interestingly, enhanced leaching of K and Mg was more prolonged in the crown zone, than in the skidder zone and the undisturbed zone. Two possible explanations are:

- the supply of mineral nutrients available for leaching diminished between 12 and 15 months after logging in the skidder zone and the undisturbed zone due to decreasing rates of decomposition and mineralisation of organic material and/or smaller quantities of organic material available for decomposition and mineralisation
- rapid growing pioneer species in the skidder zone and undisturbed zone had such a great nutrient demand and nutrient exploiting capacity of their roots that they were able to take up the majority of the nutrients in the soil solution, thereby limiting further leaching (cf. *Raaimakers*, 1994).

Although the first explanation will play a role, the rather rapid decrease of nutrient concentrations in soil water between 12 and 15 months after logging suggests that the effect of re-establishment of the vegetation was dominant (Figures 4 and 5). While the gaps were still extremely open 10 months after logging with only a few small and scattered *Cecropia obtusa* trees, 18 months after logging the central areas of the large gap were an almost impervious wilderness of secondary trees and vines. The skidder zone in the medium gap showed less vigorous re-growth and leaching rates tended to decrease more slowly to background levels than in the large gap, confirming the importance of rapid growing pioneer species in limiting leaching losses (see also *Raaimakers*, 1994).

### 5.2.3 Nitrification

The most dramatic change in soil water composition after logging concerned $NO_3$. The balance of N in the soil and its transformations into different molecules have important consequences for the soil pH, and indirectly for other soil processes (*Van Breemen* et al., 1983). When released from organic compounds, N is in the form of $NH_4$. The $NH_4$ can be taken up by plants or, in the presence of oxygen and nitrifying bacteria, can be transformed into $NO_3$, releasing two protons:

$$NH_4^+ + 40 \rightarrow NO_3^- + H_2O + 2\ H^+ \text{ (nitrification)}$$

If the $NO_3$ is not taken up by plants this will cause a drop in soil pH, which in acid soils will lead to an increased solubility of Al (*Van Breemen* et al., 1984; *Verstraten* et al., 1990). The Al ions or the protons in the soil solution displace exchangeable bases from the soil exchange complex into the soil solution, or prevent that bases released by decomposition are held by the soil exchange complex. In contrast to $NH_4$, $NO_3$ is very mobile in the soil. If excess $NO_3$ is leached downward in the soil profile it needs to be accompanied by a cation and this may therefore lead to an increased leaching of bases (*Cahn*, 1993; *Freiesleben* and *Rasmussen*, 1986).

The above described array of processes of N mineralisation, nitrification, acidification, increasing Al levels and leaching of base actions obviously occurred in the gaps after logging. Mineralisation of N was not necessarily faster in the gaps; a study in north Brazil has shown that the creation of gaps up to 2500 $m^2$ has little impact on mineralisation and nitrification rates (*Marrs* et al.1991; see also *Vitousek* and *Denslow*, 1986). Decreased N-uptake by roots already leads to a higher availability of $NH_4$, which in the presence of sufficient oxygen can be transformed rapidly into $NO_3$ (cf. the trenched plots of *Silver* and *Vogt*, 1993; *Robertson*, 1989).

Despite the much greater availability of organic debris in the crown zone, serving as potential input for mineralisation, the effects of acidification and raised Al levels were relatively minor here. This was possibly due to a neutralising effect of the relatively large quantities of Na and K that were released during the initial stages of decomposition, combined with a root system that, at least partly, remained intact in this zone.

Increased Al solubility may have some adverse effects. It is well known that Al in the soil solution can be toxic to plants, especially if the pH $< 5$ (*Gundersen* and *Rasmussen*, 1987; *Marschner*, 1991; *Wolt*, 1990). Al toxicity is generally visible in the root morphology and high Al levels in the soil often interfere with P and Ca nutrition of plants (*Lathwell* and *Grove*, 1986; *Marschner*, 1995; *Raaimakers*, 1995). Because different species have different tolerance to Al in the soil environment (*Wolt*, 1990), the soil acidification following selective logging might have a selective influence in the natural re-growth of species, and in this way influence future species composition (*Lathwell* and *Grove*, 1986). This is especially relevant in the skidder zones, where Al levels in the soil water were highest and did not return to background levels during the first 2.5 years after logging. Al may also have toxic effects on soil microorganisms, leading to slower decomposition rates (*Anderson* and *Swift*, 1983).

#### 5.2.4 Recovery to pre-logging leaching rates

The major pulse of leaching occurred during the first year after logging and for most solutes the concentrations in draining soil water clearly remained higher up to 15 months after logging. However Ca, Mg, $SO_4$ and $NO_3$ concentrations remained (marginally) higher than in closed forest up to 34 months after logging, indicating that a build-up had not started for those nutrients at that time. K seemed to be present at lower concentrations in the percolating soil water under the gaps during the last sampling date. This trend towards lower leaching losses than under pristine forests suggests that the recovering vegetation is able to reduce the leaching to such an extent that it creates the positive budget needed to build up new nutrient stocks. Obviously, the current data only suggest that this is possible, continued monitoring of the plots is necessary to supply more decisive data on this important issue.

## 6 Conclusions and recommendations for management

It was shown above that nutrient inputs by decomposition were highest in the crown zone, both due to the quantity of material and the rapid release of nutrients, and that nutrient inputs by decomposition in the skidder zone were even lower than in closed forest due to retarded decomposition. This was not reflected in the soil solution concentrations and the leached amounts of solutes, except for Na and Cl which were higher in the soil solution of the crown zone. The skidder zone with damaged top soil experienced the greatest nutrient losses by leaching.

Making a direct comparison between leaching losses in the gap areas and in closed forest, it appears that N losses were 17 times higher in the large gap than in the closed forest over a period of 1030 days. For Mg, Ca and K this was respectively 10, 5 and 4 times. In the medium gap the leaching losses were about 50% lower than in the large gap for N, Mg and K, but similar for Ca.

In an ecosystem with an extremely low content of weatherable minerals in the soil (*Brouwer*, 1996; *van Kekem* et al., 1997; see also Table 1), replenishment of nutrients lost through logging will primarily depend on atmospheric inputs, except perhaps for N. The total nutrient losses through logging (in extracted timber and due to enhanced leaching) are equal to: 88–260 (N), 18–52 (P) and 11–45 (Ca, K and Mg) years of atmospheric deposition, in case of the large gap (*Brouwer*, 1996). The ranges result from different methods to incorporate the interaction between a gap and surrounding intact forest. The ranges are wider when similar calculations are made for the medium gap: 19–260 years (N), 5–59 years (P), and 2–44 years (Ca, K, Mg). In the large gap, enhanced leaching contributes 25–50% of the total nutrient losses due to logging. In the medium gap this is 15–30%. Since leaching losses are partly avoidable and unnecessary, contrary to nutrients extracted in the timber itself, minimising these losses is important if sustainable logging is pursued. Based on the results of this study the following recommendations can be given to limit leaching losses after logging:

- Limit the range and the activity of skidders as much as possible. Skidders damage the topsoil with most of the fine roots, which is the most important cause for enhanced leaching.
- Creating a series of well distributed small gaps is to be preferred over one large clearing for two reasons: 1) leaching losses seem to increase with increasing distance to intact vegetation (i.e. with increasing gap size) and 2) a large clearing will be the centre of a zone where nutrients are depleted and it will take centuries before the vegetation biomass approaches the original biomass. In a series of small gaps, the nutrient losses are smaller and the losses are moderated by surrounding intact forest.

## Acknowledgements

We are grateful to: Joël van Brunschot and Karel de Lange who did a great job in the first months of the fieldwork; Sampurno Bruijnzeel for his comments on earlier versions of this paper and stimulating discussions; Kees Klawer and his crew of the chemistry lab for the analysis of the samples; Demarara Timbers Ltd for their kind assistance in creating the experimental gaps. This study was carried out as a part of the Tropenbos Guyana Programme.

## References

*Allen, S. E.* (1989). Chemical analysis of ecological materials. Blackwell Scientific Publications. Oxford, UK.

*Anderson, J. M.* and *Swift, M. J.* (1983). Decomposition in tropical forests. In: Tropical rain forest: Ecology and management. Sutton, S. L., Whitmore, T. C. and Chadwick, A. C. (eds.). Blackwell Scientific Publishers.

*Bauhus, J.* and *Bartsch, N.* (1995). Mechanisms for carbon and nutrient release and retention in beech forest gaps. I. Microclimate, water balance and seepage water chemistry. Plant and Soil 168–169: 579–584.

*Bormann, F. H.* and *Likens, G. E.* (1967). Nutrient cycling. Science 155: 424–429.

*Breemen, N. van, Driscoll, C. T.* and *Mulder, J.* (1984). Acidic deposition and internal proton sources in acidification of soils and waters. Nature 307: 599–604.

*Brouwer, L. C.* (1996). Nutrient cycling in pristine and logged tropical rain forest. A study in Guyana. Tropenbos Guyana Series – 1, Tropenbos Guyana, Georgetown, Guyana/PhD thesis Utrecht University, the Netherlands.

*Bruijnzeel, L. A.* (1990). Hydrology of moist tropical forests and effects of conversion: estate of knowledge review. UNESCO-IHP, Humid tropics programme, Paris, 224 pp.

*Brunschot, J. van* and *Lange, K. de* (1992). The impact of selective logging on the micro climate and the nutrient balance in two gaps of different size in tropical rain forests in Guyana. Tropenbos Guyana internal report/MSc thesis Utrecht University.

*Cahn, M. D., Bouldin, D. R., Cravo, M. S.* and *Bowen, W. T.* (1993). Cation and Nitrate Leaching in an Oxisol of the Brazilian Amazon. Agronomy Journal 85: 334–340.

*Camargo, J. L. C.* and *Kapos, V.* (1995). Complex edge effects on soil moisture and microclimate in central Amazonian forest. Journal of Tropical Ecology 11: 205–221.

*Colchester, M.* (1994). The new sultans. Asian loggers move into Guyana's forests. The Ecologist 24: 45–52.

*Cuevas, E.* and *Medina, E.* (1988). Nutrient dynamics within Amazonian forests. II. Fine root growth, nutrient availability and leaf litter decomposition. Oecologia 76: 222–235.

*Ewel, J. J.* (1977). Litter fall and leaf decomposition in a tropical forest succession in eastern Guatemala. Journal of Ecology 64: 293–309.

*Freiesleben, N. E. v.* and *Rasmussen, L.* (1986). Effects of acid rain on ion leaching in a Danish forest soil. Water, Air and Soil Pollution 31: 965–968.

*Gundersen, P.* and *Rasmussen, L.* (1987). Nitrification, acidification and aluminium release in forest soils. Laboratory of environmental sciences and ecology, technical university of Denmark, Lyngby, Denmark.

*Haines, B. L., Waide, J. B.* and *Tono, R. L.* (1982). Soil solution nutrient concentrations sampled with tension and zero-tension lysimeters: report of discrepancies. Soil. Sci. Soc. Am. J. 46: 658–661.

*Herrera, R., Jordan, C. F., Medina, E.* and *Klinge, H.* (1981). How human activities disturb the nutrient cycles of tropical rainforest in Amazonia. Ambio 10: 109–114.

*Jetten, V. G.* (1994). Modelling the effects of logging on the water balance of a tropical rain forest. A study in Guyana. Tropenbos Series 6, The Tropenbos Foundation, Wageningen, the Netherlands. 196 p. PhD thesis Utrecht University.

*Jetten, V. G.* (1994). SOAP, Soil atmosphere plant model. A one dimensional water balance model for a forest environment. Tropenbos Documents 5, Tropenbos Foundation, Wageningen, the Netherlands.

*Jordan, C. F.* (1985). Nutrient Cycling in Tropical Forest Ecosystems. Institute of Ecology, University of Georgia, Athens, Georgia 30602, USA.

*Kekem, A. J. van, Pullis, J.* and *Khan, Z.* (1996). Soils of Central Guyana. Tropenbos-Guyana series 2. Tropenbos-Guyana programme, Georgetown, Guyana.

*Lathwell, D. J.* and *Grove, T. L.* (1986). Soil-Plant Relationships in the tropics. Annual Reviews Ecological Systems 17: 1–16.

*Luxmoore, R. J.* et al. (1990). Physical and chemical controls of preferred path flow through a forested hillslope. Geoderma 46: 139–154.

*Maheswaran, J.* and *Gunatilleke, I. A. U. N.* (1988). Litter decomposition in a lowland rainforest and a deforested area in Sri Lanka. Biotropica 20: 90–99.

*Malmer, A.* (1993). Dynamics of hydrology and nutrient losses as response to establishment of forest plantation. A case study on tropical rainforest land in Sabah, Malaysia. PhD thesis, Swedish University of Agricultural Sciences, Umeå, Sweden.

*Malmer, A.* and *Grip, H.* (1993). Converting tropical rainforest to forest plantation in Sabah, Malaysia II. Effects on nutrient dynamics and net losses in stream water. Hydrological processes 8: 195–209.

*Marrs, R. H., Thompson, J., Scott, D.* and *Proctor, J.* (1991). Nitrogen mineralization and nitrification in terra firme forest and savanna soils on Ilha de Maraca, Roraima, Brazil. Journal of Tropical Ecology 7: 123–137.

*Marschner, H.* (1991). Mechanisms of adaptation of plants to acid soils. Plant and Soil 134: 1–20.

*Marschner, H.* (1995). Mineral Nutrition of higher plants. Academic Press, Harcourt Brace & Company, London.

*Nortcliff, S.* and *Thornes, J. B.* (1989). Variations in soil nutrients in relation to soil moisture status in a tropical forested ecosystem. In: Mineral Nutrients in Tropical Forest and Savanna Ecosystems, J. Proctor (ed.). Blackwell Scientific Publications.

*Parker, G. G.* (1985). The effect of disturbance on water and solute budgets of hillslope tropical rainforest in north-eastern Costa Rica. PhD thesis, University of Georgia, Athens, Ga, USA, 161 p.

*Poels, R. L. H.* (1987). Soils, water and nutrients in a forest ecosystem in Suriname. PhD thesis, Agricultural University Wageningen, the Netherlands.

*Raaimakers, D.* (1994). Growth of tropical rain forest trees as dependent on phosphorous supply. Tree saplings differing in regeneration strategy and their adaptations to a low phosphorus environment in Guyana. PhD thesis Utrecht University. Tropenbos Series 11, The Tropenbos Foundation, Wageningen, the Netherlands.

*Robertson, G. P.* (1989). Nitrification and denitrification in humid tropical ecosystems: potential controls on nitrogen retention. In: Mineral Nutrients in Tropical Forest and Savanna Ecosystems, J. Proctor (ed.). Blackwell Scientific Publications.

*Schulz, J. P.* (1960). Ecological studies on rain forest in Northern Suriname. Mededelingen van het botanisch museum en herbarium van de Rijksuniversiteit Utrecht 163, Utrecht University, the Netherlands.

*Silver, W. L.* and *Vogt, K. A.* (1993). Fine root dynamics following single and multiple disturbances in a subtropical wet forest ecosystem. Journal of Ecology 81: 729–738.

*Sollins, P.* and *Radulovich, R.* (1988). Effects of Soil Physical Structure on Solute Transport in a Weathered Tropical Soil. Soil Science Society America Journal 52: 1168–1173.

*Steege, H. ter* (1993). Patterns in tropical rain forest in Guyana. PhD thesis Utrecht University. Tropenbos Series 3. The Tropenbos Foundation, Wageningen, the Netherlands.

*Steege, H. ter* et al. (1995). Basic and applied research for sound rain forest management in Guyana. Ecological Applications 5: 904–910.

*Uhl, C.* and *Jordan, C. F.* (1984). Succession and nutrient dynamics following forest cutting and burning in Amazonia. Ecology 65: 1476–1490.

*Uhl, C.* et al. (1982). Ecosystem recovery in Amazon caatinga forest after cutting, cutting and burning, and bulldozer clearing treatments. Oikos 38: 313–320.

*Uhl, C., Clark, K., Dezzeo, N.* and *Maquirino, P.* (1988). Vegetation dynamics in Amazonian treefall gaps. Ecology 69: 751–763.

*Veenendaal, E. M.* et al. (1995). Differences in plant and soil water relations in and around a forest gap in West Africa during the dry season may influence seedling establishment and survival. Journal of Ecology 83: 83–90.

*Verstraten, J. M., Dopheide, J. C. R., Duysings, J. J. H. M., Tietema, A.* and *Bouten, W.* (1990). The proton cycle of a deciduous forest ecosystem in the Netherlands and its implications for soil acidification. Plant and Soil 127: 61–69.

*Vitousek, P. M.* and *Denslow, J. S.* (1986). Nitrogen and phosphorus availability in treefall gaps of a lowland tropical rainforest. Journal of Ecology 74: 1167–1178.

*Wolt, J. D.* (1990). Effects of acidic deposition on the chemical form and bioavailability of soil Aluminium and Manganese. In: A. A. Lucier and S. G. Haines: Mechanisms of forest response to acidic deposition, Springer Verlag, New York.

*Zanten, S. van* (1991). Structure, soil and litter of a tropical rain forest. A comparative study of three different forest types in the Mabura Hill Ecological Reserve. Tropenbos Guyana internal report/MSc thesis Wageningen Agricultural University.

*Zulkifli, Yusop* (1989). Effects of selective logging methods on dissolved nutrient exports in Berembun watershed, Peninsular Malaysis. Regional seminar on Tropical Forest Hydrology, 4–9 September, Session IV, Kuala Lumpur, organized by FRIM, IHP and Unesco.

# Logging can cause a serious lack of calcium in tropical rainforest ecosystems: An example from Sabah, Malaysia

**Nils Nykvist**

Department of Forest Ecology, Swedish University of Agricultural Sciences, S-901 83 Umeå

Postal address: Professor emeritus Nils Nykvist, Skogsstigen 1, S-144 62 Rönninge, Sweden, Telephone and telefax: +46853252631

## Abstract

Chemical analysis of total calcium in soil samples from Mendolong research area revealed that they had a very low calcium content. The content was so low that the harvesting of stemwood and stembark from all trees larger than 20 cm DBH should result in the loss of about 19 % of all calcium in the ecosystem down to 50 cm depth. This finding was not expected based on earlier results of analyses of foliar contents and contents of exchangeable calcium in the soil.

Comparable figures for total calcium contents in the soil are not available from other tropical rainforest ecosystems. They have therefore been calculated by multiplying literature values for concentrations of total calcium in areas with tropical rainforest climate by soil weights from corresponding depths in Mendolong. The total calcium contents in the mineral soil to 50 cm depth varied from more than 20 000 kg ha$^{-1}$ in Chile to less than 200 kg ha$^{-1}$ in Brazil. This great variation indicates that not all tropical rainforests have low contents of calcium in their mineral soil. However, low contents can be found in certain areas which can be rather widespread in the humid tropics. In such areas, sustainable forestry is not possible without replacing the calcium lost from forest harvesting.

## 1 Introduction

Calcium is one of the most important plant nutrients. In spite of this, no investigations have been able to show any deficiency of calcium in forest stands although many forest scientists have pointed out that calcium is one of the plant nutrients most likely to reach levels low enough to result in a deficiency in forest soils after logging, especially when branches and leaves are also harvested (*Weetman* and *Webber*, 1972; *Boyle* et al., 1973; *Silkworth* and *Grigal*, 1982; *Shortle* and *Smith*, 1988; *Federer* et al., 1989).

The aim of this paper is to show that there is a great risk for a deficiency of calcium developing in some areas when tropical forests are logged. The simplest way to determine which areas are at risk is to make a total analysis of the mineral soil.

## 2 Study area

The study area was situated in Mendolong at an altitude of 650–750 m about 35 km southeast of Sipitang, at the Brunei Bay in Sabah, Malaysia. The bedrock consisted of interbedded sandstones, siltstones and shales on top of which Orthic Acrisols or Podzols were formed depending on the type of bedrock. The lowland hill dipterocarp forest growing in the area had been lightly selectively logged about six years before sampling.

## 3 Sampling and chemical analyses

Twenty sample trees representing different diameter classes larger than 20 cm diameter of breast height were harvested. Ground vegetation and trees smaller than 20 cm DBH were harvested from ten 10 x 10 m sample plots. Forest and biomass sampling procedures are described in more detail in *Sim* and *Nykvist* (1991).

The mineral soil used for analysis of exchangeable calcium was sampled with an auger (area 13.8 cm$^2$) before planting a trial plot after clear-felling in Mendolong. For the total analysis of calcium, the soil was sampled from ten plots four years after clear-felling in the *Acacia mangium* plantation. From each plot, two cylinder samples (height 5 cm, diameter 7.2 cm) were taken from each 5-cm layer of the mineral soil down to 50 cm.

Concentrations of exchangeable calcium were determined by ICP-technique after extraction with 1 N ammonium acetate at pH 7 and filtration of the solution. A total analysis of calcium was carried out with the same technique after digesting the soil samples with perchloric and hydrofluoric acids. A more detailed description of the methods can be found in *Emteryd* (1989).

# 4 Results

## 4.1 Concentrations of calcium in biomass

### 4.1.1 Foliar calcium

Foliar analysis is commonly used, especially in agriculture, to determine whether any deficiency of plant nutrients exists in crops. The concentrations of calcium in leaves from the rainforest in Mendolong before clear-felling and in the following plantation of *Acacia mangium* are given in Table 1. In nine out of 15 investigations from tropical rainforests, the concentration of foliar calcium was lower compared with that in Mendolong (Tab. 2). *Ovington* and *Olson* (1970) reported a minimum value of 0.23 and a maximum of 2.92% for foliar calcium among trees in a montane rainforest in Puerto Rico. The lowest figure obtained for the rainforest in Mendolong was 0.48% which is not indicative of any serious lack of calcium in the ecosystem.

The concentration of calcium in the leaves of *Acacia mangium* was lower than that in the leaves of rainforest trees from the same site before clear-felling (Tab. 1). However, calcium concentrations in *Acacia mangium* leaves did not reach the critical level of 0.2%, below which growth is suppressed according to Mead, who studied *Acacia mangium* plantations in Sabah (Unpublished. From *Shrivastava*, 1993).

### 4.1.2 Concentrations of calcium in the above-ground biomass

A great amount of calcium is stored in the stems (*Sim* and *Nykvist*, 1991). Thus, the concentration of calcium in the total above-ground biomass provides a better indication of low amounts of this element in the ecosystem compared with values obtained in foliar analysis. The concentration of calcium in the above-ground biomass of the tropical rainforest in Mendolong was 0.25%. In 17 other stands for which concentrations of calcium in the above-ground biomass of tropical rainforests could be calculated based on published data, the concentration was lower in eight, most of which were situated in Brazil and Venezuela (Tab. 2).

**Table 1:** Foliar calcium in catchments W4 and W5 in Mendolong research area for trees larger and smaller than 20 cm DBH and for *Acacia mangium*. Mean figures (in per cent) with 95% confidence intervals. Sample size (n) varies from 3 to 10.

| Rainforest | | | | *Acacia mangium* | | | |
|---|---|---|---|---|---|---|---|
| Trees > 20 cm | | Trees < 20 cm | | Age 1 year | | Age 3.7 years | |
| W4 | W5 | W4 | W5 | W4 | W5 | W4 | W5 |
| 0.48 | 0.79 | 0.64 | 0.66 | 0.35 | 0.46 | 0.42 | 0.41 |
| ± 0.22 | ± 0.31 | ± 0.36 | ± 0.44 | ± 0.01 | ± 0.29 | ± 0.09 | ± 0.10 |
| n = 10 | n = 10 | n = 5 | n = 5 | n = 3 | n = 3 | n = 5 | n = 5 |

## 4.2 Amounts of calcium in the soil

### 4.2.1 Exchangeable calcium

Chemical analysis of exchangeable calcium is the most common method for determination of calcium contents in mineral soil. Amounts of exchangeable calcium in the Mendolong soil and amounts reported in 17 other investigations of tropical rainforest soils are given in Table 2. In five of the other investigations, the calcium amounts were found to be lower than that in the Mendolong soil. Thus, the Mendolong soil should not be considered as having an alarmingly low content of calcium.

### 4.2.2 Total calcium in the soil

The amount of total calcium in the mineral soil to 50 cm depth in Mendolong research area was 730 ± 110 kg $ha^{-1}$ which is about 35% of the total calcium in the ecosystem (*Nykvist*, 1997a). Corresponding amounts of mineral-soil calcium in two spruce ecosystems in Sweden were 40000 and 57000 kg $ha^{-1}$ (*Nykvist*, 1977).

The loss of calcium when harvesting stemwood from rainforest trees larger that 20 cm DBH was about 126 kg $ha^{-1}$ (*Nykvist* et al., 1994). If stemwood and stembark from all trees larger than 20 cm DBH would have been harvested, the loss of calcium would have been almost 400 kg $ha^{-1}$ which is about 19% of the total calcium in the ecosystem and almost 55% of the storage of total calcium in the soil.

Low contents of total calcium in the mineral soil were also found outside the research area in Mendolong. In a surrounding area covering about 46000 ha, the content of total calcium in the mineral soil to 100 cm depth was 790 ± 440 kg $ha^{-1}$. The soil was sampled at random from 23 sampling plots covering most of the area from which raw material for Sabah Forest Industries is obtained. Wood ash from the factory is now transported back to the forest plantations to compensate for nutrient losses associated with any further utilization of the forest resources in the area. To my knowledge, Sabah Forest Industries is the first pulp mill in the world that recycles plant nutrients on a regular basis.

# 5 Discussion

None of the most common methods for determining the availability of calcium in forest ecosystems viz. analyses of foliar contents, concentrations of calcium in aboveground biomass or exchangeable calcium in the soil, revealed alarmingly low contents of calcium in Mendolong compared with other investigations of calcium in tropical rainforest ecosystems (Tab. 2). However, when analyzing the total calcium in the soil, it was found that the amount in the soil was so low that a single harvest of logs from the rainforest in Mendolong can remove about 19% of the total amount in the whole ecosystem down to 50 cm soil depth (*Nykvist*, 1997a).

**Table 2:** Concentrations of calcium in leaves and above-ground biomass, and amounts of calcium in above- and below-ground biomass and as exchangeable in soils from different tropical rainforest areas. Limestone areas are excluded.

| | Concentrations | | Amounts | |
|---|---|---|---|---|
| | Leaves | Above-ground biomass | Biomass | Exchange-able in soil |
| | % | % | kg ha$^{-1}$ | kg ha$^{-1}$ |
| Malaysia, Mendolong<br>*Sim* and *Nykvist*, 1991<br>*Nykvist*, 1997a<br>Soil depth 0–50 cm | 0.64 | 0.25 | 1021 | 37 |
| Indonesia, Kalimantan<br>*Ruhiyat*, 1989<br>Soil depth 0–50 cm | 0.79 | 0.30 | 1744 | 60 |
| Papua New Guinea<br>Montane<br>*Edwards* and *Grubb*, 1982<br>*Grubb* and *Edwards*, 1982<br>Soil depth 0–30 cm | 1.50 | 0.41 | 1625 | 3750 |
| Ghana<br>*Greenland* and *Kowal*, 1960<br>Soil depth 0–30 cm | 1.54 | 0.75 | 2104 | 2580 |
| Ivory Coast, Banco<br>*Bernhard-Reversat* et al., 1978<br>Soil depth 0–50 cm | | 0.27 | 1500 | 150 |
| Ivory Coast, Yapo<br>*Bernhard-Reversat* et al., 1978<br>Soil depth 0–50 cm | | 0.51 | 2375 | 170 |
| Surinam<br>*Poels*, 1987<br>Soil depth 0–60 cm | 0.54 | 0.57 | 2918 | 170 |
| Brazil, Manaus<br>*Klinge*, 1976<br>Soil depth 0–30 cm | 0.21 | 0.10 | 506 | 0 |
| Brazil, Jari Florestal<br>*Russell*, 1983<br>Soil depth 0–50 cm | 0.23 | 0.23 | 1218 | 40 |
| Venezuela, San Carlos<br>Oxisol<br>Jordan, 1989<br>Soil depth 0–30 cm | 0.13 | 0.08 | 249 | 7 |
| Venezuela, San Carlos<br>Tall *caatinga* forest<br>*Herrera*, 1979. From<br>*Medina* and *Cuevas*, 1989<br>Soil depth 0–40 cm | | 0.13 | 518 | 130 |
| Venezuela, San Carlos<br>*Guaco* forest<br>*Buschbacher*, 1984. From<br>*Medina* and *Cuevas*, 1989<br>Soil depth 0–30 cm | 0.16 | 0.08 | 358 | 3 |
| Venezuela, Western Llanos<br>*Hase* and *Fölster*, 1982<br>Soil depth 0–50 cm | 1.50 | 0.85 | 3690 | 1680 |
| Venezuela, San Eusebio<br>Montane<br>*Grimm* and *Fassbender*, 1981<br>Soil depth 0–40 cm | 0.46 | 0.21 | 899 | 200 |
| Colombia, Terrace<br>*Fölster* et al., 1976<br>Soil depth 0–50 cm | 0.50 | 0.23 | 571 | 30 |
| Colombia, Slope<br>*Fölster* et al., 1976<br>Soil depth 0–50 cm | 0.50 | 0.25 | 986 | 1780 |
| Jamaica, Mull ridge<br>Montane<br>Tanner, 1985<br>Soil depth 0–40 cm | 0.93 | 0.28 | 1182 | 240 |
| Jamaica, Mor ridge<br>Montane<br>*Tanner*, 1985<br>Soil depth 0–45 cm | 0.59 | 0.17 | 615 | 30 |

Analysis of exchangeable calcium was obviously not a satisfactory method for estimating the calcium status of the soil in Mendolong. This became even more obvious when the amount of exchangeable calcium was compared with the uptake of calcium by plants.

Amounts of exchangeable calcium to 30 cm depth were 20.7 kg ha$^{-1}$ before planting a trial plot after clear-felling in Mendolong and 9.8 kg ha$^{-1}$ in the 45-month-old plantation of *Acacia mangium* in the same trial plot (*Nykvist*, 1997b). The total amount of calcium in the humus layer had decreased from 65 to 57 kg ha$^{-1}$. However, the total amount of calcium in the biomass of the 45-month-old plantation was 177 kg ha$^{-1}$. Calculations showed that the plantation had taken up and stored 47 kg ha$^{-1}$ in the biomass per year, which is much more than the decrease in calcium amounts in humus and mineral soil to 30 cm depth.

Most exchangeable calcium was found in the topsoil, with only small amounts in soil layers deeper than 30 cm (*Nykvist*, 1997a). The loss through leaching was greater than the atmospheric input of calcium (*Grip* et al., 1994), and the input of nutrients dissolved in laterally flowing water was negligible because the investigated plot was situated near the water divide. Thus, the only explanation for the great uptake of calcium by plants in relation to the loss of exchangeable amounts in the soil, could be chemical weathering of minerals in the soil.

Exchangeable calcium is only a minor part of the total storage of calcium in a soil. Most calcium is bound in organic compounds, calcium-containing silicates (e.g. augite, hornblende and anorthite) and carbonates. Thus, to estimate the total content of calcium in a soil, one must do more than just oxidize the organic substances in the soil and extract the soil samples in acids. Such treatments will bring calcium in organic substances and carbonates into solution but not that in silicates. To determine the total contents of calcim, the silicates must be completely dissolved before the chemical analyses of the different elements are carried out. The most common method is digestion in perchloric and hydrofluoric acids in which volatile $SiF_4$ is formed and lost, thus bringing all cations in solution.

Low amounts of exchangeable calcium in soils from other parts of the world (Tab. 2) indicate that the calcium content of the soil can be a critical factor influencing forest production in other rainforest areas as well.

To my knowledge, there are no other investigations of tropical rainforest ecosystems in which amounts of total calcium have been investigated in both soils and biomass. Figures for the amounts of total calcium in soils from tropical rainforest ecosystems therefore have to be obtained elsewhere.

There are such figures for concentrations of total and exchangeable calcium in mineral soil in the "FAO-UNESCO. Soil Map of the World" and in papers by *Stark* (1970, 1971) and *Waterloo* (1994). However, the concentration figures are from different soil depths and are therefore not comparable. To get them comparable, the amounts have been calculated by multiplying the concentrations by the soil weights from corresponding depths in the Mendolong research area and then summing them to a soil depth of 50 cm. Even though the bulk density varied among the different soils, this variation in mineral soils is very small compared with the great differences in amounts from different areas (Fig. 1). Amounts of calcium from Peru, Brazil and Suriname were calculated from mean figures of concentrations from only two or three soil depths multiplied by the total dry weight of the Mendolong soil to 50 cm depth. These figures are less accurate than the others; nevertheless, the very low amounts of calcium in the investigated soils from Brazil are alarming.

The very great differences in the amounts of total calcium in soils from areas with rainforest climate show that low calcium contents in soils are not a characteristic feature for tropical rainforests but restricted to certain geographical areas.

The main reason for the low content of total calcium in the Mendolong soil is that the bedrock consists of interbedded sandstones, siltstones and shales with low calcium contents. Concentrations of calcium in sandstone and shale were 0.007 and 0.022%, respectively. The soil is formed from weathering in situ; thus bedrock composition and soil age are a much better indicator of soil nutrient status in such areas than in earlier glaciated areas where the soil contains minerals from different bedrock types mixed by the inland ice. A deficiency of any plant nutrient in the bedrock would therefore be more pronounced in tropical areas than in most temperate ones. However, the low fertility of many tropical rainforest soils has long been concealed by an accumulation of plant nutrients in the biomass without any losses due to harvesting. Areas with poor soils have remained forested because the most productive soils in the humid tropics have mostly been utilized for growing cash crops.

## 6 References

*Bernhard-Reversat, F.* (1975). "Recherche sur l'ecosysteme de la foret subequatoriale de basse Cote-d'Ivoire. VI. Les cycles des macroelements". La Terre et la Vie, 29: 229–254.

*Bernhard-Reversat, F., C. Huttel* and *G. Lemée* (1978). "Structure and functioning of evergreen rainforest ecosystems of the Ivory Coast". In Tropical forest ecosystems pp. 557–574. UNESCO. Paris.

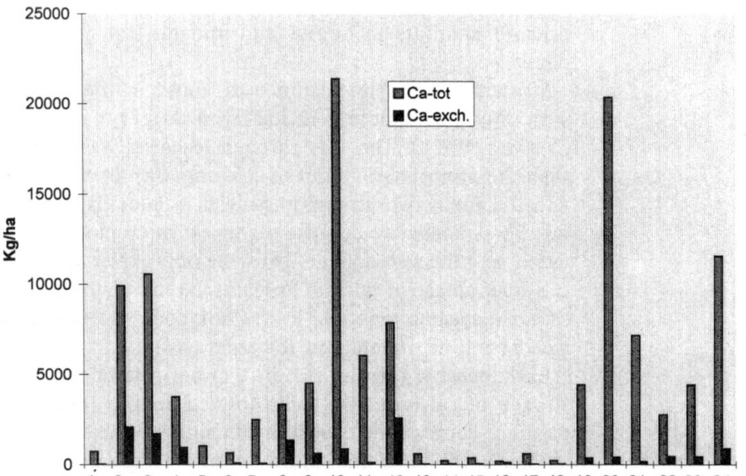


**Figure 1:** Amounts of total calcium in mineral soil to 50 cm depth. Amounts were calculated taking concentrations found in the FAO-UNESCO Soil map of the world, Stark (1970, 1971) and Waterloo (1994) and multipling them by soil weights from corresponding depths at the Mendolong research area. In kg $ha^{-1}$. Limestone areas are excluded.

1. This study.
2.–4. Fiji. From Waterloo, M. J. (1994).
5. Hawai. From FAO-UNESCO. Soil map of the world. Volume X. Australasia.
6.–8. Madagaskar. From FAO-UNESCO. Soil map of the world. Volume VI. Africa.
9. Sierra Leone. From FAO-UNESCO. Soil map of the world. Volume VI. Africa.
10. Chile. From FAO-UNESCO. Soil map of the world. Volume IV. South America.
11. India. From FAO-UNESCO. Soil map of the world. Volume VII. South Asia.
12. Peru. From Stark, N. (1971). Amounts calculated from mean figures for soil depths 0–4, 20–22 and 60–62 cm.
13.–15. Brazil. From Stark, N. (1971). Amounts calculated from mean figures for soil depths 0–4, 20–22 and 60–62 cm.
16.–18. Brazil. From Stark, N. (1971). Amounts calculated from mean figures for soil depths 0–4 and 20–22 cm.
19.–24. Surinam. From Stark, N. (1970). Amounts calculated from mean figures for soil depths 0–3 and 60–62 cm.

*Boyle, J. R., J. J. Phillips* and *A. R. Ek* (1973). "Whole-tree harvesting: Nutrient budget evaluation". Journal of Forestry, 71: 760–762.

*Buschbacher, R. J.* (1984). Changes in productivity and nutrient cycling following conversion of Amazon rainforest to pasture. Ph. D. Thesis, University of Georgia, Athens, USA.

*Emteryd, O.* (1989). Chemical and Physical Analysis of Inorganic Nutrients in Plant, Soil and Air. Swedish University of Agricultural Sciences, Department of Forest Site Research, Mimeo no 10, Umeå.

*Edwards, P. J.* and *P. J. Grubb* (1982). "Studies of mineral cycling in a montane rain forest in New Guinea. IV. Soil characteristics and the division of mineral elements between the vegetation and soil". Journal of Ecology 70: 649–666.

*FAO-UNESCO.* (1977). Soil map of the world. Volume IV, VI, VII and X.

*Federer, C. A., J. W. Hornbeck, L. M. Tritton, C. W. Martin, R. S. Pierce* and *C. T. Smith* (1989). "Long-term depletion of calcium and other nutrients in eastern US forests". Environmental Management 13: 593–601.

*Fölster, H., G. De las Salas* and *P. Khanna* (1976). "A tropical evergreen forest site with perched water table, Magdalena valley, Colombia. Biomass and bioelement inventory of primary and secondary vegetation". Oecologia Plantarum 11: 297–320.

*Greenland, D. J.* and *J. M. L. Kowal* (1960). "Nutrient content of the moist tropical forest of Ghana". Plant and Soil XII: 154–174.

*Grimm, U.* and *H. F. Fassbender* (1981). "Ciclos bioquimicos en un ecosistema forestal de los Andes Occidentales de Venezuela. I. Inventario de las reservas orgánicas y minerales". Turrialba 31: 27–37.

*Grip, H., A. Malmer* and *F. K. Wong* (1994). "Converting tropical rainforests to forest plantation in Sabah, Malaysia. I. Dynamics and net losses in streamwater". Hydrol. Proc. 8: 195–209.

*Grubb, P. J.* and *P. J. Edwards* (1982). "Studies of mineral cycling in a montane rain forest in New Guinea. III. The distribution of mineral elements in the above-ground material". J. Ecology 70: 623–648.

*Hase, H.* and *H. Fölster* (1982). "Bioelement inventory of a tropical (semi-) evergreen seasonal forest on eutrophic alluvial soils, Western Llanos, Venezuela". Acta Oecologica Plantarum 3: 331–346.

*Herrera, R.* (1979). Nutrient distribution and cycling in an Amazon caatinga forest on spodosols in southern Venezuela. Ph. D. Thesis, University of Reading, England.

*Jordan, C. F.* (1989). An Amazonian Rain Forest. Man and the Biosphere series, Volume 2. The Parthenon Publishing Group.

*Klinge, H.* (1976). "Bilanzierung von Hauptnährstoffen im Ökosystem tropischer Regenwald (Manaus)". Biogeographica 7: 56–77.

*Medina, E.* and *E. Cuevas* (1989). "Patterns of nutrient accumulation and release in Amazonian forests of the upper Rio Negro basin". Pp 217–240 in Mineral nutrients in tropical forest and savanna ecosystems. Edited by J. Proctor.

*Nykvist, N.* (1977). "Skogliga åtgärders inverkan på storlek och tillgänglighet av ekosystemets näringsförråd". Sveriges Skogsvårdsförbunds Tidskrift 75: 167–168.

*Nykvist, N.* (1997a). "Total distribution of plant nutrients in a tropical rainforest ecosystem in Sabah, Malaysia". Ambio 26: 152–157.

*Nykvist, N.* (1997b). "Uptake of nutrients in a plantation of Acacia mangium in relation to decrease in soil amounts". Journal of Sustainable Forestry 4: 131–139.

*Nykvist, N., H. Grip, B. L. Sim, A. Malmer* and *F. K. Wong* (1994). "Nutrient losses in forest plantations in Sabah, Malaysia". Ambio 23: 210–215.

*Ovington, J. D.* and *J. S. Olson* (1970). "Biomass and chemical content of El Verde lower montane rain forest plants". Pages 53–75 in Odum, H. T. (Ed.) A tropical rainforest. Div. Technical Information. U.S. Atomic Energy Commission. Washington, D.C.

*Poels, R. L. H.* (1987). Soils, water and nutrients in a forest ecosystem in Surinam. Ph. D. Thesis. Agricultural University, Wageningen, Holland.

*Ruhiyat, D.* (1989). Die Entwicklung der standörtlichen Nährstoffvorräte bei naturnaher Waldbewirtschaftung und im Plantagenbetrieb, Ostkalimantan (Indonesien). Göttinger Beiträge zur Land- und Forstwirtschaft in den Tropen and Subtropen, 35.

*Russell, C. E.* (1983). Nutrient cycling and productivity of native and plantation forests at Jari Florestal, Pará, Brazil. Ph. D. Thesis. University of Georgia, Athens, USA.

*Shortle, W. C.* and *K. T. Smith* (1988). "Aluminium-induced calcium deficiency syndrome in declining red spruce". Science 240: 1017–1018.

*Shrivastava, P. B. L.* (1993). "Soils and fertilizer requirements. Relationship between soil properties and growth of Acacia mangium". In Awang, K. and D. A. Taylor (Eds.) Acacia mangium – Growing and utilization. Winrock International and FAO, Bangkok, Thailand. MPTS Monograph Series, 3: 120–147.

*Silkworth, D. R.* and *D. F. Grigal* (1982). "Determining and evaluating nutrient losses following whole-tree harvesting of aspen". Soil Science Society of America Journal 46: 626–631.

*Sim, B. L.* and *N. Nykvist* (1991). "Impact of forest harvesting and replanting". J. Trop. For. Sci., 3: 251–284.

*Stark, N.* (1970). "The nutrient content of plants and soil from Brazil and Surinam. Biotropica 2: 51–60.

*Stark, N.* (1971). "Nutrient cycling: I. Nutrient distribution in some Amazonian soils". Tropical Ecology 12: 24–50.

*Tanner, E. V. J.* (1985). "Jamaican Montane forests: nutrient capital and cost of growth". Journal of Ecology 73: 553–568.

*Waterloo, M. J.* (1994). Water and nutrient dynamics of Pinus Caribea plantation forests on former grassland soils in Southwest Viti Levu, Fiji. PhD thesis. University of Amsterdam, The Netherlands.

*Weetman, G. F.* and *B. Webber* (1972). "The influence of wood harvesting on the nutrient status of two spruce stands". Canadian Journal of Forest Research 2: 351–369.

# Resilience of phosphorus transformations in tropical forest and derived ecosystems

**Holm Tiessen**

Department of Soil Science, University of Saskatchewan, Saskatoon, Canada S7N 5A8
E-mail: tiessen@sask.usask.ca

## Abstract

The soils associated with most tropical forests are exposed to significant leaching. This leaching has weathered the soils' mineral suite and reduced their capacity to bind nutrients, exacerbating the potential for nutrient loss from the forest or any derived ecosystem such as slash-burn agriculture. Phosphorus is often the most limiting nutrient in ecosystems on weathered tropical soils. Examples are given for P budgets of South American forests and for an estimate of P loss from a highly dystrophic rain forest. Despite a biological turnover of 6 % of the forests P every year, only about 0.001 % of the total P stocks are lost annually, indicating that biogeochemical P recycling is highly efficient. The mechanisms for such efficient cycling are internal retranslocation of P, and a litter and root mat with abundant mycotrophism which result in an effective separation of the biological P cycle from a potentially P-fixing soil. In the examples of less dystrophic forests from NE Brazil, P retention in the litter mat is less efficient and P uptake from the soil is more important. Therefore, the potential for non-forest land use is greater in NE Brazil, although P tranformations to less available forms limit the length of cultivation cycles. Few data exist on alternative land uses, such as tree crops or agroforestry, and the complex methodology of studying P has limited the number of experiments aiming at an understanding of the complete set of P transformations which determine an ecosystem's resilience to disturbance and limit net P losses.

## 1 Introduction

Most forest soils are more leached than savanna or grassland soils in similar latitudes since forests occur in regions of higher rainfall. Typically, temperate forest occur on Alfisols whereas grasslands are dominated by Mollisols. In the tropics, forests occur on Ultisols or Oxisols while savannas are associated with Alfisols. Exceptions are forests on alluvial soils that are frequently renewed or scrub or grasslands on highly leached soils that were formed where paleoclimates were wetter than the current semiarid regime that supports the low vegetation. As a result of the more intensive leaching, forest soils are commonly nutrient-poorer, more acid and sesquioxide-richer, than grassland soils on similar parent material. This has important implications for fertility and the resilience of the forest ecosystem.

The lower sorption potential of leached forest soils for inorganic nutrients, and the smaller pools of stabilised organic nutrients mean that greater proportions of nutrients are leachable than in natural grassland soils. This potential for loss affects even relatively immobile nutrients like phosphorus. In sub-humid temperate forest soils, *Frossard* et al. (1989) calculated losses of 40% of total P (3000 kg P $ha^{-1}$) and *Letkeman* et al. (1995) near 10% total P from the solum in the 10,000 y since the last glaciatiòn. *Letkeman* et al. (1995) also presented evidence that at least some of the P had moved in inorganic form. Such data indicate that P mobility may be greater than commonly assumed, and that ecosystems formed under leaching regimes can lose all nutrients, including less soluble phosphates. Resilience depends in part on avoiding conditions under which such irreversible losses of critical nutrients are exacerbated. Many grasslands are resilient after a disturbance such as fire or cultivation, and annual fires form part of the regular regime for many savannas. Forests recover less readily from such disturbance and suffer greater losses of nutrients and species diversity when the vegetation cover is disrupted.

## 2 Nutrient cycles in the forest ecosystem

Extreme examples of tropical rainforests on highly dystrophic soils exist in the northern Amazon basin, where soil parent materials are derived from highly weathered sandstone and some granite outcrops. On a Ferralsol (Oxisol) near San Carlos de Rio Negro (Venezuela), with an annual rainfall of 3600 mm, the native forest has a total biomass of 300 Mg $ha^{-1}$, 20% of which are below ground. This biomass contains 1500 kg N $ha^{-1}$ and 50 kg P $ha^{-1}$ (*Uhl* and *Jordan*, 1984). The leaf biomass is about 9 Mg $ha^{-1}y^{-1}$, 70% of which fall annually. Total litter fall (including stems and branches) is near 10 Mg $ha^{-1}$, which contain 130 kg N $ha^{-1}y^{-1}$ and 2.3 kg P $ha^{-1}y^{-1}$ (*Medina* and *Cuevas*, 1989).

Fine root turnover amounts to 11 Mg $ha^{-1}y^{-1}$, containing approximately 180 kg N and 9 kg P (*Cuevas* and *Medina*, 1983, 1988). This means that every year about 300 kg N and 11 kg P are liberated from decomposing primary producer biomass, equivalent to 20% of the total nutrient stock of the primary producers. The soil under this forest contains about 2600 kg N $ha^{-1}$ and 140 kg $ha^{-1}$ of potentially available organic and inorganic P to a depth of 36 cm, the most active rooting zone (*Tiessen* et al., 1994a, b). Annual biological turnover of plant-recycled N and P therefore amounts to near 10% of soil nutrient stocks and to 6–7% of the total P and N stocks of the entire ecosystem. In the case of P, no atmospheric fixation occurs, and P inputs from outside the region are likely to be zero (the small amounts measured in rainfall are probably derived from ash and dust generated in the region). Therefore, recycling must be very efficient if the system is not to run down its nutrient stock.

Are there indications for continuing net P losses from the native ecosystem? *Vitousek* and *Sanford* (1986) estimate that the P loss from moist tropical forest is < 1 kg $ha^{-1}y^{-1}$. This estimate, given the low nutrient stocks of the system at San Carlos, is not accurate enough to give a picture of the nutrient conservation required for maintaining this forest over thousands of years.

The total P concentration in the soil below 36 cm is 200 µg $g^{-1}$, nearly twice as high as the 110 µg $g^{-1}$ found in the first 36 cm (Fig. 1) (*Tiessen* et al., 1994a). It is common for soils to show a depletion zone of P at the depth explored by roots. In temperate soils there is usually also a net accumulation zone near the soil surface where organic P from plant uptake is recycled (Fig. 1) (*Letkeman* et al., 1995). In this Amazonian soil, the depletion zone coincides with the top layer of the mineral soil and the organic P returns accumulate above the soil in the litter layer. If the amounts of P in the above ground vegetation (40 kg $ha^{-1}$) and in the litter mat (120 kg $ha^{-1}$) are added to the amounts present in the P-depleted top soil (270 kg $ha^{-1}$) (*Tiessen* et al., 1994a), the total P concentration is this layer

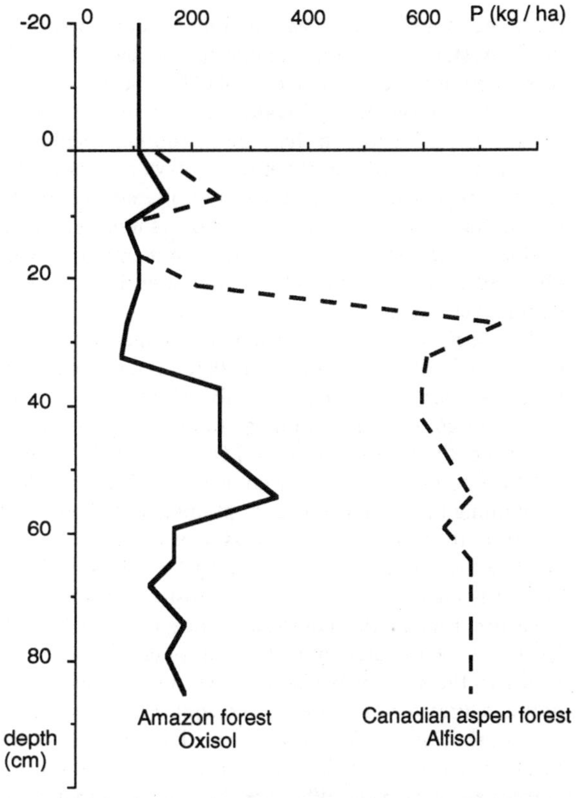


**Figure 1:** Amounts of total P with soil depth (based on 5 cm measuring increments) in an Amazonian Oxisol and a Canadian Alfisol.

is nearly doubled. As a result, the top soil shows a deficit of only 10–15 µg P $g^{-1}$ relative to the lower soil horizons.

In addition to the fine soil (sand, silt and clay) found in the top soil, the lower soil horizons also contain lateritic nodules with very high P concentrations around 800 µg $g^{-1}$, which might have accumulated because the nodules act as a sink for P (*Tiessen* et al., 1991) and might therefore have accumulated P lost from the top soil. Their high P content

**Table 1:** Organic matter and associated N and P stocks (% of total) and mean residence times for organic C of soil and litter layer to 60 cm under natural rainforest at San Carlos

| sample, depth | subsample | Carbon | Nitrogen | organic P | MRT |
|---|---|---|---|---|---|
| | | % | | | y |
| fresh litter, 0–1 cm | total | 0.4 | 0.2 | 0.1 | 1.5[(1)] |
| org. layer, 1–21 cm | total | 39 | 37 | 24 | 1.5–3.3[(1)] |
| min. soil, 21–36 cm | sand | 16 | 14 | 22 | < 30[(2)] |
| | clay + silt | 13 | 14 | 12 | 30–50[(2)] |
| min. soil, 36–60 cm | sand < 0.5 mm | 4 | 14 | 6 | < 30[(2)] |
| | clay + silt | 14 | 15 | 21 | 50–100[(2)] |
| | nodules > 0.5 mm | 14 | 6 | 15 | 4600[(2)] |
| Total to 60 cm | | 100 | 100 | 100 | |

(1) Calculated from Cuevas and Medina 1986, 1988

(2) Tiessen et al. 1994b

could therefore be an indication of inefficient P cycling in the ecosystem above. But organic carbon sequestered in these nodules was $^{14}C$-dated to 4,600 y before present (Table 1) (*Tiessen* et al., 1994b), and the intact nodules showed lower P sorption than the surrounding soil fines. So, it can be argued that the nodules have sequestered C and P some 4,600 y ago, and have not interacted with the rest of the soil since the occlusion process was completed. A similar occulusion of C, P and Fe minerals has been observed by *Tiessen* et al. (1996) in a lateritic crust from eastern Venezuela.

The C occluded in the ferruginous nodules 4600 y ago shows a $\partial\,^{13}C$ of $-25\,^{o}/_{oo}$ indicating that it is derived mainly from C-3 plants associated with forest rather than from C-4 plants common in savannah (*Volkoff* and *Cerri*, 1987). Therefore, the forest is at least 4,600 y old, and *Schubert* (1988) presents evidence for an age of at least 15,000 y. Combining these data and assumptions one arrives at a P loss from the top soil of somewhat over 10 $\mu g\ g^{-1}$ over the last five thousand years or 2–3 $\mu g\ kg^{-1}\ y^{-1}$. This excludes mass wasting of whole top soil by erosion. Erosion must have occurred in the landscape, since massive lateritic layers are encountered in the top soils of upland positions, which in the current landscape have no up-slope or up-profile source for the observed Fe concentrations. Nevertheless, the P losses from the biogeochemical cycle of the forest appear to be in the range of only 1/100,000 of the current stock per year. What are the processes that make the dystrophic tropical rain forest such an efficient nutrient recycler?

Trees are able to conserve nutrients internally by retranslocation of limiting nutrients prior to leaf abscission. Leaf litter composition at San Carlos showed adaptations related to the main nutrient limitation at three different forest sites (*Cuevas* and *Medina*, 1986). Litter produced on the hydromorphic soil had a higher P concentration, while litter on the upland soil was richer in N. Retranslocation was more extensive for the most limiting nutrient, reducing the potential for nutrient loss during litter decomposition.

The pattern of nutrient limitation indicated by litter composition was shown to reflect closely the available nutrient contents in the soil (*Tiessen* et al., 1994a). Within some 200 m, the natural forest type changed significantly, presenting very different plant adaptations suited to each soil type. The lowland site, at which retranslocation and litter composition indicate a relatively smaller P limitation was characterised by the lowest P sorption capacity of all three soils studied and by a relatively greater soil P availability. Although total soil P concentrations were very low, a greater proportion of P was maintained in available form because of the lower sorption and the frequent water saturation. Water saturation mobilises P in a manner similar to that observed in paddy soils. Despite the more common occurrence of water logging in the lowland profile, rooting depth was deeper than at the upland sites, and litter was incorporated into the topsoil to a greater extent. This reflects the greater nutrient availability and lower Al concentration in the top soil.

A superficial and dense root distribution is an important adaptation mechanism on soils that show high P fixation. At the two upland sites, a root and litter mat permits direct uptake of mineralised P by the vegetation. This root and litter mat contains one third of the soil's N and P, has a high turnover (Table 1) and maintains exceptionally high levels of resin-extractable available P of 50 $\mu g\ g^{-1}$ over a soil containing only 1 $\mu g\ g^{-1}$. *Herrera* et al. (1978) showed that $^{32}P$ from labelled leaf litter was taken up by roots via mycorrhizal associations. Applied nutrients are retained to 99% in this litter mat (*Stark* and *Jordan*, 1978). The root-litter mat above the soil is thus an important component of the forest's nutrient conservation mechanisms.

Mycorrhizae aid their host plants by exploring the soil and litter beyond the immediate rooting zone. Mycorrhizae deplete the same P sources as plants (*Bolan*, 1991), but also facilitate the utilisation of organic P through production and secretin of acid phosphatase (*Häussling* and *Marschner*, 1989). The rates of hydrolysis of organic P can exceed the P-uptake rates by mycorrhizae and plant roots, resulting in an increase of inorganic P in the mycorrhizosphere (*Häussling* and *Marschner*, 1989). At San Carlos, mycorrhizae were abundant in the nutrient-rich litter layer, but were also observed extending well into the soil below. In several cases, mycorrhizae were seen entering into the pores of lateritic nodules, where no roots were observed (Figure 2). Mycorrhizae may thus aid in the exploration of soil volumes inaccessible or inhospitable to plant roots.

In an experiment on P retention by the litter layer in the Atlantic coastal rainforest of northeastern Brazil, *Salcedo* et al. (1991) injected $^{32}P$ into the much thinner litter mat

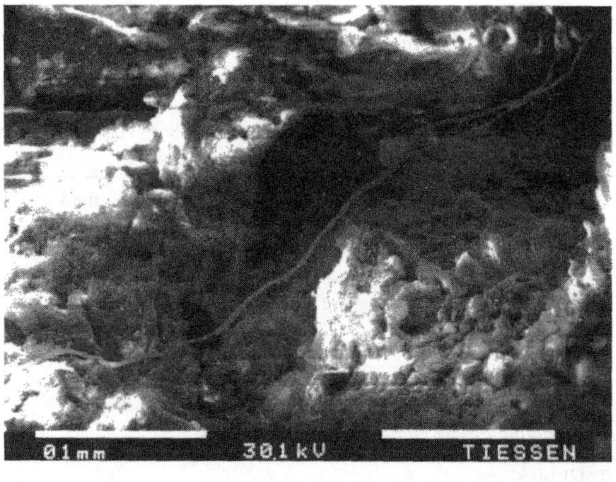


**Figure 2:** Scanning electonmicrograph of mycorrhizal hypha exploring pore spaces in a ferruginous nodule from a Ferralsol at San Carlos de Rio Negro.

typical of that forest and observed 60% break-through of $^{32}P$ into the mineral soil. While roots in the organic layer subsequently showed higher $^{32}P$ activities, more total $^{32}P$ was found in the roots growing in the mineral soil. Trenching around the injection site indicated that the roots in the organic layer were short vertical feeders connected to the horizontal web of roots in the mineral soil below. It was therefore not possible to attribute uptake to one layer or the other. Unlike the organic layer at San Carlos, the Atlantic forest layer maintained low levels of resin extractable P, and much of the P was only released upon chloroform fumigation. Chloroforming releases P from the microbial biomass and recent residues by cell lysis. Most $^{32}P$ from the injection experiment was released by chloroform from the mineral soil below the litter mat, indicating that at that site, not only roots but also decomposers are active and abundant in the mineral soil. This site is not nearly as dystrophic as the San Carlos site, the soil is less acid and contains more clay, and P uptake from the mineral soil is therefore much easier. The region has, in fact, been largely deforested and has been producing large amounts of sugar cane, albeit at yields below world standard, for centuries. The soils at San Carlos would not support agricultural crops beyond a brief 3-y period of slash-burn agriculture during which the decomposing litter mat provides most or all plant nutrients.

Comparing the fate of P released by or injected into the litter mat of the dystrophic northern Amazon and the oligotrophic Atlantic rainforest, it appears that the more nutrient-stressed system is more efficient at nutrient conservation, and that it is less suitable to be converted into any agricultural use.

## 3 Land use systems

The mechanisms evolved by natural forests to cycle nutrients effectively between the biotic and soil compartments, or – in extreme cases – to maintain nutrients within the biotic compartment with minimal loss, are disrupted by felling, burning and cultivation. The success of any agricultural or forestry production depends on this ecosystem disturbance: it requires that nutrients that are normally closely cycled are mobilised and made available for crop uptake and subsequent export. But a balance must be maintained between mobilisation and conservation. The sustainability of a production system depends on the efficiency of nutrient mobilisation for crop use, so that economic returns may be obtained, the avoidance of nutrient loss, so as to minimise required inputs and environmental degradation, and – in all but extremely eutrophic systems – the replacement of exported nutrients. Essentially, three competing processes determine the balance of nutrient transformations: biological (crop) uptake, stabilisation in the soil and leaching losses.

Relatively rapid mineralisation of organic matter and litter under tropical conditions not only supplies P for crop growth (*Adepetu* and *Corey*, 1976) but also generates a surplus, such that soil available P may be increased upon land clearing despite crop uptake (*Agboola* and *Oko*, 1976). In rain forest regions such available nutrients may be lost by leaching. Because of the large spatial variability of soils, it may be difficult to quantify leaching losses, unless they are measured in receiving water bodies (*Tiessen*, 1995). In soils with high P sorption, a surplus of available P is rapidly transformed into plant-unavailable P forms.

After clearing of a semi-arid thorn forest on a Brazilian Oxisol, six years of cultivation with minimal fertilization resulted in the mineralisation of 30% of the soil's C, N and organic P (*Tiessen* et al., 1992). The P export by cassava of only 2 to 3 $\mu g\ g^{-1}$ of soil was much less than the amounts liberated, and inorganic P stocks increased. This increase did not sustain supplies of labile P required for adequate crop production, and the site was abandoned to bush fallow because of declining production. Fractionation of soil P by a sequential extraction confirmed that the decline in P fertility was not due to net P losses but due to the transformation of the surplus mineralised P to unavailable forms. This process of P transformation led to an accumulation of 15–20 $\mu g\ g^{-1}$ P in the unextractable, residual fraction, probably due to organic P mineralisation and some fertilisation. Although soils in the region may be submitted successfully to further cultivation cycles, the transformation of organic and available to unavailable P may represent a downward spiral that twists with every cultivation cycle.

Experience form temperate grassland and forest ecosystems under cultivation indicates that soil organic matter levels stabilise after some years, and a new steady state of organic matter and nutrient cycling is established, which may maintain soil organic matter levels near 50% of the original grassland levels or even above the native values for forests converted to graminaceous crops. Commonly this steady state does not mobilise sufficient nutrients for annual crop production so that fertilisation or suitable crop – perennial cover rotations become necessary.

In tropical forested regions, a very wide range of post-slash-burn conditions have been observed. At San Carlos, the slash-burn site lost 80% of its litter layer and 30% of its soil C to 15 cm depth over three years. The end of the litter layer coincides with the end of the cultivation cycle because the nutrients liberated from the decomposition are essential for crop production (*Jordan*, 1989). Over half the total organic matter to 60 cm depth will mineralise in less than 4 years. Radiocarbon ages of the soil organic matter (Table 1) indicate that most has a short half-life under native conditions and would also be mineralised in less than 50 y. Experience of local shifting cultivators indicates that the rates of nutrient release are not sufficient to support agricultural crops beyond 3 years. Declining crop production limits C returns to the soil, and the system can not acquire a new steady state.

Humid forest soils in the Brazilian subtropics showed C turnover times of 4, 6 and 60 years in sand, silt and clay fractions, indicating a similar potential for rapid degradation (*Bonde* et al., 1992). But, on the other hand, 60% of forest-derived C was still present in a soil converted to sugar cane after 50 years (*Vitorello* et al., 1989). Since these soils are less dystrophic than the northern Amazonian soils, crops such as sugar cane grow well and maintain soil organic matter near 75% of native levels. Degradation of forest-derived C might be reduced when C inputs from the crops are abundant.

The sandy Ferralsol under thorn forest in semi-arid northeastern Brazil lost 40% of its organic C during 6 years of mechanised cultivation. When cultivated manually, i.e. with out tillage and using only planting holes, 20% of its C, N and organic P was lost during 6 years. Most of the soil organic matter losses occurred from the mineral associated fraction, which lost one third of its C, and which radiocarbon dated to near 50 y under native conditions. The carbon date of the C remaining after 6 y of cultivation was not significantly different from the native condition (actually showing a slightly greater modern C proportion), i.e. it does not have any greater stability than the C already lost at this stage. Similar to the Amazonian site, there is no indication of a new steady state, and crop production drops so low, that organic matter return would be minimal, if the site were not abandoned. It is important to note that the soils at that site were Feralsols (Oxisols) with their typical strong micro-aggregation (*Bui* et al., 1989). It might therefore be expected that C held in association with these Fe-rich microaggregates would show high radiocarbon ages and would be resistant to mineralisation under cultivation. Both field observation and $^{14}C$ dating contradicted this expectation.

Where leaching is not excessive, production can be maintained by substantial fertilisation, possibly in combination with liming (to reduce P sorption). Soils with moderate or high retention capacity for nutrients and organic matter, i.e. those which have moderate clay content and some proportion of high activity clays, can be managed through a combination of organic and mineral nutrition. Soils at Yurimaguas (upper Amazon basin, Peru) on less weathered sediments have been successfully maintained fertile through liming, fertilization and the judicious maintenance of organic matter levels and soil cover (*Sanchez* et al., 1983). No cropping system with inherent resilience or sustainability was found in these experiments. Instead, the production system was empirically optimised with interventions that counteracted fertility decline, acidification, erosion etc. when such problems became apparent. The management system thus requires a level of expertise difficult to achieve for local subsistence farmers. Distance from markets limits the availability of critical inputs and the marketing opportunities for produce that would support such an intensive, fine-tuned management system. Because of these socio-economic constraints, the system has found little acceptance amongst farmers (*Fearnside*, 1987). Given the high rainfall regime at San Carlos (> 3500 mm $y^{-1}$) and limited stabilisation of organic and other materials by the soil's silts and clays, fertilisation is unlikely to be an option even for cash crops, unless a successful litter-organic matter cycle can be established. In a semi-arid environment were crop production is potentially limited by moisture availability, fertilisation and other inputs are usually neither economical nor available to subsistence farmers. It is therefore important that the constraints on the availability of natural soil nutrients under shifting cultivation be understood, in order to design viable regimes of management and supplementary fertilisation.

Alley cropping and tree crop plantations maintain components of forest during the cropping phase, but in the few cases studied, nutrient budgets and availabilities have still shown considerable degradation, both through the transformation of available P to unextractable forms and through the relatively high P exports (up to 25 kg $ha^{-1}$ for oil palm) from some plantation crops (*Palm* et al., 1991). Relatively little is known about the detailed P transformation and losses of these systems, since most studies of alley cropping have concentrated on N budgets and other simple indicators such as soil pH.

## 4 Forest regrowth

Land clearing and cultivation induce C, N and organic P mineralisation in both the humid (*Mueller-Harvey* et al., 1985) and the semi-arid (*Tiessen* et al., 1992) tropics. The attendant decline in fertility frequently results in land abandonment, particularly by subsistence farmers who cannot afford the inputs required to stabilise the production system. At the time of abandonment, when the soil's nutrient content is too low to support adequate crop production, the forest usually re-invades the site, unless erosion, extreme nutrient loss or acidification cause the establishment of shrub or savanna.

At the time of forest re-establishment, the soil releases nutrients in sufficient quantities and at a rate adequate for the successful re-establishment of secondary successions. In humid environments such as at San Carlos, this regrowth often shows higher net productivity than the native forest and can reestablish relatively high soil organic matter levels (*Uhl*, 1987). During regrowth, aboveground live biomass accumulates steadily over 60 to 80 years, relying on mineral-associated nutrients which did not sustain the annual crops, possibly because of their higher acquisition rates. Despite the relative success of secondary successions, basal area and total biomass level of the mature forest may only reach native levels after some 200 years (*Saldarriaga* et al., 1988) when internal nutrient cycles are fully re-established.

The rate and success of regeneration of the forest after abandonment depends, among other factors, on the size of the clearing, pointing to the importance of complex biotic linkages such as seed propagation, symbioses etc. in the system's resilience. Soil analyses of inorganic nutrient availability, or soil organic matter levels are unable to predict the ecosystem fertility at this point, since fertility is largely determined by biotic processes and pools.

At the semi-arid site in NE Brazil, 40% of soil organic matter lost during cultivation were largely recovered after approximately 10 years of bush fallow, even though the fallow vegetation had not nearly attained the height, density or species composition of the native vegetation. Accretions of total C during fallow are probably the result of slow but continuously increasing inputs of organic matter from litter fall and turnover of root biomass under fallow vegetation. The relatively slow rates of additions of organic matter as well as the slow rates of nutrient uptake by the reestablishing native vegetaion are more effective in building up and maintaining soil organic matter levels than the periodic additions of crop residues under cultivation without fertilisation. Several native leguminous species are involved in the recovery of N levels during the bush fallow period. The increased amounts of organic P were supplied by the soil's inorganic P, but no specific pools could be identified as the source.

While the general processes of transformations between soil organic and inorganic P have been described, few specifics have been elucidated. There is the potential for leaching loss during the mineralisation of organic P, and some movement of P into resistant fractions, of possibly low plant availability, has been observed. No clear indicator pools that could be used as measures of ecosystem resilience and fertility have been identified. This is due to methodological problems associated with the study of P.

## 5 Methods of study

It has been suggested that C:P ratios may serve as indicators of net mineralisation or immobilisation potential (*Blair* and *Boland,* 1978). *Tiessen* et al. (1992) reported losses of both C and organic P from a cultivated thorn forest amounting to near 30%, which left the C:P ratio almost unmodified at 430:1. This indicated that C mineralisation was the main driving force for organic P mineralisation, although at such a high C:P ratio, immobilisation might be predicted. When the determination of organic P was improved with a new method, the amounts determined increased, and the C:P ratio dropped to about 250. So, even the relatively simple problem of distinguishing organic and inorganic forms of P still faces methodological obstacles. Organic P is invariably determined by difference between inorganic and total P, and no absolute reference method has yet been developed.

Some organic P fractions have been found to change predictably under manipulations such as incubations, fertilisations, manuring, exhaustive cropping etc. To fully understand the organic P cycle, though, organic matter transformations must be considered (*Tiessen,* 1991). Many active biological structures are resistant to extraction (*Stevenson* and *Elliot* 1989), and therefore, some unextractable organic P is biologically active as a result of organic matter turnover. Separation of recent litter, floatable organic matter, clay associated or mineral stabilised organic matter (*Feller* et al. 1983), and analysis of the associated P (Table 1) are methods that have shown some success, since they separate organic matter fractions of different stability. The determination of the C/N/P ratio in different extracts might add some information about the nature and the availability of the organic P extracted. The extensive analytical labour involved is daunting, and the interpretation of results demands a combination with other chemical soil characteristics (*Tiessen* et al., 1994c).

In ecosystems where biotic P cycling through close recycling of plant litter is important, such as in nutrient limited tropical rain forests, chemical P transformations in the soil are of limited relevance. In such cases, critical ecosystem components such as leaf P, litter P, plant P, decomposition rates, root growth and root distribution and the role of soil fauna, bacteria and fungi must be evaluated. In the absence of a long-lived isotopic tracer for P, only the use of repeated elemental budgets in different ecosystem components along seasons has provided insight into the P cycle in such ecosystems (*Medina* and *Cuevas*, 1989), and an understanding of P transformations invariably hinges on an understanding of the associated carbon cycles.

## 6 Conclusions

Tropical rainforests, through long-term biological adaptation have evolved systems of extremely efficient nutrient recycling, particularly when they grow on poor soils. It will be nearly impossible to devise land use systems that match this level of adaptation. On richer soils, where biotic adaptations are less essential for nutrient conservation, permanent agricultural and forestry uses have been more successful, although many low-input systems are suffering from internal transformations of P to low-availability forms. In attempts to achieve sustainable, resilient land use systems, it is important to evaluate the functioning of the native forest, and to emulate its conservation mechanisms, or to match nutrient replacements to loss rates. This is still hampered by the complex methodologies required for studying P transformations on soils with considerable P sorption, low available P and important biotic transformation mechanisms.

## 7 References

*Adepetu, J. H.* and *Corey, R. B.* (1976): Organic phosphorus as a predictor of plant available phosphorus in southern Nigeria. Soil Sci. 19, 65—80.

*Agboola, A. A.* and *Oko, B.* (1976): An attempt to evaluate plant available P in western Nigerian soils under shifting cultivation. Agronomy Journal 68, 798–801.

*Blair, G. J.* and *Boland, O. W.* (1978): The release of phosphorus from plant material added to soil. Aust. j. soil Res. 16, 101–111.

*Bolan, N. S.* (1991): A critical review on the role of mycorrhizal fungi in the uptake of phosphorus by plants. Plant. and Soil. 134, 189–207.

*Bonde, T. B., Christensen, B. T.* and *Cerri, C. C.* (1992): Dynamics of soil organic matter as reflected by natural 13C abundance in particle size fractions of forested and cultivated soils. Soil Biol. Biochem. 24, 275–277.

*Bui, E. N., Mermut, A. R.* and *Santos, M. C. D.* (1989): Microscopic and ultramicroscopic porosity of an Oxisol as determined by image analysis and water retention. Soil Sci. Soc. Am. J. 53, 661–665.

*Cuevas, E.* and *Medina, E.* (1983): Root production and organic matter decomposition in a Tierra Firme forest of the Upper Rio Negro Basin. In: (W. Bohm, L. Kutschera and E. Lichtenegger, eds.) Root Ecology and its Practical Applications. International Symposium. Verlag Gumpenstein, Austria. pp. 653–666.

*Cuevas, E.* and *Medina, E.* (1986): Nutrient dynamics within Amazonian forests I. Nutrient flux in fine litter fall and efficiency of nutrient utilization. Oecologia 68, 466–472.

*Cuevas, E.* and *Medina, E.* (1988): Nutrient dynamics within Amazonian forests II. Fine root growth, nutrient availability and leaf litter decomposition. Oecologia 76, 222–235.

*Feller, C., Bernhardt-Reversat, F., Garcia, J. L., Pantier, J. J.* and *Roussos, S.* (1983): Études de la matière organique de différentes fractions granulométriques d'un sol sableux tropical. Effet d'un amendment organique (compost). Cahiers ORSTOM ser. Pedol. 20, 223–238.

*Fernside, P. (1987): Rethinking continuous cultivation in Amazonia. BioScience 37, 209–214.*

*Frossard, E., Stewart, J. W. B.* and *St. Arnaud, R. J. (1989): Distribution and mobility of P in grassland and forest soils of Saskatchewan. Can. J. Soil Sci. 69, 401–416.*

*Häussling, M.* and *Marschner, H.* (1989) Organic and inorganic soil phosphates and acid phosphatase activity in the rhizosphere of 80-year-old Norway spruce [Picea abies (L.) Karst.] tress. Biol. Fertil. Soils 8: 128–133.

*Herrera, R., Mérida, T., Stark, N. and Jordan, C. F.* (1978): Direct phosphorus transfer from dead litter to roots. Naturwissenschaften 65, 208–209.

*Jordan, C. F.* (1989): An Amazonian rain forest. Man and the Biosphere Series – UNESCO, Parthenon Publishing Group, Park Ridge, N.J. USA. 176 pp.

*Letkeman, L. P., Tiessen, H.* and *Campbell, C. A.* (1996): Phosphorus Transformation and Redistribution During Pedogenesis of Western Canadian Soils. Geoderma 71, 201–218.

*Medina, E.* and *Cuevas, E.* (1989): Patterns of nutrient accumulation and release in Amazonian forests of the upper Rio Negro basin. In: J. Proctor (ed.) Mineral nutrients in tropical forest and savanna ecosystems. Blackwell scientific Publications, Oxford, U.K. p. 217–240.

*Mueller-Harvey, I., Juo, A. S. R.* and *Wild, A.* (1985): Soil organic C, N, S and P after forest clearance in Nigeria: mineralisation rates and spatial variability. J. Soil Science 36, 586–591.

*Palm, C. A., McKerrow, A. J., Glaesner, K. M.* and *Szott, L. M.* (1991): Agroforestry systems in lowland tropics: is phosphorus important? in: H. Tiessen, D. López-Hernández and I. H. Salcedo (eds.), Phosphorus Cycles in Terrestrial and Aquatic ecosystems Regional Workshop 3: South and Central America. Saskatoon: Saskatchewan Institute of Pedology, University of Slaskatchewan. pp. 134–141.

*Saldarriaga, J. G., West, D. C., Tharp, M. L.* and *Uhl, C.* (1988): Long-term chronosequence of forest succession in the upper Rio Negro of Colombia and Venezuela. J. Ecology 76: 938–958.

*Sanchez, P. A., Villachica, J. H.* and *Bandy, D. E.* (1983): Soil fertility dynamics after clearing tropical rainforest in Peru. Soil Sci. Soc. Am. J. 47, 1171–1178.

*Salcedo, i. H., Elliott, E. T.* and *Sampaio, E. V. S. B.* (1991): Mechanisms controlling phosphorus retention in the litter mat of Atlantic coastal forests. in: H. Tiessen, D. López-Hernández and I. H. Salcedo (eds.), Phosphorus Cycles in Terrestrial and Aquatic ecosystems Regional Workshop 3: South and Central America. Saskatoon: Saskatchewan Institute of Pedology, University of Saskatchewan. pp. 109–120.

*Schubert, C.* (1988): Climatic changes during the last glacial maximum in northern South America and the Caribean: a review. Interciencia 13, 128–137.

*Stark, N.* and *Jordan, C.* (1978): Nutrient Retention by the root mat of an Amazonian rain forest. Ecology 59, 434–437.

*Stevenson, F. J.* and *Elliot, E. T.* (1989): Methodologies for assessing the quantity and quality of soil organic matter. In: Dynamics of soil organic matter in tropical ecosystems. Eds. pp. 173–199. University of Hawaii, Honolulu.

*Tiessen, H.* (1991): Characterization of soil phosphorus and its availability in different ecosystems. Trends in Soil Sci. 1, 83–99.

*Tiessen, H.* (1995): Introduction and synthesis. In: H. Tiessen (ed.) Phosphorus in the global environment. Transfers, cycles and management. John Wiley & Sons, Chichester U.K. pp. 1–6.

*Tiessen, H., Lo Monaco, S., Ramirez, A., Santos, M. C. D.* and *Shang, C.* 1996. Phosphate minerals in a lateritic crust from Venezuela. Biogeochemistry 34, 1–17.

*Tiessen, H., Salcedo, I. H.* and *Sampaio, E. V. S. B.* (1992): Nutrient and soil organic matter dynamics under shifting cultivation in semi-arid northeastern Brazil. Agric. Ecosys. Env. 38, 139–151.

*Tiessen, H., Chacon, P.* and *Cuevas, E.* (1994a): Phosphorus and nitrogen status in soils and vegetation along a toposequence of dystrophic rainforests on the upper Rio Negro. Oecologia 99, 145–150.

*Tiessen, H., Cuevas, E.* and *Chacon, P.* (1994b): The role of soil organic matter in sustaining soil fertility. Nature 371, 783–785.

*Tiessen, H., Frossard, E., Mermut, A. R.* and *Nyamekye, A. L.* (1991): Phosphorus Sorption, and Properties of Ferruginous Nodules from Semi-arid Soils from Ghana and Brazil. Geoderma 48, 373–390.

*Tiessen, H. Stewart, J. W. B.* and *Oberson, A.* (1994c): Innovative phosphorus availability indices: assessing organic phosphorus. In: Soil Testing: Prospects for Improving Nutrient Recommendations. SSSA special publication 40 pp. 143–162.

*Uhl, C.* (1987): Factors controlling succession following slash-and-burn agriculture in Amazonia. J. Ecol. 75, 377–407.

*Uhl, C.* and *Jordan, C. F.* (1984): Succession and nutrient dynamics following forest cutting and burning in Amazonia. Ecology 65, 1476–1490.

*Vitorello, V. A., Cerri, C. C., Andreux, F., Feller, C.* and *Victoria, R. L.* (1989): Organic matter and natural carbon-13 distribution in forested and cultivated Oxisols. Soil Sci. Soc. Am. J. 53, 773–778.

*Vitousek, P. M.* and *Sanford, R. L., Jr.* (1986): Nutrient cycling in moist tropical forest. Ann. Rev. Ecol. Syste. 17, 137–167.

*Volkoff, B.* and *Cerri, C. C.* (1987): Carbon isotope fractionation in subtropical Brazilian grassland soils. Comparison with tropical forest soils. Plant and soil 102, 27–31.

# Nutrient disorders and nutrient management in fast growing plantations

**Wolfgang Zech** and **Pay Drechsel**

Institute of Soil Science and Soil Geography, University of Bayreuth, D-95440 Bayreuth, Germany

## Abstract

The negative impact of fast-growing tree plantations, especially when grown in short rotations, on nutrient reserves of poor tropical soils is often described. The intensity of soil and forest degradation may be assessed by different approaches, including detailed observation of deficiency symptoms and yield development, foliar and soil analyses, as well as diagnostic fertilizer experiments. Input/output analyses, taking into consideration water and nutrient fluxes between the different plant and soil compartments are means to evaluate the sustainability of tropical tree plantations. Only a small amount of data is available up to now, indicating poor sustainability. Soil and tree nutrient status can be improved by the application of mineral fertilizer, which is usually rapidly effective. In contrast, amelioration by management practices is only slowly effective, nevertheless important in view of sustainability, environmental protection and biodiversity. A big gap exists in reliable data on water and element dynamics in forest plantations with different tree species under different ecological conditions, and under different management practices.

## 1 Introduction

Soils of tropical forest ecosystems are in most cases poor in nutrients. Despite this fact, undisturbed natural forests seldom reveal symptoms of mineral deficiencies which are typical indicators of degradation. It seems that nutrient balances in undisturbed systems are in a "steady state", and losses – if they occur – are more or less "neutralized" by input rates (*Proctor*, 1987). In contrast to undisturbed natural forests, man-made systems with annual or perennial plantations often show signs of degradation, such as symptoms of mineral deficiency or yield decline, thus indicating that plantation systems are highly susceptible to nutrient imbalances or other forms of degradation, such as phytopathological attacks. Soil related imbalances are mainly due to the following reasons:

- Impacts of clearing of natural forests, site preparation and establishment of the plantation,
- Impoverishment of soil fertility due to nutrient export by (multiple) harvesting (*Chijioke*, 1980).

In the following overview we will focus on decline phenomena caused by degradation of the soil and tree nutrient status with special emphasis on the early identification of nutrient imbalances, and on the sustainability of tropical tree plantations taking into consideration changes in nutrient cycles in connection with forest clearing and tree harvesting. In a final section aspects of ameliorating soil and tree nutrient status without and with mineral fertilizer will be discussed.

## 2 Identification of nutrient disorders in tropical forest plantations

For the reliable diagnosis of nutrient disorders in plants several approaches are available, including detailed observation of symptoms, laboratory analyses, and diagnostic fertilizer experiments (*Mead*, 1984; *Drechsel* and *Zech*, 1993).

### 2.1 Detailed observation of degradation symptoms

Generally, plants with adequate nutrient supply look healthy, develop green leaves and grow vigorously. Latent deficiencies primarily reduce growth, but do not induce deficiency symptoms. Only severe disturbances of the tree nutrient status due to deficiency or toxicity induce symptoms like leaf discoloration, die back and growth deformities or decline of the stand. If only one nutrient is lacking (or toxic) an identification often is possible simply by detailed observation of the symptoms, taking into consideration site conditions, age of foliage, symptom development during the season, etc. In Table 1 some typical symptoms of mono-causal deficiencies are summarized using the example of tropical conifers. Special guidelines for identifying deficiency symptoms of, e.g., eucalypts and teak are reported by *IPI* (1962), *Drechsel* and *Zech* (1993) and *Evans* (1992).

However, several deficiency symptoms are very similar. In addition, under field conditions often not only one nu-

**Table 1:** Typical symptoms of monocausal nutrient deficiencies in conifers (*Drechsel* and *Zech*, 1993; *Weetmann* and *Wells*, 1990; modified).

| | |
|---|---|
| **N:** | Markedly reduced shoot growth and uniformly light-green to yellow needles, occurring first in the older foliage. Partly stunted needles. |
| **S:** | In general similar to those of N-deficiency, but more on younger foliage |
| **P:** | Untypical symptoms on older foliage: yellow (later dead) needle tips, or purple-brown tinged needles or thin and small needles |
| **Mg:** | Yellow needle tips on older foliage with distinct borderlines to the rest of the needle |
| **K:** | Firstly, older needles develop yellow tips which turn to brown-reddish, no distinct borderlines to the rest of the needle |
| **B:** | Terminal and leader die back often with orange-red discolouration. Resinous buds often fail to flush – main stem forks (multiple leader) and becomes deformed. Black or dark brown pith, shortened young needles. The higher the moisture stress, the higher the foliar levels at which the tree develops deficiency symptoms. |
| **Cu:** | Often dark blue-green foliage, distorted (snake-tailed) shoots and branches, bushiness. Some needle tip burn. Prostrate growth in extreme cases. Application of N(PK)-fertilizer may accentuate the deficiency. |
| **Fe:** | Older needles green. Increasing chlorosis towards younger foliage. White colours in extreme cases (only common on calcareous soils). |
| **Mn:** | Yellow tipped young needles, sparse, light green foliage, stunting (especially on dry, alkaline soils; e.g. Vertisols, but also on highly weathered and acid Oxisols) |
| **Zn:** | Terminal needles very short. Premature needle fall and rosetting, tip dieback and yellowing. Partly multiple leadering. |
| **Mo:** | Blue-green needles, partly purple-brown tips with distinct borderlines to the rest of the needle, may occur on older or younger foliage. |

trient is lacking but multiple deficiencies occur. The reason could be a general low soil fertility or a synergism between a mono-deficient nutrient and physiologically related elements. Then it is difficult or impossible to identify nutrient disorders just by detailed observations. In these cases analytical or experimental approaches are necessary, like foliar and soil analyses in combination with diagnostic fertilizer trials.

## 2.2 Soil analysis and site evaluation

Without doubt the majority of growth disturbances and decline phenomena is closely related to soil and site properties. Though, a thorough evaluation of these properties will contribute significantly to the identification and prevention of growth and health disorders. In contrast to foliar analysis soil and site studies can be carried out before the establishment of a plantation. They should refer to the evaluation of soil fertility and soil water supply considering first of all key parameters like root available soil depth, slope position, occurrence (or depth) of stagnating water, texture (soil water capacity), acidity (pH, Al-saturation), organic matter content (CEC) and salinity. These parameters do not require great efforts to study and usually show quite good correlations with tree growth, as illustrated in the example of Figure 1. In this case study a young teak plantation was studied in a Vertisol landscape (10 000 ha) in West Africa. Tree growth was significantly correlated with water stagnation. Decline phenomena were most pronounced in basins where the clay was covered by a nutrient poor sand sheet of about 30 cm, inducing typical pseudogley conditions (*Drechsel* and *Zech*, 1994; *Drechsel* et al., 1995).

Soil nutrient analysis only gives limited information on tree availability of soil nutrients because most chemical extraction methods were developed for agricultural crops. In addition, there are no data about the influence of mycorrhiza and very seldom an idea about the lower boundary of the forest root network. In fact, there exist nearly no verified guidelines for the interpretation of "available" soil nutrient levels with respect to tropical trees. Nevertheless, linear and multiple regression analyses between tree growth and foliar or soil/site data can give valuable information, especially if a distinct nutrient is lacking or if complex parameters, like "topsoil depth", "amount of organic matter" or "pH" are used. Multiple regression analysis allows the quantification of the influence of more than one factor on tree growth. It should be tried to concentrate the calculations on key parameters, which allow a transfer of the results to soil mapping and site classification (e.g. *Zech* and *Cepel*, 1970; *Drechsel* et al., 1995). In view of soil fertility, the storage capacity for nutrients in kaolinitic soils highly depends on soil organic matter contents. This fact explains why a soil parameter such as the thickness of the Ah-horizon, often used in site evaluation, is frequently well correlated with tree growth, soil nutrient reserves (especially N, P, cations) and foliar nutrient contents (e.g. *Stuhrmann* et al., 1994; *Bergmann* et al., 1994).

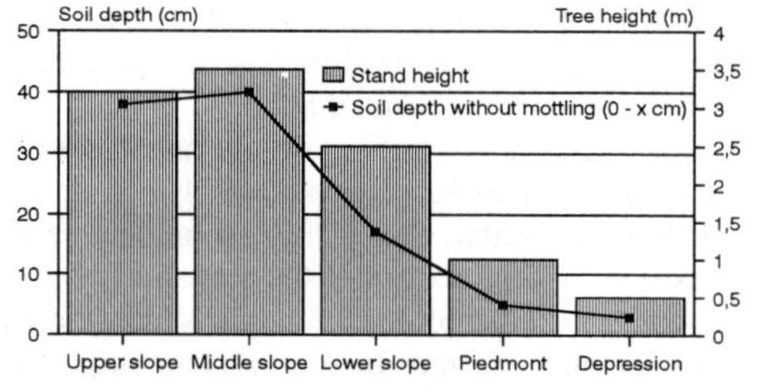


**Figure 1:** Relationship between topography, depth of water stagnation in the soil, and average height of teak in a Vertisol landscape of Southern Benin (*Drechsel* et al., 1995, modified).

### 2.3 Foliar analysis

In contrast to soil nutrient analysis, foliar analysis has been found to be very suitable in the identification of nutritional stress, even in the case of multiple deficiency (*Bowen* and *Nambiar*, 1984; *Brunck*, 1987; *Drechsel* and *Zech*, 1993; *Van den Driessche*, 1974; *Westerman*, 1990). Most (reference) foliar data is available for pines and eucalypts especially from Australia and South Africa (e.g. *Lambert*, 1984; *Grant*, 1991; *Herbert* and *Schönau*, 1991), and there are only a few studies on tropical broad-leaved trees, especially tropical hard woods (*Evans*, 1992). Recently, *Drechsel* and *Zech* (1991) published a tabular review on foliar nutrient levels of broad-leaved tropical and subtropical trees in order to support the identification of ranges in concentrations indicating deficient, marginal (low), adequate (intermediate), high and toxic foliar mineral-nutrient levels. However, this work has to be considered as a first approximation since most reviewed papers did not mention when (during the wet season) the foliage was sampled. Foliar analyses only can be a valuable tool in characterizing the nutrient status of trees if the principal factors that influence foliar nutrient contents (age and position of the foliage, sampling time, number of trees to sample etc.) are taken into consideration. For example, the N concentration of one and the same leaf may decrease over the wet season by more than 50% while the Ca concentrations may increase by up to 100% (*Walworth* and *Sumner*, 1987).

Several possibilities are available to analyse nutrient disorders:

- the direct comparison of data of healthy and declining trees,
- the comparison of analysed data with already published reference values ("critical levels"),
- the Diagnosis and Recommendation Integrated System (DRIS).

DRIS (*Beaufils*, 1973) has the advantage of considering all nutrient interactions like N/P, N/K, N/Ca etc. simultaneously and to analyse the nutrient balance. Under multiple deficiency DRIS is able to rank deficient nutrients (i.e., to identify the most deficient nutrient and best fertilizer composition) and to detect over-supply, like Mn-toxicity (*Walworth* and *Sumner*, 1987, 1989; *Drechsel*, 1994). Up to now only few DRIS-related studies have been carried out with tropical trees for instance with Anacardium occidentale, Eucalyptus grandis, E. saligna, E. viminalis, Pinus taeda, P. patula, and P. radiata (for sources see *Schutz* and *De Villiers*, 1987; *Drechsel*, 1994). Recent DRIS studies deal with Cunninghamia lanceolata (*Zhong* and *Hsiung*, 1993) and Tectona grandis (*Drechsel* and *Zech*, 1994).

Taking again the example of teak on Vertisols in West Africa, we observed in some regions interveinal chloroses (of middle age foliage). The symptoms indicated according to the symptom key several different possibilities of mono-nutrient deficiency, but also multiple deficiency was possible. The comparison of foliar data with reference values showed that indeed a broad range of nutrient concentrations (N, S, P, K, Mn, Zn, Cu, ..) have to be classified as deficient (Table 2).

To get additional information we analysed the nutrient balance using DRIS. Fig. 2 shows that DRIS allows to identify the order of deficient nutrients: Mn > Fe > N > Cu etc. This agreed with the occurrence of lime in the soil. The example shows that the combination of detailed field observations plus foliar and soil analyses can significantly contribute to the identification of nutrient disorders.

It shows in addition that a soil check before the establishment of tree plantations in tropical countries can avoid failure, but such a pre-evaluation is up to now practiced very little. This is an avoidable risk which contributes in many cases to unsatisfying results of tropical reforestation programmes (*Zech* et al., 1989).

**Table 2:** Foliar nutrient levels of 2 years old teak (site 21/Togo) in comparison with concentration ranges of adequate nutrient supply according to *Drechsel* (1992). Several nutrient concentrations have to be classified as deficient.

| Element | | Reference values | Site 21 |
|---|---|---|---|
| N | % | 3.16 ± 0.50 | 1.39 ± 0.22 |
| S | mg g$^{-1}$ | 1.94 ± 0.34 | 1.19 ± 0.18 |
| P | mg g$^{-1}$ | 2.65 ± 0.71 | 1.58 ± 0.21 |
| K | % | 2.00 ± 0.36 | 1.59 ± 0.11 |
| Ca | % | 0.96 ± 0.30 | 0.78 ± 0.23 |
| Mg | % | 0.33 ± 0.08 | 0.25 ± 0.04 |
| Si | % | 9.63 ± 3.42 | 16.1 ± 2.1 |
| Fe | mg kg$^{-1}$ | 112 ± 50 | 55 ± 12 |
| Mn | mg kg$^{-1}$ | 45 ± 14 | 19 ± 8 |
| Zn | mg kg$^{-1}$ | 22 ± 4 | 13 ± 5 |
| Cu | mg kg$^{-1}$ | 17 ± 3 | 9 ± 2 |

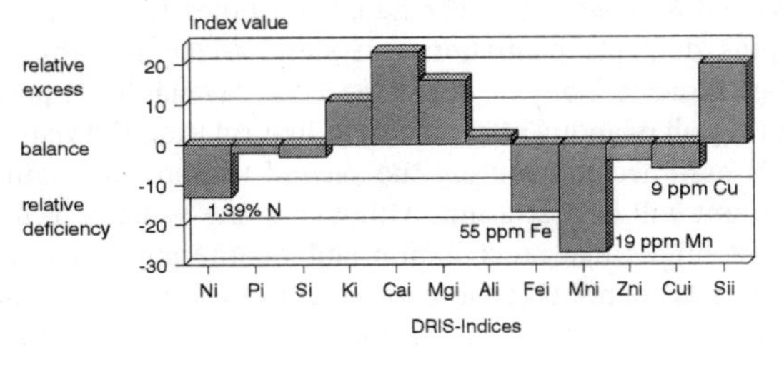


**Figure 2:** DRIS indices of Tectona grandis with intercostal chloroses on site 21/Togo (*Drechsel*, unpubl.). DRIS allows to rank the deficient nutrients (see Table 2 for comparison).

## 3 Evaluation of sustainability of tropical tree plantations

To estimate the sustainability of tree plantations nutrient stores and nutrient dynamics in the different compartments of the system have to be taken into consideration. Tropical tree plantations often are established on marginal soils with fast growing species like Eucalyptus (39%), Pines (28%), Tectona grandis (17%) and Acacia spp. (5%) (*Zech* et al., 1989). Rotation time in these monocultures usually is very short, and each harvest is accompanied by nutrient exports in the form of wood, but also by soil erosion, nutrient leaching and volatilization due to the destruction of the protecting canopy, burning of residues and higher microbial activity. These losses may reduce soil and biomass nutrient stores and thus sustainability. However, whether nutrient reserves are exhausted after two or five rotations depends on tree species, site conditions and management. But reliable data which allow modelling of nutrient dynamics and yields in tropical plantations are scarce (*Sanchez* et al., 1985; *Bruijnzeel*, 1992). Basic figures on nutrient storage in tropical tree plantations (and hence on potential nutrient losses upon harvesting) are reported from a couple of countries. Examples are Tanzania (*Lundgren*, 1978), Nigeria (e.g. *Kadeba* and *Aduayi*, 1986; *Egunjobi* and *Bada*, 1979; Nwoboshi, 1983, 1984), India (e.g. *Singh*, 1982), and Indonesia (e.g. *Bruijnzeel*, 1983, 1992). Some studies focus as well on losses due to accelerated mineralization, leaching and erosion during clearing or the first harvest (e.g. *Nykvist* et al., 1994; *Bruijnzeel*, 1992).

As there are in general no long-term observations, most studies compare different stands of different age or cleared with still uncleared sites to get an idea about development ("false time series"). As there are usually not enough repetitions the "temporal changes" may be confounded by spatial variability in soil conditions (*Bruijnzeel*, 1992). Nevertheless, as all studies show the same tendency, it seems to be significant that clearing of a natural forest by logging and burning is accompanied by significant nutrient losses, as shown in the example of Table 3 (*Ruhiyat*, 1989; *Fölster* and *Ruhiyat*, 1991).

The situation becomes worse after the first rotation (*Fölster* and *Ruhiyat*, 1991). The figures for potassium in Eucalyptus deglupta plantations are summarized in Table 4. They indicate a drastic decrease of the "available" K pool in the soil of about 34% within the first rotation (20 years). It is assumed that during the second rotation the same amount will be lost again. Although some sources of nutrient inputs like precipitation and weathering of parent materials are not taken into consideration, the results indicate that in contrast to the majority of natural tropical forests which reveal "closed" nutrient cycles (input = output), frequently disturbed systems like short term tree plantations are highly susceptible to degradation of at least one nutrient pool. Similar results from Nigeria were published by *Egunjobi* and *Bada* (1979) or *Kadeba* and *Aduayi* (1986) for Pinus caribaea. *Hase* and *Fölster* (1983) compared nutrient stores and cycles in the different compartments of natural forests, teak plantations and secondary forests growing on Fluvisols in Venezuela. As in Indonesia, it was mainly K besides Ca and Mg being primarily lost in the plantation due to hardwood exploitation (Ca) and leaching (K), especially on sandy soils. N and P were less affected.

For Eucalyptus uvo-grandis plantations in Eastern Amazonia *Spangenberg* (1994) analysed average nutrient losses due to exportation of wood and bark to amount 65% N, 54% P, 76% Ca, 57% K, and 61% Mg of the corresponding stores in the whole tree biomass. Assuming the same nutrient removal during the second rotation, losses would correspond to about 100% K, more than 100% Ca, about 30% Mg of the actual "available" soil stores in these elements. The bark of Eucalyptus in particular is very rich in Ca (*Turner* and *Lambert*, 1983). Figures like these explain why with rising numbers of rotations base saturation and e.g. Ca soil stores sharply decrease. For compensation an application of up to 250 kg Ca $ha^{-1}$ appeared to be necessary in the case of the Amazonia study.

In a case study in Sabah, Malaysia, *Nykvist* et al. (1994) found that wood extraction using crawler tractors followed by slash burning and planting resulted in much

**Table 3:** Total nutrient stores (kg $ha^{-1}$) in a Dipterocarpacae forest of East-Kalimantan, and average nutrient losses due to clearing (*Ruhiyat*, 1989; *Fölster* and *Ruhiyat*, 1991).

| | Total reserves in biomass, litter and soil (50 cm) before clearing the natural forest | Reserves in the biomass only (natural forest) | Mean losses during clearing the natural forest |
|---|---|---|---|
| N | 9000 (100%) | 1180 (13%) | 2000 (22%) |
| K | 1150 (100%) | 800 (70%) | 400 (35%) |
| Ca | 2000 (100%) | 1400 (70%) | 700 (35%) |
| Mg | 900 (100%) | 230 (25%) | 90 (10%) |

**Table 4:** Decrease of K stores (kg $ha^{-1}$) in Eucalyptus deglupta plantations on clayey Ultisols in mid slope position due to wood exploitation and accelerated losses by erosion, mineralization and leaching in the context with harvesting (*Fölster* and *Ruhiyat* 1991; *Ruhiyat*, 1989)

| | |
|---|---|
| Total K stores at the beginning of the 1st generation | 1013 (100%) |
| K loss due to wood exploitation at the end of the 1st generation (20 years) | 231 (23%) |
| K loss due to accelerated mineralization, leaching and erosion during the 1st harvest | 114 (11%) |
| Total K stores at the beginning of the 2nd generation | 668 (66%) |
| K loss due to wood exploitation a the end of the 2nd generation | 231 (23%) |
| K loss due to accelerated mineralization, leaching and erosion during the 2nd harvest | 114 (11%) |
| Total K stores at the beginning of the 3rd generation | 323 (32%) |

**Table 5:** Losses (kg $ha^{-1}$) of inorganic nutrients connected with the removal of stem wood after clear-felling a tropical rainforest in Sabah, Malaysia, and treatment effects on dissolved nutrient losses during the first 2.8 years after clear-felling (*Nykvist* et al., 1994).

| | Treat-ment[1] | Losses caused by removing stem wood | Dissolved nutrient losses after clear-felling | Total losses | Dissolved nutrient losses in % of total losses |
|---|---|---|---|---|---|
| N | A | 100 | 27 | 127 | 21% |
| | B | 80 | 40 | 120 | 33% |
| P | A | 2.3 | 0.8 | 3.1 | 26% |
| | B | 1.8 | 1.3 | 3.1 | 42% |
| K | A | 71 | 106 | 177 | 60% |
| | B | 57 | 189 | 246 | 77% |
| Ca | A | 139 | 25 | 164 | 15% |
| | B | 112 | 27 | 139 | 19% |
| Mg | A | 39 | 8 | 47 | 17% |
| | B | 32 | 16 | 48 | 33% |
| Zn | A | 0.4 | 1.1 | 1.5 | 73% |
| | B | 0.3 | 0.4 | 0.7 | 57% |
| Fe | A | 9 | 22 | 31 | 71% |
| | B | 7 | 38 | 45 | 84% |

[1] A: Manual felling, wood extraction and planting row preparation without burning
B: Manual felling, wood extraction using crawler tractors, slash burnt.

higher nutrient losses compared with manual wood extraction and no burning (Table 5). A considerable proportion of the total losses can be attributed to the leaching/runoff of dissolved nutrients. For K, Zn and Fe, dissolved nutrient losses exceeded losses by the removal of stemwood.

In summary, the following statements can be given:

a) To evaluate sustainability of forest plantations water and nutrient balances have to be studied using an ecosystemary approach. Clearly, there is a need for more quantitative information. There is still a great deficiency in our knowledge of nutrient budgets (input – output) under different management and fertilizer practices, considering e.g. root necromass decomposition and below-ground biomass turnover, inputs due to dry deposition, chemical weathering of the parent material, and interflow.
b) However, we already have a quite good idea about the general dimensions of nutrient inputs and outputs, and the available data indicate poor sustainability. No successful stable crop production system can be based on "mining" marginal soils without returning a proportionate amount of what was removed by harvesting and leaching.

Optimizing plantation management offers some possibilities to keep nutrient cycles in a better equilibrium or even to ameliorate soil and tree nutrient status. This subject will be discussed in the last section.

## 4 Amelioration of soil and tree nutrient status without and with mineral fertilizer

Improving and maintaining fertility of tropical soils are big challenges in view of increasing population and increasing demands of people, especially in developing countries. Degradation and amelioration of the soil nutrient status both relates to forestry and agriculture. Since mineral fertilizers usually are expensive, alternatives for improving and maintaining soil fertility and reducing the risk of nutrient disturbances are of great importance.

In a case study in Sabah, Malaysia, nutrient losses in stream water during clear-fellings, harvest and site preparation were reduced by 50% and growth of the planted forests increased by 100% if soil disturbances were minimized and burning not used. In addition, weeds were more abundant after burning and the extra weeding needed increased costs for plantation establishment (*Nykvist* et al., 1994). Slash burning should be avoided as well between rotations as it could accelerate decline in the following generation (*Wood*, 1990). In Central Java, *Bruijnzeel* (1992) and *Bruijnzeel* and *Wiersum* (1985) demonstrated that proper land management systems excluding whole tree harvesting may avoid soil degradation on moderately fertile Andosols. If besides boles also bark, twigs and leaves were exported, P, Ca and again K stores were rapidly exhausted even on these fertile soils. Data of bark, branch, twig and foliage nutrient contents indicate quite clearly that there are possibilities to reduce nutrient losses and soil degradation by debarking or other ways of avoiding whole-tree harvests (*Nykvist* et al., 1994). Other elements of nutrient management, starting with site evaluation before planting and site adapted tree species selection, through careful litter management, up to optimal (long) rotation periods, were summarized by *Drechsel* and *Zech* (1993) and *Pancel* (1993ab). The effects of these methods are of course less obvious than of mineral fertilizer but probably more lasting. As there exists only a small amount of data on their contribution to nutrient balance, future research should be focussed on these methods. In view of fertilizer prices and long-term sustainability nutrient management will become more and more important not only in cropping systems but also in plantation forestry.

To obtain high yields within a short time, mineral fertilizers usually seem to be most promising. Proper fertilizer application improves mineral disorders, supports the establishment of plantations on infertile sites, stimulates growth and may increase resistance against diseases, frost and drought (*Evans*, 1992). Many different methods of applying fertilizers especially to eucalypts and pines have been studied all over the world. The results were variable and partly confusing. They depend mostly on the specific site conditions (e.g. soil texture, clay type, water regime, pH), weed competition, weather conditions, tree species, time and kind of application as well as type and quantity of fertilizer. General conclusions valid for different regions, soils and species are not possible. Recommendations have always to be based on local fertilizer trials and experiences. Only the establishment of such experiments by applying the identified deficient elements in different quantitites and mixtures allows a definitive verification of the diagnosis (*Cellier* and *Correll*, 1984). Most data on the fertilization of (sub)tropical tree plantations has been gathered from South Africa (e.g. *Schönau*, 1984; *Schutz*, 1976; *Schönau* and *Herbert*, 1989), Australia (e.g. *Crane*, 1984; *Grant*, 1991; *Turner* and *Lambert*, 1983, 1986), New Zealand (e.g. *Will*, 1985), Brazil (*Bellote*, 1990) and e.g. Malaysia (*Thajib* and *Pushparadjah*, 1984), where fertilizer application in forest, rubber or oil palm plantations is widely used. In contrast to the silviculture practices in these countries, little fertilizer is used in most developing countries due to high costs despite poverty of the soils. Since several reviews on fertilization of (sub)tropical tree plantations, especially with eucalypts and pines, have been published (e.g. *Ballard*, 1984; *Herbert* and *Schönau*, 1991; *Drechsel* and *Zech*, 1993) we will focus in this paper on Terminalia superba as hardwood, Gmelina arborea as pulp and paper source and Azadirachta indica for pole and firewood production (Table 6). The examples cover different causes (low soil fertility and high acidity) and purposes of fertilization (amelioration of tree growth and wood production as well as of wood quality).

In the first example from Southern Nigeria different amounts of lime were tested to enhance growth of Terminalia superba on acid Ultisols and to study changes in soil and foliar nutrient levels. Best growth was obtained using 6–9 g $Ca(OH)_2$ per seedling, increasing the soil pH-$H_2O$ from 4.6 to about 6.0. At this pH highest foliar P levels were measured. However, soil tests for available P (Bray method) were not adapted to these conditions and showed no increase in soil P (Tab. 6a). The problem of non-sensitive soil tests for the analysis of tree available nutrients has already been mentioned above.

In the second example different kinds of fertilizer were used to increase yields of Neem on sandy Luvisols in Northern Nigeria. On the three stations under study, the application of farmyard manure usually gave the best results in comparison with comparable amounts of NPK (Table 6b). Similar results were found for Eucalpytus camaldulensis in the same region showing the importance of organic material under tropical conditions.

Nutritional disorders do not only decrease tree yield, but also wood quality, e.g. with regard to the production of pulp and paper. In this last example the effects of different kinds of N-fertilizer ($NO_3$-N, $NH_4$-N, $NH_4NO_3$-N, urea-N) on Gmelina growth and wood properties were studied. Due to the central role of nitrogen in plant metabolism significant differences were found, e.g. in tree growth, specific gravity of the wood and size of fibre cells (Table 6b).

**Table 6a:** Case studies of fertilizer experiments carried out in Nigeria.

| Example | Problem | Species | Kind of trial | Treatments | Lime rate | Tree height (cm) | Results: pH $H_2O$ | Soil-P (ppm) | Foliar-P (%) |
|---|---|---|---|---|---|---|---|---|---|
| 1 SW-Nigeria Ultisol (*Aluko*, 1990) | soil acidity | Terminalia superba (seedlings) | greenhouse trials over 6 months | different lime quantities: 0/3/6/9/12/15 g $Ca(OH)_2$ per seedling | 0 | 18 | 4.6 | 6.0 | 0.14 |
| | | | | | 3 | 21 | 5.7 | 5.8 | 0.17 |
| | | | | | 6–9 | 26 | 6.0 | 5.7 | 0.18 |
| | | | | | 12 | 18 | 6.5 | 5.6 | 0.15 |
| | | | | | 15 | 11 | 7.2 | 5.4 | 0.14 |

**Table 6b:** Case studies of fertilizer experiments carried out in Nigeria.

| Example | Problem | Species | Kind of trial | Treatments | Results | | | |
|---|---|---|---|---|---|---|---|---|
| 2 N-Nigeria Ferric Luvisol (*Kadeba*, 1990/91) | low fertility (2% clay) | Azadirachta indica (2.5 yrs old) | field trials over 2 years | different fertilizer: 2 kg farmyard manure and 84 g NPK (15-15-15) per tree | Mean tree height (m) at 3 research stations | 1 | 2 | 3 |
| | | | | | Manure | 4.2 | 4.5 | 4.9 |
| | | | | | NPK | 4.3 | 4.4 | 4.6 |
| | | | | | Control | 3.7 | 3.8 | 4.0 |
| 3 S-Nigeria "Latosol" (*Ogbonnaya*, 1993) | low wood quality and growth | Gmelina arborea (seedlings) | greenhouse trials over 5 months | different N-sources: $NO_3$-N as $KNO_3$; $NH_4$-N as $(NH_4)_2SO_4$; $NH_4NO_3$-N as $CaNH_4(NO_3)_3$; Urea-N | 2.5 g N per seedling | Tree growth | Specific gravity | Fibre length |
| | | | | | $NO_3$-N | 210% | 0.62 | 0.59 |
| | | | | | $NH_4$-N | 200% | 0.57 | 0.42 |
| | | | | | $NH_4NO_3$-N | 300% | 0.64 | 0.57 |
| | | | | | Urea-N | 240% | 0.61 | 0.62 |
| | | | | | Control | 100% | 0.56 | 0.55 |

The authors found that the application of $NH_4$-N fertilizer should not be recommended.

Even this data make clear that the kind of fertilizer as well as application rates of nutrients depend on the objective of the plantation, the tree species selected, site and soil properties, stand age and density, and several other parameters including of course the financial resources.

## 5 Conclusions

In contrast to the sustainability of natural ecosystems, man-made sustainability is until now only possible with external inputs. Due to nutrient export tropical tree plantations, especially in short rotations, will exhaust the nutrient reserves of the system. The time span until the system will become unproductive depends on initial soil fertility, number of rotations and a range of parameters which can be influenced by man. A careful nutrient mangement starting at clear-felling as well as the diagnosis and treatment of appearing nutrient disorders are important tools with respect to the amelioration and stabilization of the nutrient balance.

## References


*Aluko, A. P.* (1990): Effect of liming an Ultisol for the establishment of a tropical hardwood in Southern Nigeria. Journal of Tropical Forest Science 2, 187–194.

*Ballard, R.* (1984): Fertilization of plantations. In: G. D. Bowen and E. K. S. Nambiar (Eds.): Nutrition of plantation forests. p. 327–360. Academic press, London.

*Beaufils, E. R.* (1973): Diagnosis and Recommendation Integrated Systems (DRIS). Soil Sci. Bull. 1, University of Natal, Pietermaritzburg, 132 pp.

*Bellote, A. F. J.* (1990): Nährelementversorgung und Wuchsleistung von gedüngten Eucalyptus grandis-Plantagen im Cerrado von Sao Paulo (Brasilien). Freiburger Bodenkundliche Abhandlungen 26, Universität Freiburg.

*Bergmann, Ch., Stuhrmann, M.*, and *W. Zech* (1994): Site factors, foliar nutrient levels and growth of Cordia alliodora plantations in the humid lowlands of Northern Costa Rica. Plant and Soil 166, 193–202.

*Bowen, G. D.*, and *E. K. S. Nambiar* (1984): Nutrition of plantation forests. Academic Press, London.

*Bruijnzeel, L. A.* (1983): The chemical mass-balance of a small basin in a wet monsoonal environment and the effect of fast-growing plantation forest. Intern. Assoc. Hydrol. Science Publ. 141, 229–239.

*Bruijnzeel, L. A.* (1992): Sustainability of fast-growing plantation forests in the humid tropics with particular reference to nutrients. in: Jordan, C. F. et al.: Taungya: Forest plantations with agriculture in Southeast Asia. CAB International, Wallingford, p. 51–67.

*Bruijnzeel, L. A.*, and *K. F. Wiersum* (1985): A nutrient balance sheet for Agathis dammara Warb. plantation forest under various management conditions in central Java, Indonesia. Forest Ecology and Management 10, 195–208.

*Brunck, F.* (1987): Tropical forest species, in P. Martin-Prével, J. Gagnard, P. Gautier: Plant analysis as a guide for the nutrient requirements of temperate and tropical crops. Lavoisier, New York, p. 479–486.

*Cellier, K. M.*, and *R. L. Correll* (1984): Design and analysis of tree nutrient experiments, in G. D. Bowen, E. K. S. Nambiar: Nutrition of plantation forests. Academic Press, London, p. 439–461.

*Chijioke, E. O.* (1980): Impact on soils of fast-growing species in lowland humid tropics. FAO Forestry Paper 21, FAO, Rome, 111 pp.

*Crane, W. J. B.* (1984): Fertilization of fast-growing conifers, in D. C. Grey, A. P. G. Schönau, C. J. Schutz: Symposium on site and productivity of fast-growing plantations. IUFRO 1, 238–251.

*Drechsel, P.* (1992): Beziehungen zwischen Standort, Ernährungszustand und Wuchsleistung von Teak (Tectona grandis L.f.) im humiden Westafrika. Bayreuther Bodenkundliche Berichte 31, 264 pp.

*Drechsel, P.* (1994): DRIS – Diagnosis and Recommendation Integrated System – in Theorie und Praxis. Eine SPSS-orientierte Anleitung. Bayreuther Bodenkundl. Berichte 34, 72 pp.

*Drechsel, P.*, and *W. Zech* (1991): Foliar nutrient levels of broad-leaved tropical trees: A tabular review. Plant and Soil 131, 29–46.

*Drechsel, P.*, and *W. Zech* (1993): Mineral nutrition of tropical trees, in L. Pancel: Tropical Forestry Handbook, Springer, p. 515–567.

*Drechsel, P.*, and *W. Zech* (1994). DRIS evaluation of teak (Tectona grandis L. F.) mineral nutrition and effects of nutrition and site quality on teak growth in West Africa. Forest Ecology and Management 70, 121–133.

*Drechsel, P., D. Hincourt*, and *W. Zech* (1995): Entwicklung und Bewertung von Vertisolen im semi-humiden West-Afrika. Mitt. deutsche Bodenkdl. Ges. 76: 1009–1012.

*Egunjobi, J. K.*, and *S. O. Bada* (1979): Biomass and nutrient distribution in stands of Pinus caribaea L. in the dry forest zone of Nigeria. Biotropica 11 (2), 130–135.

*Evans, J.* (1992): Plantation forestry in the tropics. Clarendon Press, Oxford, 403 pp.

*Fölster, H.*, and *D. Ruhiyat* (1991): Die Entwicklung der Nährstoffvorräte in Holzplantagen nach Kahlschlag des Regenwaldes, in W. Zech: Mineralstoffversorgung tropischer Waldbäume. Bayreuther Bodenkundliche Berichte, 21, 63–77.

*Grant, M. J.* (1991): Recent state of knowledge concerning the mineral nutrition of conifer plantations in Queensland, Australia, in W. Zech: Mineralstoffversorgung tropischer Waldbäume. Bayreuther Bodenkundliche Berichte 21, 45–61.

*Hase, H.*, and *R. H. Fölster* (1983): Impact of plantation forestry with teak (Tectona grandis) on the nutrient status of young alluvial soils in West Venezuela. Forest Ecology and Management 6, 33–57.

*Herbert, M. A.*, and *A. P. G. Schönau* (1991): Fertilizing commercial forest species in Southern Africa: research progress and problems, in W. Zech: Mineralstoffversorgung tropischer Waldbäume. Bayreuther Bodenkundliche Berichte 21, 7–43.

IPI (International Potash Institute) (Ed.). (1962): On the mineral nutrition of some tropical crops. IPI, Berne.

*Kadeba, O.* (1990/91): Fertilizer application in aid of plantation establishment in the savanna areas of Nigeria. Water, Air, and Soil Pollution 54, 641–649.

*Kadeba, O.*, and *E. A. Aduayi* (1986): Dry matter production and nutrient distribution in a Pinus caribaea stand planted in a subhumid tropical savanna site. OIKOS 46, 237–242.

*Lambert, M. J.* (1984): The use of foliar analysis in fertilizer research, in D. C. Grey, A. P. G., Schönau, C. J. Schutz: IUFRO Symposium on site and productivity of fast growing plantations. Pretoria and Pietermaritzburg, Vol. 1, p. 269–291.

*Lundgren, B.* (1978): Soil conditions and nutrient cycling under natural and plantation forests in Tanzania highlands. Reports in Forest Ecology and Forest Soils, 31, 1–426. Swedish University of Agricultural Sciences. Uppsala.

*Mead, D. J.* (1984): Diagnosis of nutrient deficiencies in plantations, in G. D. Bowen, E. K. S. Nambiar: Nutrition of plantation forests. Academic press, London.

*Nwoboshi, L. C.* (1983): Growth and nutrient requirements in a Teak plantation age series in Nigeria. I. Linear growth and biomass production. Forest Science 29, 159–165.

*Nwoboshi, L. C.* (1984): Growth and nutrient requirements in a Teak plantation age series in Nigeria. II. Nutrient accumulation and minimum annual requirements. Forest Science 30, 35–40.

*Nykvist, N., H. Grip, B. L. Sim, A. Malmer* and *F. K. Wong* (1994): Nutrient losses in forest plantations in Sabah, Malaysia. AMBIO 23, 210–215.

*Ogbonnaya, C. I.* (1993): Effects of nitrogen sources on the wood properties of Gmelina arborea relevant to pulp and paper production. Forest Ecology and Management 56, 211–223.

*Pancel, L.* (1993a): Species selection, in L. Pancel: Tropical Forestry Handbook. Springer, p. 463–514.

*Pancel, L.* (1993b): Forestation, in L. Pancel: Tropical Forestry Handbook. Springer, p. 645–725.

*Proctor, J.* (1987): Nutrient cycling in old secondary rainforests. Applied Geography 7, 135–152.

*Ruhiyat, D.* (1989): Die Entwicklung der standörtlichen Nährstoffvorräte bei naturnaher Waldbewirtschaftung und im Plantagenbetrieb, Ostkalimantan (Indonesien). Göttinger Beiträge zur Land- und Forstwirtschaft in den Tropen und Subtropen 35, 1–206.

*Sanchez, P. A., C. A. Palm, C. B. Davey, L. T. Szott*, and *C. E. Russell* (1985): Tree crops as soil improvers in the humid tropics? in: Cannell, M. G. R. and J. E. Jackson: Attributes of trees as crop plants. Institute of Terrestrial Ecology, Huntingdon, U. K. p. 327–358.

*Schönau, A. P. G.* (1984): Silvicultural considerations for high productivity of Eucalyptus grandis. Forest Ecology and Management 9, 295–314.

*Schönau, A. P. G.*, and *M. A. Herbert* (1989): Fertilizing eucalypts at plantation establishment. Forest Ecology and Management 29, 221–244.

*Schutz, C. J.* (1976): A review of fertilizer research on some of the more important conifers and eucalypts planted in subtropical and tropical countries, with special reference to South Africa. Dep. For. Pretoria, 53, 89 pp.

*Schutz, C. J.*, and *J. M. De Villiers* (1987): Foliar diagnosis and fertilizer prescription in forestry – the DRIS System and its potential. South Afr. Forestry Journal 141, 6–12.

*Singh, B.* (1982): Nutrient content of standing crop and biological cycling in Pinus patula ecosystem. Forest Ecology and Management 4, 317–332.

*Spangenberg, A.* (1994): Nährstoffvorräte und -exporte von Eucalyptus uvo-grandis-Plantagen in Ostamazonien (Jari), Brasilien. Göttinger Beiträge zur Land- und Forstwirtschaft in den Tropen und Subtropen 93, 114 pp.

*Stuhrmann, M., Ch. Bergmann*, and *W. Zech* (1994): Mineral nutrition, soil factors and growth rates of Gmelina arborea plantations in the humid lowlands of northern Costa Rica. Forest Ecology and Management 70, 135–145.

*Thajib, A.*, and *E. Pushparadjah* (1984): Soils and nutrition of perennial crops. Malaysian Soil Science Society, Kuala Lumpur, 453 pp.

*Turner, J.*, and *M. J. Lambert* (1983): Nutrient cycling within a 27-year-old *Eucalyptus grandis* plantation in New South Wales. Forest Ecology and Management 6, 155–168.

*Turner, J.*, and M. J. Lambert, (1986): Fate of applied nutrients in a long term superphosphate trial in Pinus radiata. Plant and Soil 93, 373–382.

*Van den Driessche, R.* (1974): Prediction of mineral nutrient status of trees by foliar analysis. Bot. Rev. 40 (3), 347–394.

*Walworth, J. L.*, and *M. E. Sumner* (1987): The diagnosis and recommendation integrated system (DRIS). Advances in Soil Science 6, 149–188.

*Walworth, J. L.*, and *M. E. Sumner* (1989): Foliar analysis: A review. Advances in Plant Nutrition 3, 193–241.

*Weetmann, G. F.*, and *C. G. Wells* (1990): Plant analyses as an aid in fertilizing forests, in R. L. Westerman: Soil testing and plant analysis 3rd edn., SSSA Book Series 3, Madison, p. 659–690.

*Westerman, R. L.* (Ed.) (1990): Soil testing and plant analysis. 3rd edn. SSSA Book Series 3, Madison.

*Will, G.* (1985): Nutrient deficiencies and fertilizer use in New Zealand exotic forests. New Zealand Forest Service, FRI Bulletin 97, 53 pp.

*Wood, R. W.* (1990): Second rotation decline in Pinus radiata plantations in South Australia has been corrected. Water, Air, and Soil Pollution 54, 607–619.

*Zech, W.*, and *N. Cepel* (1970): Beziehungen zwischen der geschätzten nutzbaren Wasserkapazität von Waldböden und dem Höhenwachstum von Pinus brutia – Beständen in den mediterranen Subtropen Anatoliens. Z. Pflanzenernährung und Bodenkunde 127 (1), 41–49.

*Zech, W., D. Elz, L. Pancel*, and *P. Drechsel* (1989): Auswirkungen und Erfolgsbedingungen in Aufforstungsvorhaben in Entwicklungsländern. Report of the Ministry for Economic Cooperation 93. Weltforum Verlag, München, 348 pp.

*Zhong, A.-L.* and *W-Y. Hsiung* (1993): Evaluation and diagnosis of tree nutritional status in Chinese-fir (Cunninghamia lanceolata (Lamb) Hook) plantations, Jiangxi, China. Forest Ecology and Management 62, 245–270.

# Growth response of rainforest cabinet timber species to fertiliser application in North Queensland plantations

**Rodney Keenan[1], Alison Hambleton[1], Ken Robson[1]** and **Michael Webb[2]**

[1]Queensland Forest Research Institute, DPI-Forestry, P.O. Box 1138, Atherton, Australia
e-mail: keenanr@dpi.qld.gov.au
[2]CSIRO land and water, PMB, Aitkenvale, Australia

## Abstract

Declining supplies of cabinet-timber from natural stands of tropical rainforest combined with an increasing global demand for all classes of timber has given rise to a significant interest in developing plantations of high-value cabinet timber species in many parts of the tropics. In many cases the plantations are being established on land with low inherent fertility or sites that have been degraded as a result of poor landuse practices. The use of fertilisers has the potential to increase the production on these sites, but our understanding of the rates, timing and types of fertiliser to use with different species on different soil or site types is poor. This paper describes results of experiments at three sites in the wet tropics of north Queensland with similar soil types to sites being used for plantation establishment in South East Asia. The experiments aimed to determine the response to varying rates of P fertiliser in three high-value rainforest cabinet timber species, the Australian natives: black bean (*Castanospermum australe* Cunn. ex C. Fraser), and kauri pine (*Agathis robusta* (C. Moore ex F. Muell.) Bailey) and the exotic West Indian cedar (*Cedrela odorata* L.). Responses to fertilisation varied between sites and between species. There was a more pronounced response to fertilisation on coarse-textured soils derived from granite parent materials than on finer extured soils derived from metamorphic or basalt substrates. We discuss these results in light of their implications for plantation management and for the design and implementation of further nutritional research.

## 1 Introduction

Tropical rainforests are an important economic, environmental, and social resource. Historically, natural rainforests in South East Asia and north Queensland formed the basis of major timber industries. In north Queensland sought after cabinet timber species included red cedar (*Toona ciliata*), Queensland maple (*Flindersia brayleyana*), northern silky oak (*Cardwellia sublimis*) and kauri pine (*Agathis robusta*). In South East Asia species in the family Dipterocarpaceae and teak (*Tectona grandis*) are the most commonly harvested high-value species. However, in much of the region the area of natural rainforest has diminished substantially due to conversion to other landuses or degradation following uncontrolled harvesting and there are increasing pressures to include remaining natural forests in conservation reserves. In Australia tropical rainforests receive a relatively high level of legal protection, in large part because of the inclusion of remaining state-owned rainforests on the UNESCO World Heritage List in 1988, but historically they suffered similar impacts to those now being experienced in the rainforests of South East Asia. Clearing for agricultural and urban development began with European settlement in the mid 1800's and clearing of small areas of privately-owned rainforest still continues.

Because of their timber value there have been many attempts to grow high-value native species in plantations (*Anonymous* 1983, *Keenan* et al. 1995, *Appanah* and *Weinland* 1996) but for a number of reasons, except for teak, the area planted has been relatively small. With the reduction in area of natural forest available for timber production, improvements in silvicultural knowledge gained from other species, and anticipated price increases for many timber species (*Lyke* 1995) there is a renewed interest in plantations of cabinet timber species.

However, the land base available for plantation establishment in much of the region is often on soil types that are too poor to sustain agricultural production or land was left in a degraded condition from past clearing (*Jordan* 1993). This is particularly so in Australia where government policy restricts the clearance of native vegetation for plantation establishment. Consequently, there is a pressing need to develop sustainable forest plantation production systems for poorer quality or degraded lands.

Fertilisers are now commonly used in plantation silviculture to correct identified nutritional deficiencies, to facilitate plantation establishment, or to increase timber yields (*Evans* 1992). Although even in areas with major plantation estates the use of fertilisers to increase yields has been inconsistent due to the variability of response across a planted area, uncertainty over the financial benefits and operational problems with fertiliser application (*Hunter* and *Smith* 1996). Given that most plantations in the tropics are being established on poorer quality or degraded soils, addition of fertiliser and other forms of nutritional enhancement generally have the potential to improve establishment success or increase timber production (*Davidson* 1996). However, knowledge of the nutrient requirements of different high value timber species on different soil types and responses to fertilisation at different stages of stand development is scant.

Most studies of fertilisation responses in the South East Asian and Australasian tropics have been undertaken with exotic species being used in various plantation programs. Substantial growth responses have been reported for tropical and subtropical pines (*Hunter* et al. 1992, *Simpson* 1995), and many eucalypts have been quite responsive to fertilisation. For example, 3 to 4 year old *Eucalyptus deglupta* plantations exhibited an increase in mean annual volume increment of 4 $m^3$/ha three years after fertilising with 400 kg/ha of P on sandstone and mudstone derived soils in Sabah (*Cromer* et al. 1992). *Pegg* (1991) reported increases in volume production of 37 to 127 percent at 5.5 years of age for a number of tropical and subtropical eucalypt species planted on eroded acid latosols derived from sedimentary parent material near Nanning in southern China when fertilised at planting with a complete mix of 100 kg/ha N, 50 kg/ha P, 50 kg/ha K plus trace elements.

For high value timber species, Queensland maple and red cedar exhibited strong early height growth response to slow release fertiliser (Osmocote 14:14:4, N:P:K) planted on recent volcanic rocklands low in P, K, Ca and Mg in Hawaii (*Walters* 1972). Height growth of the dipterocarps *Dryobalanops lanceolata* and *Shorea leprosula* planted on degraded log landings in Sabah was up to three times greater than non-fertilised treatments six months after planting when seedlings were fertilised with 100 g of rock phosphate and 40 g of 12:12:7, N:P:K (*Nussbaum* and *Hoe* 1996).

The objective of this study was to determine the response to varying rates of P fertiliser of three high-value rainforest cabinet timber species, the Australian natives: black bean (*Castanospermum australe* Cunn. ex C. Fraser), and kauri pine (*Agathis robusta* (C. Moore ex F. Muell.) Bailey, and the exotic West Indian cedar (*Cedrela odorata* L.). Results are presented from experiments undertaken on three different soil types in north Queensland, that are analagous to soils being used for plantation establishment in South East Asia. P has been identified as a limiting nutrient in glasshouse nutrient omission studies at one of the three sites (*Webb* et al. 1991).

The study was part of a larger project funded by the Australian Centre for International Agricultural Research (ACIAR) that aimed to characterise soils, identify nutritional deficiencies and mycorrhizal associations for a range of sites and tree species in the Western Pacific region (*Poa* and *Mazza* 1995).

## 2 Methods

### 2.1 Study Areas

The study was undertaken at three sites near the towns of Feluga, South Johnstone, and Atherton (between 17° and 18° S) in North Queensland. Altitude and climatic

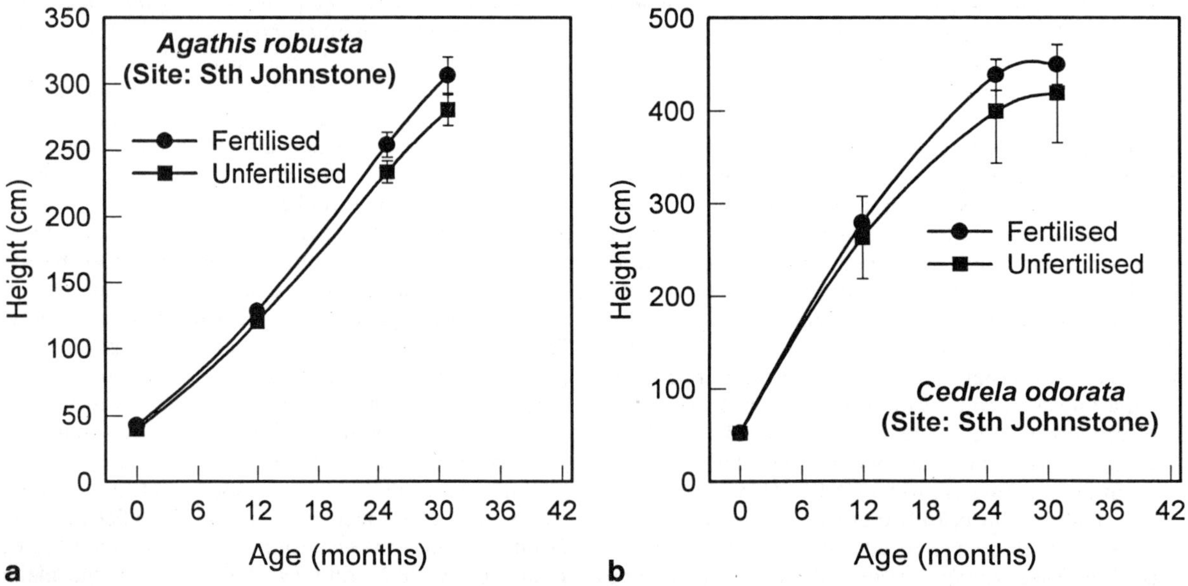


**Figure 1:** Height growth of (a) kauri pine (*Agathis robusta*) and (b) West Indian cedar (*Cedrela odorata*) for the first 30 months after treatment with varying levels of P fertilisation on metamorphic soils at South Johnstone in north Queensland. All fertilised treatments have been combined for comparison with the nil treatment. See text for details of fertiliser treatments. Error bars are one standard error of the mean.

**Table 1:** Location and climatic variables for the three study sites.

| Location | Latitude (S) | Longitude (E) | Altitude (m) | Mean Annual Rainfall (mm) | Mean minimum temperature in coldest month (°C) | Mean maximum temperature in warmest month (°C) |
|---|---|---|---|---|---|---|
| Atherton | 17°16′ | 145°29′ | 760 | 1425 | 10.0 | 28.5 |
| South Johnstone | 17°36′ | 146°00′ | 18 | 3625 | 14.0 | 31.5 |
| Feluga | 17°56′ | 145°55′ | 10 | 4321 | 15.0 | 30.0 |

**Table 2:** Soil characteristics for the three study sites.

| Location | Geology | Slope (%) | Depth (m) | Texture | N% | P (mg.g$^{-1}$) | K | CEC (cmol(c).kg$^{-1}$) |
|---|---|---|---|---|---|---|---|---|
| Atherton | Basalt | 3.5 | > 2 | clay loam | 0.08 | 28 | 0.21 | 6.2 |
| South Johnstone | Metamorphic | 9 | 1.2 | clay loam-light clay | 0.06 | 2 | 0.07 | 1.6 |
| Feluga | Granite | 7 | 1.5 | sandy loam | 0.05 | 19 | 0.11 | 2.4 |

conditions for the three areas are shown in Table 1 and soil characteristics in Table 2. All sites had been cleared for over 50 years. Feluga and South Johnstone are situated on the coastal lowlands and Atherton is at the northern end of a broad plateau about 50 km from the coast.

The higher rainfall site at Feluga had previously been used for sugar and more recently for paw paw (*Carica papaya*) farming. Soils are sandy-loam yellow earths (Thorpe series, *Cannon* et al. 1992; Dystropeptic tropudult, *Soil Survey Staff* 1975) which developed on a granite fan.

At South Johnstone soils are red podzolics (Galmara series, *Murtha* 1986; Tropeptic haplorthox, *Soil Survey Staff* 1975) derived from metamorphic rocks with light clay to clay loam texture. The site had been unused for over 30 years and contained a high proportion of the grass *Imperata cylindrica* and other weeds prior to site preparation.

The Atherton site is at a higher altitude (760 m) and received a lower rainfall than the other two sites. Soils are krasnozems of the Pin Gin series (*Laffan* 1988, Tropeptic haplorthox *Soil Survey Staff*, 1975) derived from deeply weathered basalt with clay-loam to clay texture. The site was being used for maize cropping prior to the experiment and had a history of fertilisation for cropping, and pasture with legume incorporation.

The soils chosen for this study are also represented in South East Asia (*Isbell* 1978; *Isbell* and *Edwards* 1988) and two of these (Thorpe and Galmara) are found in areas that would be available for forestry activities (R.F. ISBELL, PERS. COMM.).

## 2.2 Experimental design

The experimental design differed in each location because of the area available and the species used. Two species were used at South Johnstone (kauri pine and West Indian cedar) and Feluga (Kauri pine and black bean), and one species at Atherton (black bean).

At Atherton the design was a randomised complete block with four replicates each of six fertiliser treatments (24 plots in total). Plots consisted of 30 trees (three rows with 10 trees in each row). Spacing was 3 m between rows and 1.8 m along the row (1833 stems per ha). Measurements were undertaken on the eight internal trees in each plot with a buffer of one row around the measured plot that received the same treatment.

At Feluga and South Johnstone the design was a randomised complete block of six treatment plots in four blocks (24 plots in total) with split plots for each of the two species used. Plots consisted of 40 trees (four rows of 10 trees). The two internal rows were planted with the target species while the outer buffer rows were planted with *Flindersia pimentaliana* at South Johnstone and *Araucaria cunninghamii* at Feluga and treated in the same way as the measured plot.

## 2.3 Site preparation and planting

Site preparation varied with location. At Atherton the site was disc-ploughed and sown with the grass *Brachyaria*, and planting lines were ripped and sprayed with a residual herbicide. At Feluga the site was slashed, raked and planting lines cultivated. At South Johnstone the site was burnt and sown with *Brachyaria*, ploughed and the planting lines cultivated with a rotary hoe. At all sites a 1 m strip was kept free of weeds using the knockdown herbicide, glyphosphate and the residual herbicide, simizine. Site preparation was completed by January 1992 and seedlings were planted during the wet season in January and February 1992. Seedlings used were propagated from seed collected off trees in the natural forest, or from locally planted West Indian cedar. Seedlings were grown in the the Queensland Forest Service nursery at Walkamin, near Atherton, in a potting mix of peat, vermiculite and slow-release fertiliser, and were about 1 yr old at the time of planting.

### 2.4 Fertilizer treatments

Details of fertilizer treatments are shown in Table 3. Six fertilizer treatments were used at each site. These consisted of a control (no fertiliser) treatment, four different levels of P with basal application of macronutrients (N, K, S, Ca, Mg) and trace elements (Zn, Cu, Mn, fe, Mo, and B), and a treatment with the highest level of P with macronutrients but no trace elements. At the Atherton site P levels were 0, 100, 250 and 500 kg $ha^{-1}$. At Feluga and South Johnstone P levels were 0, 50, 100, and 250 kg $ha^{-1}$. Fertilizers were applied in a split application, with half of the amount applied about 1 month after planting, and the remainder about 6 months after planting. Fertilizers were applied in measured amounts to individual trees and inserted in slits made with a spade about 20 cm from the base of each tree.

To investigate potential response to later-age fertilisation, treated plots in the Feluga experiment were re-fertilised at 24 months after planting to bring treatments up to a total level of P of 500 kg/ha and to test three different N fertilisers: urea (200 kg N/ha), diammonium phosphate (200 kg N/ha), and IBDU (iso-butylidene di-urea, 104 kg/ha).

### 2.5 Measurements

Seedling heights were measured immediately after planting and height, diameter at ground level (dgl), and diameter at breast height (when trees had reached sufficient size) were measured at approximately 6 monthly intervals for up to 41 months after planting.

### 2.6 Statistical analysis

Plot means for the four replicates of each treatment were used for statistical comparisons between treatments and for comparisons between sites. A stem volume index ($dgl^{2}$*height) was used to integrate diameter and height data. For each species mean height, dgl, and volume index were calculated for each plot. To compare the influence of fertilisation as a whole the means of non-fertilised and fertilised plots were calculated.

Results were initially compared using ANOVA, and where significant differences were identified individual treatments were compared using Tukey's test. In some cases individual orthogonal contrasts were made. Significance level was $p <= 0.05$ in all cases. Analyses were carried out using SYSTAT (*Wilkinson* 1990).

## 3 Results

### 3.1 Height growth

There were significant differences between sites in the growth of each species. *Agathis robusta* grew faster at South Johnstone (Fig. 1a) than at Feluga (Fig. 2a). *Castanospermum australe* grew faster at Feluga (Fig. 2b) than at Atherton (Fig. 3). The best performing species was the exotic *Cedrela odorata* planted at South Johnstone with an average height of over 4 m at 31 months.

In the first two years, there were no significant difference in height growth between individual fertiliser treatments for *Castanospermum australe* at Atherton, or either species at Feluga. For the 20 month measurement at South Johnstone, in both *Cedrela odorata* and *Agathis robusta*, trees receiving N, K, 250 kg/ha of P, but no micronutrients were significantly taller than those receiving the complete − P treatment.

With time, however, differences in height growth between fertilised and unfertilised treatments began to emerge. These differences were small at the South Johnstone and Atherton sites (Figs. 1, 3) but were quite pronounced at the Feluga site with relatively coarse-textured soils derived from granite (Fig. 2). The Feluga experiment was refertilised at 24 months. By this time the average height of *Agathis robusta* in all fertiliser treatments combined was almost twice that in unfertilised treatments

**Table 3:** Fertiliser treatment formulations applied at to plantations of rainforest cabinet timber species at three sites in North Queensland.

| Treatment | Nutrient and form | Elemental rate kg/ha | | | Fertiliser rate kg/ha | | |
|---|---|---|---|---|---|---|---|
| | | Total | 0 months | 6 months | Total | 0 months | 6 months |
| Complete | N as Urea (46% N) | 100 | 50 | 50 | 217 | 108.5 | 108.5 |
| | K as KCL (50% K) | 160 | 80 | 80 | 320 | 160 | 160 |
| | S as Sulfur (99.5% S) | 120 | 60 | 60 | 120 | 60 | 60 |
| | Ca as Dolomite (19% Ca) | 108 | 54 | 54 | 571 | 285.5 | 285.5 |
| | Mg as Dolomite (7.8% Mg) | 44 | 22 | 22 | | | |
| | Trace elements* | | | | 129 | 129 | |
| P Treatments | | | | | | | |
| P50 | P as Trifos (20.7% P) | 50 | 12.5 | 12.5 | 241.5 | 120.75 | 120.75 |
| P100 | P as Trifos (20.7% P) | 100 | 25 | 25 | 483 | 241.5 | 241.5 |
| P250 | P as Trifos (20.7% P) | 250 | 125 | 125 | 1207 | 603.5 | 603.5 |
| P500 | P as Trifos (20.7% P) | 500 | 250 | 250 | 2415 | 1207.5 | 1207.5 |

* Trace elements mix composed of the following by weight: $ZnSO_4$, $7H_2O$ 17%; $CuSO_4$, $5H_2O$ 11%; $MnSO_4$, $5H_2O$ 19%; Fe $SO_4$, $7H_2O$ 35%; $Na_2MoO_4$, $2H_2O$ 1%; $Na_2B_4O_7$, $10H_2O$ 17%.

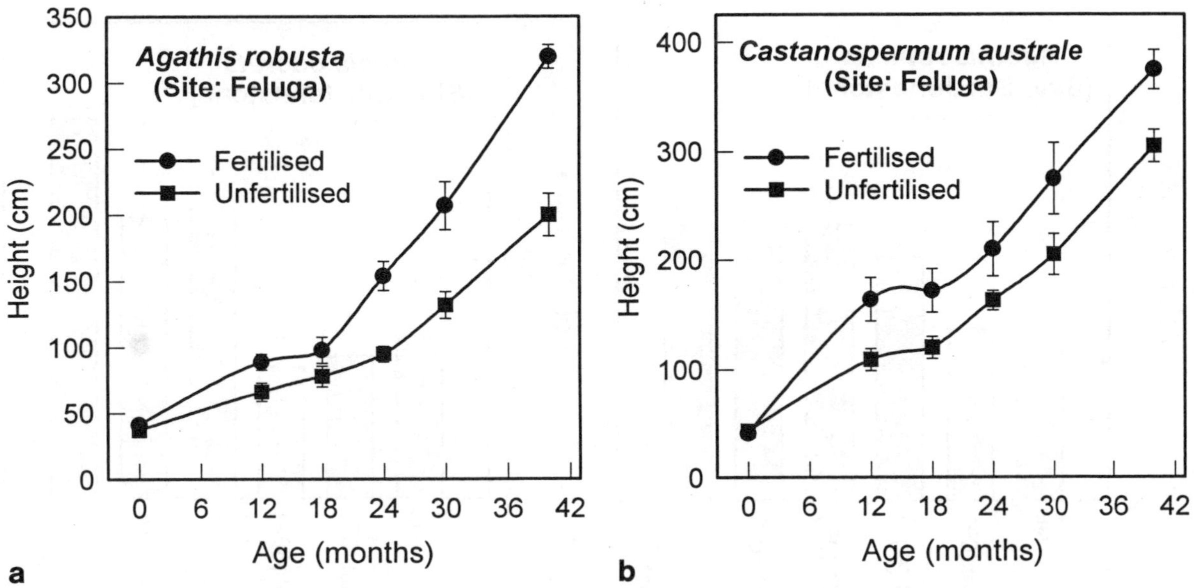


**Figure 2:** Height growth of (a) kauri pine (*Agathis robusta*) and (b) black bean (*Castanospermum australe*) for the first 40 months after treatment with varying levels of P fertilisation on granite derived soils at the Feluga site in north Queensland. All fertilised treatments have been combined for comparison with the nil treatment. See text for details of fertiliser treatments. Error bars are one standard error of the mean.

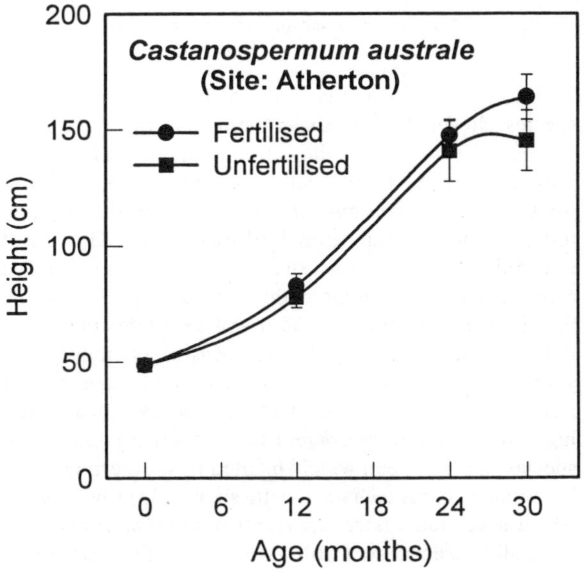


**Figure 3:** Height growth of black bean (*Castanospermum australe*) for the first 30 months after treatment with varying levels of P fertilisation on basalt-derived soils at the Atherton site in north Queensland. All fertilised treatments have been combined for comparison with the nil treatment. See text for details of fertiliser treatments. Error bars are one standard error of the mean.

(Fig. 2a). Although this response began before refertilisation it appears to have been enhanced by this second fertiliser addition, with height growth accelerating more rapidly to 41 months. *Castanospermum australe* at the same site demonstrated a lesser response (but more quickly) with an approximately 50% greater height in fertilised treatments after 12 months (Fig. 2b). However, this difference in height growth did not increase and growth curves for the fertilised and unfertilised treatments remained parallel.

### 3.2 Volume index

Results for the volume index (dgl$^2$*height) were qualitatively no different to the height responses except at the South Johnstone site. While variability in volume between replicates was high, there was an indication that the addition of P resulted in an increased stem volume in *Agathis robusta* at 31 months (Fig. 4a). At the higher rates of P, stem volume was almost double that of the nil fertiliser and nil P treatments. This effect was even more pronounced with *Cedrela odorata* grown on the same site (Fig. 4b).

Interestingly, while the stem volume of *Agathis robusta* gradually increased with increasing nutrient addition up to 100 kg P/ha, volume growth of *Cedrela odorata* was slightly depressed compared to the nil treatment when other nutrients were added but P was omitted or supplied at only a low rate, but exhibited a major response when 100 kg/ha P was added (Fig. 4).

By contrast, even though *Agathis robusta* showed a height response to a complete fertiliser application at 24 months (before refertilisation), this effect was the same irrespective of the level of P in that complete fertiliser (data not shown).

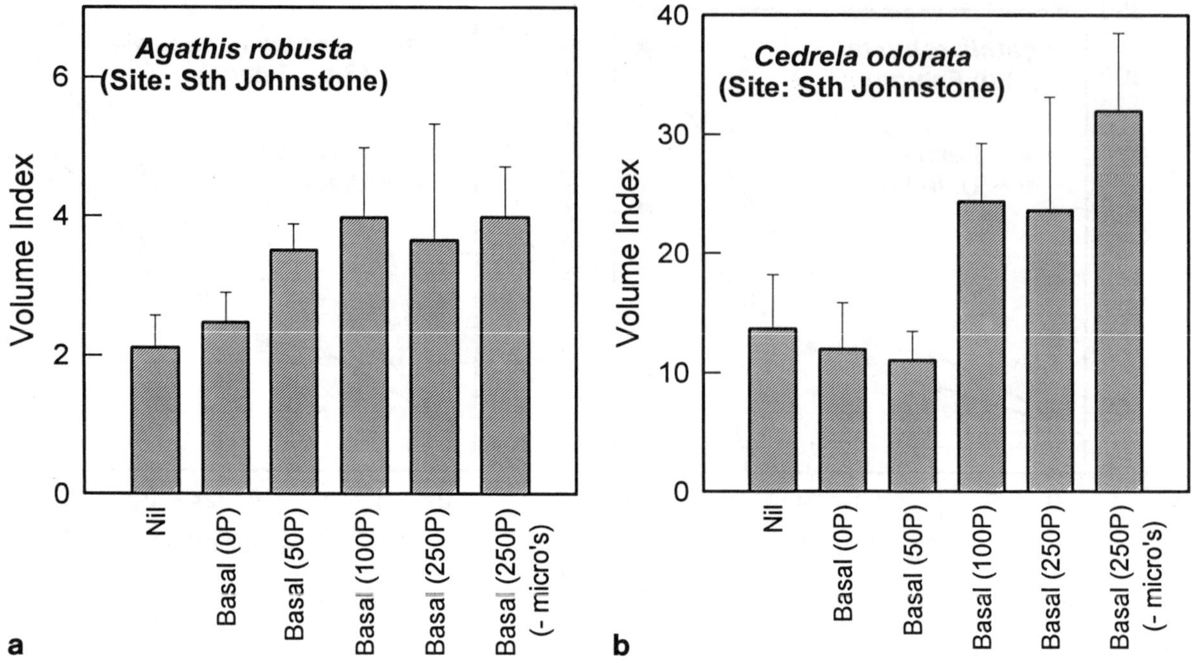


**Figure 4:** Volume index (diameter$^2$ x height) for (a) kauri pine (*Agathis robusta*), and (b) West Indian cedar (*Cedrela odorata*) 30 months after treatment with either no fertiliser, or a basal application of N, K and micronutrients plus varying levels of P on metamorphic-derived soils at the South Johnstone site in north Queensland. See text for details of fertiliser treatments. Error bars are one standard error of the mean.

## 4 Discussion

Survival in all treatments at all sites, including the nil fertiliser treatment, was relatively high and there was little evidence through visual symptoms or early growth response to indicate that major nutritional deficiencies limited establishment or growth in the first 6 months of the tree species used on these soil types. This may have been due to sufficient release of nutrients during the site preparation process to satisfy early nutritional demands of the species, carry over of residual nutrition from fertilising associated with farming practices, carry over of residual nutrients from the nursery phase (either within the plant or the media surrounding the roots), and the generally slow early growth characteristics of the species used in the experiments. All sites were occupied by productive rainforest prior to their clearance for agriculture, and the degradation associated with agricultural practices has not been particularly severe.

However, in spite of this, it is clear that nutritional factors are limiting on some sites as growth begins to accelerate. *Castanospermum australe* planted on granite soils at the Feluga site demonstrated a response to fertilisation after 12 months and *Agathis robusta* has begun to respond strongly after 24 months.

Lack of specific visual symptoms of nutrient limitation, even though growth has been affected, has been observed in other rainforest species (*Webb* et al. in press).

Although it is clear that nutrient limitations exist, the species chosen for the experiment may not have been the best for discerning the full consequence of nutrient limitations. The Atherton site was probably not optimum for *Castanospermum australe* in that it was too dry and too cool for optimum growth of this species. Growth of unfertilised trees at this site was slower over the last 6 months of measurement than the average for all the fertilised trees combined and it will be interesting to see if this trend continues. *Agathis robusta* generally grows slowly early in stand development, and demand for nutrients in the first two or three years is probably not high. *Araucaria cunninghamii* is a closely related native species that has been widely-planted in southern Queensland that also has a characteristic slow early growth phase which accelerates later. *Araucaria cunninghamii* does not generally respond to fertilisation in the first two years, although substantial responses may occur with the addition of N, and sometimes P, 3 to 6 years after planting (*Xu* et al., 1995). The response at 24 months in *Agathis robusta* indicates some of the initial fertiliser has been retained in the system, and available to satisfy later growth demand.

Although these species may not have been the best for rapidly detecting nutritional constraints, the differential responses between species show how important it is to conduct experiments on species that are likely to be widely planted.

Although *Cedrela odorata* had more rapid initial growth and may have demonstrated potential nutrient limitations more effectively at these sites, variability in the growth of this species did not allow the statistical discrimination of treatment effects on height or volume growth. In spite of this variability, there were substantial responses between particular treatments indicating potential growth increases if fertiliser delivery can be improved. The results with this species also suggest that a balanced formulation is essential as the application of a complete mix without, or with a small amount of, P had a slight depressing effect on growth. This effect has been described by *Hunter* and *Smith* (1996).

For fertiliser to be effective and elicit a response it must reach the root system. The method of application used in this study may not have been the most effective way of applying fertiliser, and other methods that concentrate nutrients near the roots may be more effective.

## 5 Conclusions

This study has identified soil types with substantial nutritional limitations where fertilisation can increase productivity of different tree species. It is not possible at this stage to make fertiliser application prescriptions for different species on different sites as the upper limit for improvement has not yet been realised, and the increased value of timber growth associated with responsed to fertiliser need to be weighed against the costs associated with fertilisation.

The experience of this study also indicate that in designing nutritional experiments it is important to consider:

- Matching the species used carefully to sites in relation to climatic, soil drainage, or topographic conditions.
- The growth pattern of the species. Most tree species used in tropical timber plantations, such as pines, acacias and eucalypts, have rapid early growth, but other high value species have different growth patterns. It is important to determine the time of maximum nutritional demand for each species and design and apply fertiliser treatments accordingly. If the objective is to rapidly determine general nutritional limitations for a particular site or soil type, a range of species with rapid early growth that quickly place nutritional demands on the site should be used, but with the knowledge that other species may respond differently.
- Maintaining an adequate balance of nutrients is likely to be as important as rectifying a shortage of one particular nutrient especially where multiple nutrient deficiencies are likely or known.

Further experiments have been established on these sites and on other important soil types to investigate fertilisation responses in a wider range of cabinetimber species. The extent of nutrient uptake by trees, the fate of applied fertiliser, and examination of whether fertilisation can increase biomass accumulation in later stages of stand development will be investigated in future studies.

## 6 Acknowledgements

Experiments were established with financial support from the Australian Centre for International Agricultural Research, in a joint project (ACIAR project no. 9114) in collaboration with CSIRO land and water. Ron Blanch, Nick Kelly, Arthur Wright, Mila Bristow and other staff from the QDPI-Forestry and CSIRO assisted with site preparation, planting, fertilisation and weed control. Grahame Applegate and Paul Reddell (SCIRO) selected the sites and contributed to the experimental design.

## 7 References

*Anonymous* (1983): Rainforest research in north Queensland, Position Paper, Queensland Department of Forestry, Brisbane.

*Appanah, S.* and *G. Weinland* (1996): Experience with planting dipterocarps in Peninsula Malaysia. In: Shulte, A. and D. Schone (eds.) Dipterocarp Forest Ecosystems, Towards Sustainable Management. World Scientific, Singapore. pp. 411–435.

*Evans, J.* (1992): Plantation Forestry in the Tropics. Clarendon Press, Oxford, 403p.

*Cannon, M. G., C. D. Smith*, and *G. G. Murtha* (1992): Soils of the Cardwell-Feluga area, North Queensland. Divisional Report No. 115, CSIRO, Division of Soils.

*Cromer, R. N., K. C. Tan, E. R. Williams*, and *W. H. M. Rawlins* (1992): Response of *Eucalyptus deglupta* to phosphate fertiliser. J. Trop. For. Sci. 5: 74–89.

*Davidson, J.* (1996): Off site and out of sight! How bad cultural practices are offsetting genetic gains in forestry. In: Dieters, M. J., A. C. Matheson, D. G. Nikles, C. E. Harwood, and S. M. Walker (eds.) *Tree Improvement for Sustainable Tropical Forestry*. Proc. QFRI-IUFRO Conf. Caloundra, Queensland Australia, 27 October–1 November, 1996. pp. 288–294.

*Hunter, I. R.* and *W. Smith* (1996): Principles of forest fertilisation – illustrated by the New Zealand experience. Fert. Res. 43: 21–29.

*Hunter, I. R., R. Sandberg*, and *J. Prince* (1992): A comparison of the nutrient requirements and growth of young radiata pine and Carribean pine on 5 New Zealand soils. Trop. Sci. 32: 53–62.

*Isbell, R. F.* (1978): Soils of the tropics and sub-tropics. In: Andrew, C. S. and E. J. Kamprath (eds.) *Mineral Nutrition of Legumes in Tropical and Subtropical Soils*. CSIRO, Melbourne. pp. 1–20.

*Isbell, R. F.* and *D. G. Edwards* (1988): Soils and their management in the Australian wet tropics. In: Loveday, J. (ed.) National Soils Conference 1988: review papers. Australian Society of Soil Science Inc. pp. 152–180.

*Lyke, J.* (1995): World supply and demand for forest products. J. For. 93: 22–26.

*Jordan, C. F.* (1994): Ecology of tropical forests. in L. Pancel: Tropical Forestry Handbook. Springer-Verlag, Berlin. p. 165–198.

*Keenan, R. J., D. Lamb*, and *G. Sexton* (1995): Experience with mixed species plantations in North Queensland. Comm. For. Rev. 74: 315–321.

*Laffan, M. D.* (1988): Soils and land use on the Atherton Tableland, North Queensland. Soils and Land Use Series No. 61, CSIRO, Division of Soils.

*Murtha, G. G.* (1986): Soils of the Tully-Innisfail area, north Queensland. Div. Rep. No. 82 CSIRO Division of Soils, Adelaide.

*Nussbaum, R.* and *A. L. Hoe* (1996): Rehabilitation of degraded sites in logged over forest using Dipterocarps. In: Shulte, A. and D. Schone (eds.) Dipterocarp Forest Ecosystems, Towards Sustainable Management. World Scientific, Singapore. pp. 446–463.

*Poa, D.* and *G. Mazza* (1995): Nutrition trials promise gains in Pacific timber production. ACIAR Forestry Newsletter Feb. 1995 p. 1,3.

*Pegg, R. E.* (1991): Eucalyptus species selection and improvement strategies at Dongmen State Forest Farm, Peoples Republic of China. In: Abod, S. A., P. M. Tahir, L. M. Tsai, N, A. A. Shukor, A. S. Sajap, and D. Manikam (eds.) Proc. Reg. Symp. on Recent Developments in Tree Plantations of Humid/Subhumid tropics of Asia. Faculty of Forestry, Selangor, Malaysia. pp. 156–170.

*Soil Survey Staff* (1975): Soil taxonomy: a basic system of soil classification for making and interpreting soil surveys. US Dept. Agriculture handbook No. 436. Govt. Printer Washington.

*Simpson, J.* (1995): Site-specific fertilizer requirements of exotic pine plantations in Queensland, Australia. In Schulte, A. and D. Ruhiyat (eds.) *Proc. Int. Congress on Soils of Tropical Forest Ecosystems.* 3rd Conference on Forest Soils (ISSS – AISS – IBG), vol. 5 Soil Fertility and Fertilization. Mulawarman University Press, Samarinda, Indonesia. pp. 90–106.

*Walters, G. A.* (1972): Slow-release fertilizer aids early growth of Australian toon and Queensland-maple in Hawaii. Tree Planters Notes: 26: 12–13.

*Webb, M. J., P. Reddell, A. Hambleton,* and *G. M. Mazza* (1991): Nutritional Constraints to Growth of Australian Red Cedar (*Toona ciliata*) in Five North Queensland Soils. Australian Forestry 60: 46–62.

*Wilkinson, L.* (1990): SYSTAT: The System for Statistics. SYSTAT, Inc. Evanston, IL. 677p.

*Xu, Z. H., J. A. Simpson, D. O. Osborne,* and *K. A. Bubb* (1995): The role of fertilisation in improving the productivity of hoop pine plantations. Paper presented to the Hoop Pine Establishment Workshop, Gympie, 11–14 September 1995. Internal document Queensland DPI-Forestry, Brisbane.

# Site specific fertiliser requirements of tropical Pine plantations

**John Simpson**

Queensland Department of Primary Industries-Forestry, Forestry Research Institute, M/S 483 Fraser Road, Gympie 4570 Queensland, Australia

## Abstract

Tropical pine plantations are an important resource in helping to meet the future timber needs of southeast Asian countries. Information on the soil resource and nutrient management of these plantations is scant and given that a large proportion of these industrial plantations are on degraded soils, there is a huge potential for increasing sustainable productivity through effective site specific use fertilisers. An Australian example of tropical pine management is detailed to illustrate intensive plantation management based on understanding of sites and specific fertiliser requirements of these sites.

The Queensland Department of Primary Industries (QDPI) Forestry manages in excess of 130 000 hectares of exotic pines, predominantly slash pine (*Pinus elliottii* Engelm. var. *elliottii* Little and Dorman) and Honduras Caribbean pine (*P. caribaea* Mor. var. *hondurensis* Bar. and Golf.). Since 1991 the hybrid between these two species has been planted extensively in the south of the State. The plantations are established essentially on the coastal lowlands on infertile, light textured, acid soils derived from granitic outwash or sedimentary materials. Drainage varies from good on the low ridges to poor in the lower lying areas. The success of the plantation program is dependant upon the widescale use of fertilisers especially phosphorus. Lesser responses to nitrogen applied at planting occur on a range of sites under particular conditions and severe deficiencies of potassium and copper are common on the podzols. Fertiliser prescriptions based on soils and site history are described. Current fertiliser prescriptions have evolved over a long period as research information has become available, species and sites planted changed and fertiliser cost and availability altered. Major nutrient problems have been effectively solved by the efficient use of fertiliser. More sophisticated nutrient management on a site specific basis is being pursued to maximise productivity at minimum cost and to maintain or improve site productivity in the long term while minimising any adverse consequences of fertiliser addition or nutrient manipulation.

There is a dearth of information on the soils resource and the scientific basis for effective nutrient management of *Pinus* plantations in southeast Asia. Many of the plantations have been established on degraded sites and there is a huge potential to increase sustainable productivity by the judicious use of fertilisers.

---

## 1 Introduction

In order to help meet anticipated future demands for timber in tropical countries in the southeast Asian region there has been a large increase in plantation forestry. *Evans* (1992) estimated that in the tropics and hotter subtropics there was 14.7 x $10^6$ hectares of plantation in southeast Asia and a further 0.4 x $10^6$ hectares in Australia and the Pacific Islands. Estimates of plantation areas established vary in their accuracy from one country to another (*Persson*, 1974). The annual planting programs have increased enormously over the last decade. From 1980 data, *Evans* (*ibid*) estimated that ca. 34% of the tropical plantations consisted of *Pinus* species. Based on these data it is conservatively estimated that there is now in excess of 5 x $10^6$ hectares of *Pinus* species in tropical reigons of southeast Asia, Australia and the Pacific Islands. The main species include: *P. caribaea, P. elliottii, P. kesiya, P. massoniana, P. merkusiana, P. oocarpa* and *P. tecunumanii.* There is an increasing interest in exploiting the advantages available from the development of hybrid pines.

The soil resource information of tropical forests is scanty at best and in most developing countries soil management has not been an important component of forest management (*Eswaran* et al., 1995). Tropical soils vary considerably from site to site and there are dangers in making too many generalisations. Soils of tropical plantation forests are often described as fragile and infertile but there are many exceptions to this popular belief. The soils of tropical southeast Asia most commonly planted with *Pinus* species are well drained, acid and base unsaturated, and are Oxisols (Ferralsols) or Ultisols (Acrisols) (Soil Survey Staff, 1975; FAO, 1974). The major edaphic constraints to sustainable plantation production are low fertility, soil acidity, aluminium and other associated toxicities, and unfavourable physical and/or biological factors. These features can be either inherent or induced by past management history. Soils deteriorate quickly under exploitative management practices. The majority of *Pinus* plantations in some of the countries of southeast Asia are established on degraded lands often dominated by *Imperata. Vanderbeldt* (1993) estimated that there are 20 x $10^6$ hectares of such grasslands in southeast Asia alone.

In addition to an increasing size of the plantation estate there is greater emphasis being placed on site manipulation and intensification of forest management in order to maximise productivity and profitability. More appropriate species selection, coupled with the use of genetically improved stock, is now common and more intensive silvicultural regimes are being imposed. Intensive site preparation, control of competing vegetation, use of fertilisers and management of within stand competition are among the modern silvicultural techniques used in intensively managed plantations. With the intensification of plantation management there is an accompanying need to tailor silvicultural operations to specific sites. Maximising the benefits achievable through site specific silviculture involves the need for a detailed understanding of the edaphic resource and a knowledge of the site by silviculture interactions. The Queensland Department of Primary Industries (QDPI) Forestry has been active in intensifying management of the *Pinus* plantation estate through the pursuit of site specific silviculture.

This paper describes in some detail the land resources for and development of site specific fertiliser prescriptions currently used in establishing and managing the exotic pine (*Pinus*) plantations in tropical and subtropical Australia and discusses this situation in relation to tropical countries of southeast Asia.

## 2 Land Resources for *Pinus* species in Tropical Australia

In excess of 98% of the plantings of pines in tropical Australia (defined as north of latitude 28° S) are in Queensland and are, in the main, managed by QDPI Forestry. Plantings commenced in 1925 to provide utility timber to meet anticipated demands. Little expansion of the planted area is currently envisaged and increases in productivity and profitability will rely on improving silviculture and management efficiency.

The major exotic pine plantation centers are located on the infertile coastal lowlands of southeast Queensland with lesser plantings near the major centers of population in the northern parts of the State (Figure 1). The climates range from humid subtropical to humid tropical with mean annual rainfall in the range 1337 mm at Tuan to 2138 mm at Cardwell. There is a strong summer incidence at all centres. Mean annual temperatures vary from 20.2 °C (Toolara) to 24.2 °C (Cardwell).

The total exotic pine estate as at 30th June 1994 was 130 669 hectares. Slash pine (*Pinus elliottii* Engelm. var. *elliottii* Little and Dorman) comprised 45% of the plantings and Honduras Caribbean pine (*P. caribaea* Mor. var. *hondurensis* Bar. and Golf.) comprised 42% (Queensland Department of Primary Industries Forest Service, 1995). Since 1991 the hybrid between slash pine and Honduras Caribbean pine has been planted exclusively on the

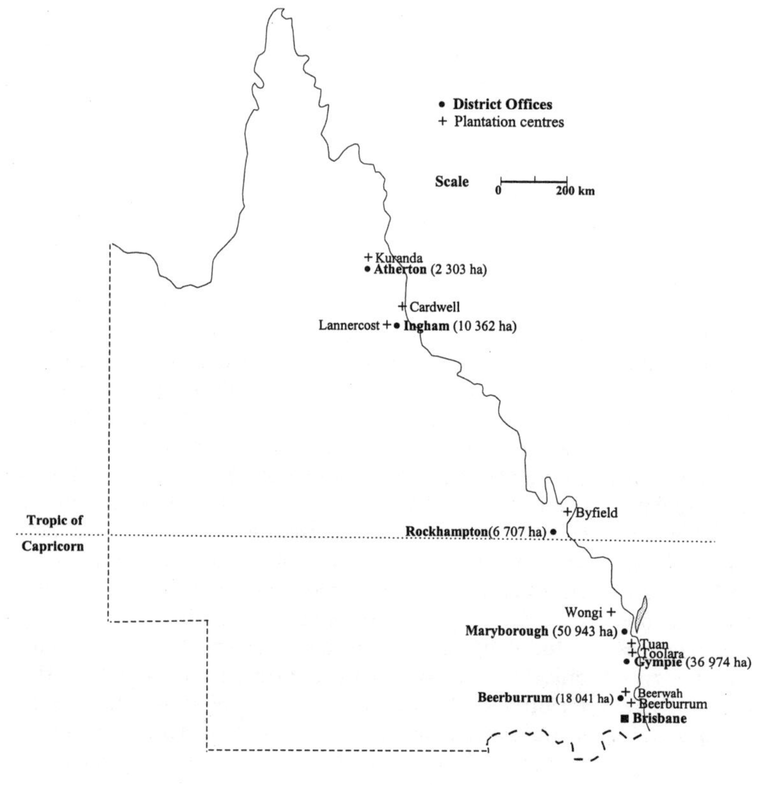


**Figure 1:** Map of Queensland showing location of the QDPI Forestry's major exotic pine plantations. Area of plantation (as at 30th June 1994) shown by district in brackets.

southern coastal lowlands and forms 9% of the exotic pine estate. Less than 2% of the estate is loblolly pine (*P. taeda* L.). Lesser areas of radiata pine (*P. radiata* D. Don.) and patula pine (*P. patula* Schl. and Cham.) have been established on the southern highlands. The potential of other taxa eg. the hybrid between Honduras Caribbean pine and Tecun Uman pine (*P. tecunumanii* P. Equiluz and J. P. Perry), are currently being investigated.

Soils of the major planting areas on the south-east coastal lowlands are derived chiefly from Mesozoic sandstones and Quaternary sands. North of the Tropic of Capricorn, plantation soils are chiefly derived from granitic outwash material. Most soils are coarse textured, acid to very acid, and have low nutrient status. Drainage varies from good on the broad low ridges, to poor in lower lying areas. Clay B horizons underlie most soils at depths ranging from 20 cm to more than 100 cm.

Early plantings were confined to the better drained sites, where depth to clay was the bais for determining plantability and species to be used (*Rogers,* 1957). As a result of work by *Pegg* (1967) in relating site index* to site variables, and with the advent of improved site preparation techniques, plantings were extended in 1970 to include the generally less fertile soils with poor drainage. Prior to 1983, virtually all sites carried low quality native forest before being cleared for plantation establishment. Since then, sites formerly developed to pastures for cattle grazing have been acquired and planted with pines. General descriptions of the major exotic pine planting areas as well as aspects of management of the plantations have been described by *Hawkins* and *Muir* (1968) – updated 1987 (Queensland Department of Forestry).

The land resource established with exotic pines in tropical Australia can be broadly classified into four main categories: well drained or poorly drained sites derived from either granitic outwash materials or sedimentary materials (Mesozoic sandstone or Quaternary sands). The major soil types associated with each of these classes, proportion of the estate and approximate equivalents in terms of Soil Taxonomy and the World Soil Map are summarise in Table 1.

Superimposed on the land resource classes are site histories, *ie* first rotation ex native forest sites, first rotation ex pasture sites or second rotation sites. It is the combination of land resource and site history that determines the fertiliser schedule for a particular species.

A summary of the plantation resources by site history and by period of establishment is presented in Table 2. The land resource established with exotic pines has changed over time. During the period 1927–70 predominantly well drained soils carrying native forests were cleared for planting but after this period a wider range of sites including poorly drained sites were established. Sites initially developed for cattle grazing and subsequently acquired for exotic pine planting now comprise 12 percent of the estate. The rapidly increasing proportion of annual plantings established on second rotation areas is a reflection of the increasing areas of plantation approaching maturity and on the increasing dependency of industry on plantation material.

Drainage conditions of the soil as reflected by the soil type is of prime importance as it predetermines many of the silvicultural requirements of the site in particular species selection, site preparation and fertiliser requirements. Slash pine was planted on the poorly drained sites as it tolerates waterlogging. Honduras Caribbean pine was confined to the better drained ridge sites because of its intolerance of waterlogging and its susceptibility to wind damage particularly on wet sites. The hybrid, now planted exclusively in the south east, combines the continuous growth habit of Caribbean pine and the tolerance of wet sites from slash pine. *Foster* and *Constantini* (1991a, b, c) described the field survey for site preparation, planning and site design, site preparation classes and design for the establishment of exotic pines.

## 3 Fertiliser Responses of *Pinus* species in Tropical Australia

Each year fertilisers are applied to in excess of 3000 hectares of exotic pine forests in Queensland to increase forest growth and value. Biological responsiveness and economic factors make fertilising an attractive investment. Although the benefits of fertilisation are clearly demonstrable, implementing the optimum prescription on site specific basis to maximise benefits remains a difficult challenge. Experimentation in Queensland over a long period has tested many soils for limiting elements and examined many combinations of forms, rates, timing of applications and combination of elements with the view of providing the background data necessary to enable preparation of specific fertiliser schedules for particular sites.

Most of the experimental work underpinning the current fertiliser prescriptions was carried out on first rotation sites. The results of these investigations are very relevant to second rotation sites currently being planted as it has been shown that although some residual benefits from first rotation fertilising (essentially to phosphorus additions) can be detected in the second rotation, carry over effects are small and do not substantially alter fertiliser regimes. With attention to nutrient conservation practices particularly residue management, nutrient depletion of the site is minimised and the fertiliser regimes designed for first rotation sites should not need to be supplemented for the second rotation.

Considerable effort has gone into investigation of the genotype (at species, variety, provenance, family and more recently clone levels) by fertiliser interactions. An understanding of these interactions enables maximum benefits

* Site index is defined as the average height (in metres) of the tallest 50 trees per hectare at age 25 years.

to be gained from early work which used test species now of lesser commercial interest. Optimum fertiliser regimes are now better tailored to specific genetic material on particular sites.

**Table 1:** Land resource classes, extent, and soil classification for exotic pine plantations in Queensland.

| Land Resource Class | Proportion of Estate (%) | Australian Great Soil Groups (Stace *et al.* 1968) | Approximate Equivalent: Soil Taxonomy (Soil Survey Staff 1975) | World Soil Map (FAO 1974) |
|---|---|---|---|---|
| Well drained Soils of Sedimentary Origin | 49 | Red Earth | *Paleustalf, Haplustalf,* Torrox. Haplargid. Haploxeralf. Paleustult. Paleudult | *Eutric Nitrosol Ferric Luvisol,* Orthic Feralsol, Luvic Yermosol, Dystric Nitosol |
| | | Yellow Earth | *Haplustalf,* Paleustult. Haplargid, Paleustalf, Plinthustult. Palexeralf | *Ferri,* Albic and Chromic *Luvisol* Ferric Acrisol, Luvic Xerosol, Plinthic Luvisol |
| | | Red Podzolic Yellow Podzolic | *Haplustalf, Haplustult,* Paleustalf, Haploxeralf, Paleustult | *Albic Luvisol,* Orthic and Ferric Acrisol |
| | | Lateric podzolic | *Palexeralf,* Plinthoxeralf, Paleustalf, Paleustult, Plinthustult, Haploxerult, Albaqualf | *Albic and Plinthic Luvisol,* Ferric and Plinthic Acrisol, Eutric Planosol |
| Poorly Drained Soils of Sedimentary Origin | 33 | Gleyed Podzolic | *Albaquult,* Paleustalf, Paleaquult, Ochraqualf | *Dystric, Eutric and Solodic Planosol,* Gleyic Acrisol |
| | | Humic Gley | *Paleaquult,* Albaquult, Humaquult, Haplaquept | *Gleyic Acrisol,* Humic and Dystric Gleysol |
| | | Podzol | Troporthod, Haplorthod, Placorthod, Quartzipsamment | *Orthic and Placic* Podzol |
| | | Soloth | *Natrustalf, Natrixeralf, Paleustalf,* Haplustalf | *Orthic Solonetz Albic Luvisol, Solodic Planosol* |
| Well Drained Soils of Granitic Origin | 8 | Red Earth | *Paleustalf, Haplustalf,* Torrox. Haplargid. Haploxeralf. Paleustult. Paleudult | *Eutric Nitrosol Ferric Luvisol,* Orthic Feralsol, Luvic Yermosol, Dystric Nitosol |
| | | Yellow Earth | *Haplustalf,* Paleustult. Haplargid, Paleustalf, Plinthustult. Palexeralf | *Ferri,* Albic and Chromic *Luvisol.* Ferric Acrisol, Luvic Yerosol, Plinthic Luvisol |
| | | Yellow Podzolic | *Haplustalf, Haplustult,* Paleustalf, Haploxeralf, Paleustult | Albic Luvisol, Orthic and Ferric Acrisol |
| Poorly Drained Soils of Granitic Origin | 5 | Gleyd Podzolic | *Albaquult,* Paleustalf, Paleaquult, Ochraqualf | Dystric, Eutric and Solodic Planosol, Gleyic Acrisol |
| | | Soloth | *Natrustalf, Natrixeralf, Paleustalf,* Haplustalf | Orthic Solonetz, Albic Luvisol, Solodic Planosol |

Approximate equivalent groups from *Moore* et al. (1983).

**Table 2:** QDPI Forestry exotic pine plantation programme by site type (areas established in hectares).

| Establishment Period | First rotation sites | | | Second rotation sites | Total |
|---|---|---|---|---|---|
| | ex native forest | | ex pasture | | |
| | Well drained | Poorly drained | | | |
| Pre 1950 | 6364 | – | – | – | 6364 |
| 1951–60 | 11676 | – | – | – | 11676 |
| 1961–70 | 20928 | – | – | – | 20928 |
| 1971–80 | 18219 | 32255 | – | – | 50475 |
| 1981–90 | 16174 | 10783 | 11564 | 1857 | 40378 |
| 1991–2000* | 1336 | 891 | 1419 | 15532 | 19178 |
| 2000–2010* | – | – | – | 27790 | 27790 |
| Total | 74769 | 43929 | 12983 | 45179 | 176789 |

* Figures are based on projections from *Robinson* (1990).

A summary of nutrient responses of exotic pines for the major site history classes is outlined below.

### 3.1 Establishment fertiliser responses of first rotation exotic pines on ex native forest sites.

#### 3.1.1 *Phosphorous*

In the early 1930's, phosphorus (P) was recognised as the major element affecting the growth of exotic pines (*Young*, 1935). The viability of the exotic pine plantation program is dependent upon the broad-scale use of phosphate fertiliser. Initially, (in the late 1930's) soil sampling to determine the total P content of the surface 10 cm was carried out in conjunction with plantability surveys and Nauru rock phosphate was broadcast to bring the soil P to the required level, namely 48 mg/kg for slash pine and 57 mg/kg for loblolly pine. With an increasing planting programme covering a wider range of sites, this approach proved unreliable and was supplanted in the early 1950's by a blanket application of Nauru rock phosphate to all sites in south-east Queensland. The application, up to three years after planting, supplied 50 kg/ha P. From 1970 to 1987 superphosphate or triple superphosphate replaced Nauru rock phosphate as the major source of P used; and aerial application was introduced to cater for the expanded planting programme. The rate of application was increased to 60 kg/ha P to compensate for uneven distribution associated with the use of aircraft (Queensland Department of Forestry 1987). The importance of P for soils derived from both sedimentary or granite outwash materials is illustrated in the rate/response curves in Graph 1.

This graph summarises data from a range of experiments carried out since the early 1950's. In all situations there has been a major response to the addition of P applied at planting. The magnitude of the response varies with the edaphic conditions, with the largest responses occurring on the less fertile sites ie. the sedimentary derived soils in the case of Honduras Caribbean pine, or the poorly drained soils in the case of slash pine. It is interesting to note that the optimum rate of P applied at planting (ie. 50 kg/ha P) is similar for each situation.

Plantations of Honduras Caribbean pine in coastal central Queensland on granite derived soils were only fertilised where site index, estimated at 7 to 11 years of age, fell below 30. A mixture of superphosphate and Nauru rock P (1:3) was broadcast over these areas to supply 73 kg/ha P. Since 1970, plantings at Byfield have extended to wetter sites and to poorer soils derived from mudstones, shales or sandstones and the fertiliser prescription changed so that all sites received 60 kg/ha P as triple superphosphate at planting.

The early plantings of Honduras Carribean pines on ridge sites in the north of the state were not fertilised. From 1982 the ridge sites received 60 kg/ha P while the poorly drained sites received a mixed fertiliser containing a similar amount of P.

#### 3.1.2 *Nitrogen*

In the 1960's research using slash and loblolly pines as the test species demonstrated that an ephemeral growth response to nitrogen (N) fertilisers could be obtained on a range of sites under certain conditions. Optimum response was obtained on cultivated, weed free sites when the N was applied to individual trees at planting with an available source of P (*Richards* and *Bevege*, 1967). *Francis* and *Bacon* (1983) reported improved growth and root/shoot ratios and as a consequence, better wind stability in cyclonic winds for young slash pine where N fertiliser had been applied together with P. From 1982 N has been applied operationally on podzol sites to stimulate early growth and keep the pines above the vigorous health regrowth on these sites. With the improved availability and competitive pricing of the ammonium phosphates, MAP (mono ammonium phosphate) has, since 1987, been widely used as the major source of N and P applied to exotic pines in Queensland.

#### 3.1.3 *Potassium*

The potassium (K) status of many soils derived from Mesozoic sandstone or Quaternary sands is marginal and deficiencies of K in exotic pines are known to occur on

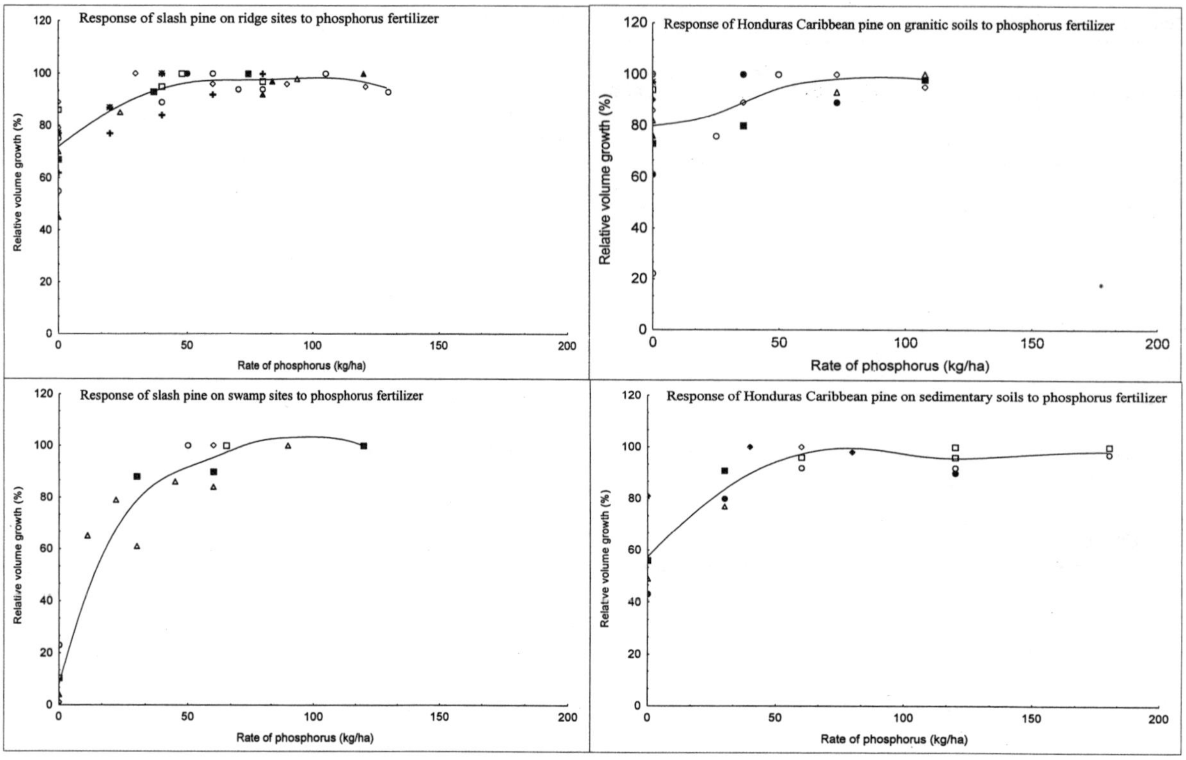


**Figure 1:** Response of Slash and Honduras Pines on Different Sites to Phosphorus Fertiliser.

many of the podzols. With the increased areas of podzols planted since 1970, investigations of K deficiency intensified. It was demonstrated that once the gross deficiencies of P on these sites were corrected, significant growth responses to K were achieved. Volume growth has been increased by between 10 and 110% by the addition of 50 kg/ha K (*Simpson* and *Grant*, 1991).

#### 3.1.4 *Other Elements*

In 1975 deficiencies of copper (Cu), manifested by gross stem and branch malformation and in severe cases, very poor growth, were identified in young slash pine plantations established on poorly drained humus podzol soils. This finding led to the inclusion of 5 kg/ha Cu in the fertiliser mix applied to all podzols from 1975. Deficiencies of Cu are most pronounced during the first three to five years after planting and heavy dressings of N fertiliser exacerbate the deficiency. Of the species planted, Honduras Caribbean pine is particularly sensitive to low Cu levels. Cu deficiency appears to be under strong genetic control.

Low foliar zinc (Zn) levels have been recorded in some plantations but in the absence of any growth response to added Zn in any of the field experiments, this element is not included in any operational fertiliser schedules.

Boron (B), calcium (Ca), magnesium (Mg), manganese (Mn), molybdenum (Mo) or sulphur (S), deficiencies have not been confirmed as growth limiting factors in any of the commercial plantings.

### 3.2 Current establishment fertiliser practices

The current fertiliser schedules have been derived from a wide range of fertiliser trials and for non pasture sites, are summarised in Table 3. Improvements in the schedules have been based on research findings, in particular long term growth responses from multi-site-trials and foliar nutrient data, and fertiliser cost and availability.

### 3.3 Later age fertilise responses of first rotation exotic pines on ex native forest sites

The need to re-apply fertilisers to established plantations in continually being assessed. The initial P dressing applied at planting is essentially designed to ensure satisfactory establishment. Given the range in quality of sites and species planted, it was not expected that the initial dressing would be adequate for all stands for the full rotation. Foliar analysis techniques have been developed to identify stands that are nutritionally stressed. Critical or acceptable foliar nutrient levels have been established for the major exotic pine taxa planted in Queensland and are used as a guide to the nutritional status of a stand (*Simpson* and *Osborne*, 1993). Numerous attempts using foliar analysis

**Table 3:** Exotic pine fertiliser prescriptions for first rotation, non-pasture sites.

| Location (Forestry District and State Forest) | Latitude | Area of exotic pine planted 1993/94 (ha) | Major species currently planted⁺ | Predominant soil parent material | Fertiliser prescription codes⁺⁺ |
|---|---|---|---|---|---|
| Atherton | | | | | |
| Kuranda | 16° 00′ | – | PCH | Metamorphics | 1 |
| Ingham | | | | | |
| Cardwell/ | 18° 16′/40′ | 326 | PCH | Granitic outwash | 1 |
| Lannercost | | | | Soloths | 2 |
| Rockhampton | | | | | |
| Byfield | 22° 50′ | 100 | PCH | Granite, | 1 |
| | | | PCH | Sedimentary | 1 |
| | | | PCH | Podzols | 2 |
| Maryborough | | | | | |
| Wongi/Tuan | 25° 27′/40′ | 678 | F1 hybrid | Sedimentary | 1 |
| | | | | Podzols | 3 |
| Gympie | | | | | |
| Toolara | 26° 00′ | 647* | F1 hybrid | Sedimentary | 1 |
| | | | | Podzols | 3 |
| Beerburrum | | | | | |
| Beerburrum/ | 26° 57′ | 613* | F1 hybrid | Sedimentary | 1 |
| Beerwah | | | | Podzols | 3 |
| Yarraman | | | | | |
| Pechy | 27° 20′ | – | RDP | Basalt | Not fertilised** |
| | | | PTP | | |
| Dalby | | | | | |
| Gambubal | 28° 20′ | – | RDP | Basalt | Not fertilised |
| Paschendale | 28° 35′ | 16 | RDP | Granite | 1 |

(Adapted from Simpson and Grant 1991)

⁺ Species Codes
PCH Honduras Caribbean Pine
F1 hybrid = Hybrid between slash pine and PCH
RDP = Radiata pine
PTP = Patula pine

⁺⁺ Fertiliser Prescription Codes

(kg/ha of element)

| | | N | P | K | Cu |
|---|---|---|---|---|---|
| 1 | = | 23 | 50 | – | – |
| 2 | = | 23 | 50 | – | 5 |
| 3 | = | 23 | 50 | 50 | 5 |

Potassium and copper are applied on podzol sites only.
Higher rates of potassium (100 kg/ha K) are recommended for severely deficient podzol sites.
* Includes ex-pastures sites – see Table 4 for fertiliser schedule
** To be reviewed

have been made at identifying areas requiring refertilising. Where intensive foliar sampling has been undertaken, the approach has been useful, however, for extensive plantations, the technique has met with only limited success.

Data collected over a period of up to 14 years from 110 plots of slash pine refertilised with P (30 to 500 kg/ha) indicated that basal area increment was increased by 20% by refertilising with P. Growth simulation suggested that wood yields at the end of the rotation are increased by 9% to 14%; and there were real rates of return of between 9% to 16% on money invested in refertilising. All slash pine plantations over 10 years of age have now been refertilised. Similar information is being investigated for Honduras Caribbean pine experiments, but to date there is no evidence to suggest that refertilising of this species is necessary.

A series of experiments has been conducted to investigate the N response applied after the first commercial thinning of established slash and Honduras Caribbean pine stands. Only in one situation has the application of N (200 kg/ha) stimulated a significant growth response. Because of the cost of N fertiliser and its limited response no widescale use of N fertiliser is undertaken in established stands.

Potassium is a very mobile element in the soil and plant, and is now applied operationally at planting to all podzol sites. All established stands on podzols were fertilised with either 50 or 100 kg/ha K in the late 1980's. The maintenance of adequate K status is being monitored.

Until recently no evidence existed to suggest that it is necessary to apply other elements to establish plantations. The recent occurrence of stem malformation in the crowns of established fast growing Honduras Caribbean pine on some sites suggests that Cu nutrition may be implicated and further investigations are current.

### 3.4 Fertiliser response of first rotation exotic pines on ex pasture sites

QDPI Forestry has, since 1977, acquired large areas of improved pasture land for conversion to exotic pine plantations (*Simpson*, 1991). Typically, such sites had received heavy applications of P (100 to 250 kg/ ha), K (100 to 380 kg/ha), and smaller (or nil) applications of N. As these quantities are well in excess of those normally used in forestry it was expected that additional P fertilising would not be generally required at the time of plantation establishment. However, since the residual value of P is dependant *inter alia* upon the quantity and timing of application, allowance has been made for pastures that have received only limited dressings well before plantation establishment. Recommended fertiliser rate is 50 kg/ha P. While the guidelines may overestimate the area requiring early fertilising, they allow for small errors in farmer application and for the fact that the later applications to pastures may have been minor.

### 3.5 Fertilising responses of second rotation exotic pines

Virtually all future plantings will be on second rotation sites. Design of second rotation fertiliser schedules must take into account any changes in site fertility from resulting the first rotation. Responses of second rotation crops to P applied more than 30 years earlier to the first rotation have been detected but these responses are of insufficient magnitude to eliminate the need for P fertilising for the establishment of second rotation stands.

The response of exotic pines on second rotation sites to elements apart from P is currently being researched. Early indications are that, given adequate attention to nutrient conservation measures in management and in the conversion of rotations, the fertiliser schedules developed for the first rotation sites will be appropriate for second and subsequent rotations. The exception is for ex pasture sites where it is expected that phosphorous fertiliser will be required to optimise establishment of the second rotations.

### 3.6 Fertiliser requirements of exotic pines re-established on fire killed plantation areas

Severe wildfires in the exotic pine plantations in 1991 and 1994 resulted in excess of 6000 hectares of plantation of various ages (2–35+ years) being destroyed. Redevelopment plans for these areas are in place and fertiliser schedules have been tailored to specific situations (Table 5). These schedules have taken into account site characteristics, the quantity and timing of previous nutrient additions and the residual value of fertiliser applied to the burnt stand.

## 4 Fertiliser Use in *Pinus* Plantations in Southeast Asia

Use of fertiliser is soil dependant so it is necessary to firstly understand the soil resource. Tropical soils vary considerably from site to site and the soil resource information for many plantation areas is at best scant. Few detailed soil information reports have been published and the FAO-UNESCO soil map of the world is frequently the only assessment of soils for major industrial plantations. Despite the potential benefits that an understanding of the soil resource can afford, unfortunately soil management is not generally regarded as high priority by plantation managers. While generalisations about tropical soils have led to many misconceptions such generalisations have often formed the basis for soil management for specific sites.

Despite the size of the pine plantation estate in tropical southeast Asian region there is a paucity of information in the literature concerning the use of fertilisers. The fact that many of these plantations have been established on

**Table 4:** Guidelines for fertilising exotic pines on pasture sites.

| Total quantity of phosphorus applied to pasture (kg/ha) | Years since last fertiliser application 0–4 | 5–9 | 10+ |
|---|---|---|---|
| 0–60 | Yes | Yes | Yes |
| 60–100 | No | Yes | Yes |
| 100–140 | No | No | Yes |
| 140+ | No | No | No |

**Table 5:** Fertiliser recommendations for re-establishing fire killed areas.

| Original Stand | Age Class | Site Type: All sites except Podzols | | Podzols | | | |
|---|---|---|---|---|---|---|---|
| Young | < 5yrs | $N_{22}$[a] | $P_{25}$[a] | $N_{23}$[b] | $P_{50}$[b] | $K_{50}$[c] | $Cu_5$[d] |
| Semi Mature/ Mature | > 5yrs | $N_{23}$[b] | $P_{50}$[b] | $N_{23}$[b] | $P_{50}$[b] | $K_{50}$[c] | $Cu_5$[d] |

Notes: Subscript numerals denote the quantity of element to be applied in kg/ha. All fertilisers to be applied as either individual tree of banded application at or soon after planting.

a $N_{22}$ $P_{25}$ Apply as DAP (18% N, 20% P) at the rate of 125 kg/ha
b $N_{23}$ $P_{50}$ Apply as MAP (Starterfos – 10% N, 21.9% P) at the rate of 228 kg/ha
c $K_{50}$ Apply as Muriate of Potash (50% K) at a rate of 100 kg/ha
d $Cu_5$ Apply as Copper Sulphate

previously degraded soils has meant that fertilising is frequently necessary and is commonly used during the establishment phase to ensure successful plantation establishment. Little reference to fertiliser use in established stands was found. With the increasing size of the tropical *Pinus* plantation estate and the intensity of management, effective nutrient management is critical to the improvement of productivity and the sustainability of these artificial ecosystems.

Most tropical soils have low P contents, usually below 200 ppm. In highly weathered Oxisols and Ultisols (Ferralsols and Acrisols), the two most important groups of acid upland soils, up to 80% of all P can be in the organic form and thus highly concentrated in the surface soil horizon (von *Uexkull*, 1986). As such it is very vulnerable to loss through site mismanagement. Phosphorus is commonly regarded as the major limiting nutrient of *Pinus* species and the management of this nutrient is frequently crucial to plantation succss. Deficiencies of P in *Pinus* plantations have been diagnosed from results of field fertiliser trials or through foliar diagnosis in southern China (*Simpson*, 1986), in the Philippines (*Zech*, 1990), Vietnam (*Simpson*, 1990), and Indonesia (*Simpson*, 1992). Correction of P deficiencies is commonly, simply, efficiently and economically effected by the addition of the readily available phosphatic fertilisers. Single dressings at planting often stimulate long term growth responses. Attention should be paid to defining, for each representative situation, the most efficient form, rate and method of application of P if responses are to be optimised. Phosphorus forms the bais of many operational fertiliser prescriptions.

After correction of the primary P deficiency other nutrients may then limit growth. Nitrogen fertilisers, if applied under optimum conditions, often promote a lesser ephemeral response in tropical *Pinus* species. Nitrogen is a highly mobile element in the plant and soil and carefully timed applications to weed free situations of appropriate fertilisers are necessary to optimise responses and economic returns. Nitrogen fertilisers are sometimes applied at relatively low rates at the time of planting of *Pinus* species to encourage establishment and rapid early growth. Because of the limited and ephemeral nature of the response, careful analysis of the economics of using nitrogenous fertiliser in operational plantings is recommended.

Potassium deficiency occurs in soils which have a low cation exchange capacity and low base saturation. These soils are often derived from either siliceous materials or from parent materials containing primary minerals low in K. Potassium deficiencies have been reported in *Pinus* plantations in China (*Simpson*, 1989) and although suspected to be a problem in other countries, no references have been located. Potassium is often included as a component in mixed fertilisers sometimes with little scientific justification.

Deficiencies of trace elements, Cu, Zn, and B are known to occur on particular sites in *Pinus* plantations in the tropics. By way of examples severe malformation of the stem and branches of slash pine on degraded sites at Dongmen in central southern China is consistent with the visual symptoms of Cu deficiency. Low foliar Zn concentrations have been reported in *P. oocarpa* in south Kalimantan, Indonesia (*Simpson*, 1992). Boron deficiencies have been diagnosed in Honduras Caribbean pine in the Philippines and in north Vietnam (*Simpson*, 1990).

No southeast Asian references for *Pinus* species were found reporting the inclusion of Ca, Mg or S in operational fertiliser schedules to correct deficiencies of these elements. Liming is commonly used in agriculture to manipulate soil pH and availability of nutrients. *Pinus* species are in general acidophyllic and the cost of liming at rates high enough to elicit responses becomes prohibitive in a low cash flow, long term forestry environment.

There is a serious lack of information on the soils and fertiliser responses in plantations of tropical pines for sourtheast Asian countries. Given the enormous potential to improve productivity of plantations of tropical pines by site specific fertiliser recommendations, that has been demonstrated by the Australian example, it is surprising that greater advantage is not being taken of this opportunity.

## 5 Conclusions and Future Directions

Locally grown tropical pines are an important resource in helping to meet the future timber demands of southeast Asia. The lack of appropriate soil resource and nutrient management information is a major constraint to productivity of these plantations. The development of site specific silviculture by QDPI Forestry provides an example of a modern silvicultural system.

The success of the tropical pine plantation program in Australia is inextricably linked to the use of fertilisers. The current fertiliser schedules have evolved over a long period and are based on a sound research background. While the major nutritional problems have been resolved by the effective use of fertilisers and efficient operational strategies put in place, there is an increasing need for more sophisticated nutritional management to maximise productivity at minimum cost, to maintain or improve site productivity in the long term and to ensure that there are minimum adverse off site consequences of any fertiliser (or nutrient manipulation) operations. Greater attention is being paid to site specific silviculture and given the existing site fertility variation across the exotic pine estate, there remains a lot of investigation to fine tune the nutrient management for specific sites or silvicultural regimes. The wide scale use of P fertiliser in the exotic pine program will continue but a greater understanding of N dynamics and the complex interaction governing the N response is required. The duration of the K response on podzols is being followed with interest. Conditions under which symptoms of Cu deficiency are expressed are not fully understood and will continue to be investigated. Large local data bases are now available which contain soils and stand nutritional and fertiliser response data. It is now timely to review these bases and to use the information to develop mathematical models which can be incorporated into decision support systems. Being able to model both cost and yield responses for various fertiliser scenarios will enable identification of more sophisticated fertiliser regimes to maximise productivity at minimal cost.

Despite the size of the *Pinus* plantation estate in southeast Asia there is a dearth of information on plantation soils and appropriate site and species specific nutrient management regimes. The fact that many of the plantations are established on degraded soils accentuates the need to better understand the nutritional management of these areas if productivity is to be optimised and the plantations managed on a sustainable basis. Phosphorus is the major limiting nutrient, but responses to N, K, Zn and B are known/suspected as being important on specific sites. This serious lack of information on soils and nutrient management regimes is a severe impediment to effective plantation management, maintenance of long term productivity and minimisation of adverse off site effects.

## Acknowledgement

Many people of the QDPI Forestry, formally the Queensland Forest Service and Department of Forestry, have contributed to the work reported in this paper. The valuable comments on the earlier drafts of this paper made by Mr. John Comrie Greig and Dr. Zhihong Xu are specifically acknowledged.

## References

*Eswaran, H., Rich, P.* and *Van Den Berg, E.* (1995). Soil resources of global tropical forest ecosystems. State-of-the-Art-Lecture. Proceedings of the International Congress (Third Conference on Forest Soils), "Soils of Tropical Forest Ecosystems" – Schulte, A. and Ruhiyat, D. (eds) – Balikpapan, Indonesia, 29 October–3 November 1995. Mulawarman University Press, Samarinda, Indonesia. 22 pp.

*Evans, J.* (1992). Plantation forestry in the tropics – Tree planting for industrial, social, environmental, and agroforestry purposes. Second Edition. Clarendon Press, Oxford. 403 pp.

*FAO* (1974). FAO-Unesco soils map of the world 1:5 000 000. Vol. 1: Legend (Unesco: Paris).

*Foster, P. G.* and *Costantini, A.* (1991a). *Pinus* plantation establishment in Queensland: I. Field surveys for site preparation planning and site design. Aust. For. 54 1/2: 75–82.

*Foster, P. G.* and *Costantini, A.* (1991b). *Pinus* plantation establishment in Queensland: II. Site preparation classes. Aust. For. 54 1/2: 83–89.

*Foster, P. G.* and *Costantini, A.* (1991c). *Pinus* plantation establishment in Queensland: III. Site preparation design. Aust. For. 54 1/2: 90–94.

*Francis, P. J.* and *Bacon, G. J.* (1983). Ripping trials in coastal south Queensland *Pinus* plantations. Qld. Dept. For. Res. Paper No. 8: 16 pp.

*Hawkins, P. J.* and *Muir, J. D.* (1968). Aspects of management of plantation in tropical and subtropical Queensland. Paper from 9th Cwth For. Conf. New Delhi. 36 pp.

*Moore, A. W., Isabel, R. F.* and *Northcote, K. H.* (1983). Classification of Australian soils. In Soils an Australian Viewpoint CSIRO/Academic Press, Melbourne, Australia. p 253–266.

*Pegg, R. E.* (1967). Relation of slash pine site index to soil, vegetation and climate in south-east Queensland. Qld. Dept. For. Res. Note 19: 53 pp.

*Persson, R.* (1974). World forest resources. Royal College of Forestry, Stockholm. 261 pp.

Queensland Department of Forestry (1987). Aspects of management of plantation in tropical and subtropical Queensland. 66 pp. Based on Hawkins, P. J. and Muir, J. D. (1968). Paper from 9th Cwth. For. Conf., New Delhi.

Queensland Department of Primary Industries Forest Service (1995). DPI Forest Service Yearbook 1993–1994. Brisbane: 57 pp.

*Richards, B. N.* (1967). Effect of cultivating and fertilising on potential yield of pulpwood from loblolly pine. Aust. For. 31 (3): 202–210.

*Robinson, M. J.* (1990). Seed production of the F1 (Pee x Pch) hybrid in Queensland. Paper from 'Status of the PEE X PCH F1 hybrid in Queensland and strategies for further development' seminar Qld. For. Service, Gympie, Australia. October 1990. 15 pp.

*Rogers, L. J.* (1957). Exotic Forest Trees in Queensland. Paper from 7th British Cwth. For. Conf., Aust. and N.Z: 27 pp.

*Simpson, J. A.* (1986). Fertiliser responses of forest species at Dongmen. Paper presented at the Third Technical Exchange Seminar, China-Australia Afforestation Project, Dongmen, P. R. of China. December 1986. 14 pp.

*Simpson, J. A.* (1989). Potassium fertilising of pines. Paper presented at the Fourth Technical Exchange Seminar, China-Australia Afforestation Project, Nanning, P. R. of China. October 1989. 11 pp.

*Simpson, J. A.* (1990). An assessment of the soils and nutrition of eucalypt and acacia plantations for the Vinh Phu project in the Socialist Republic of Vietnam. Report prepared for Interforest AB, Lidingo, Sweden. 51 pp.

*Simpson, J. A.* (1992). An assessment of the soils and nutrient status of trial plantations in the Indonesian-Finland project at Riam Kiwa, South Kalimantan, Indonesia. Report prepared for Enso Forest Development Ltd, Finland. 115 pp.

*Simpson, J. A.* (1991). Changing land resources, taxa preference and fertiliser use in exotic pine plantations in south east Queensland. Paper from The Third Australian Forest Soils and Nutrition Conference 'Productivity in Perspective', Melbourne, Australia. October 1991. p 31–32.

*Simpson, J. A.* and *Grant, M. J.* (1991). Exotic pine fertiliser practice and its development in Queensland. Qld. For. Serv. Tech. Paper No. 49. 17 pp.

*Simpson, J. A.* and *Osborne, D. O.* (1993). Relative fertiliser requirements and foliar nutrient levels of slash pine, Honduras Caribbean pine and the hybrid in Queensland. Cwth. For. Rev. 72 (2): 105–113.

Soil Survey Staff (1975). Soil Taxonomy – a basic system of soil classification for making and interpreting soil surveys. USDA: Washington, D. C.

*Stace, H. C. T., Hubble, G. D., Brewer, R., Northcote, K. H., Sleeman, J. R., Mulcahy, M. J.* and *Halloworth, E. G.* (1968). A Handbook of Australian Soils. Relliem Tech. Publs: Glenside, South Australia.

*Vanderbeldt, R. J.* (1993). *Imperata* grasslands in southeast Asia: summary reports from the Philippines, Malaysia and Indonesia. MPTS Report No. 18, Winrock International and USAID, Bangkok, Thailand. 3 pp.

*von Uexkull, H. R.* (1986). Efficient fertilizer use in acid upland soils of the humid tropics. FAO Fertilizer and Plant Nutrition Bulletin No. 10. FAO Rome. 59 pp.

*Young, H. E.* (1935). Fused needle of species of *Pinus*. Qld. Agric. J. XLIV: 286–298.

*Zech, W.* (1990). Mineral deficiencies in forest plantations of North-Luzon, Philippines. Tropical Ecology 31 (1): 22–31.

# Above-ground nutrient dynamics of Caribbean Pine (*Pinus caribaea*) plantation ecosystems

**Olatunji Kadeba**

Department of Forestry and Wood Technology, Federal University of Technology, P.M.B. 704, Akure, Nigeria

## Abstract

Growth, biomass and nutrient accumulation by 14-year-old stands of Caribbean pine (*Pinus caribaea* Morelet var. *hondurensis* Barr. and Golf.) on three subhumid savanna sites in Nigeria were investigated. Soil organic carbon, total N, extractable P and exchangeable K, Ca and Mg showed significant decline in the upper 10 cm of the mineral soil as a result of reafforestation with the pine.

Estimates of the mean residence time of organic matter and nutrients in the litter layer of the ecosystems were 2 to 4 years thus indicating a slower rate of litter decomposition compared to litter of mixed natural savanna vegetation and broad-leaved plantations in which the residence time of organic matter rarely exceeds 6 months. Nutrient inventory in the different compartments of the pine ecosystem showed that a sizeable proportion of P and K was held in the tree biomass. Nutrient budget analysis showed that total biomass removal in above-ground tree harvest could lead to problem of P and K nutrition in future rotations on some sites. Estimates of nutrient fluxes and their relative contributions to tree nutrition were made. It was concluded based on the relative contributions of the fluxes that the soil remains the dominant source of the nutrients taken up by the tree and consequently tree nutrition as well as tree growth were soil fertility-dependent. Sites with low inherent nutrient quality were considered to be at a risk of nutrient depletion.

## 1 Introduction

Managed plantations of exotic tree species began in the tropics as a compensation for the diminishing wood resources from the natural forests (*Evans*, 1986). Over 90 different tree species have been tried in Nigeria (*Kemp*, 1970). The Caribbean pine (*Pinus caribaea* Morelet var. *hondurensis* Barr. and Golf.) is one of such promising introduced species being planted to meet increasing demands for long fibred pulpwood.

In the tropics, the replacement of native vegetation by any other type of ecosystem is often associated with a more or less pronounced drop in organic matter and nutrients (*Nye*, 1961). This has been documented for timber monocultures (*Cornforth*, 1970 and *Lundgren*, 1978). Thus it is important to investigate the nutrient-economy of these fast-growing monoculture plantations in view of the concern about the capacity of fragile tropical soils to sustain multiple short rotations (*Cuevas* et al., 1991; *Jorgensen* and *Wells*, 1986) and the role of nutrient cycling in maintaining forest productivity (*Bowen* and *Nambiar*, 1984).

Little is known about the nutrition and productivity of the Caribbean pine in the Nigerian savanna. Elsewhere in West Africa, its poor adaptation to sites of low nutrient quality has been reported (*Zech* and *Drechsel*, 1992). This paper reports the investigation of tree nutrition and nutrient cycling through the synthesis of data on nutrient pools in the soil, litter and vegetation of young stands of *Pinus caribaea* planted at three sub-humid savanna sites in northern Nigeria.

## 2 Materials and methods

### 2.1 Description of experimental sites

The savanna area of Nigeria covers an area of 684 000 km$^2$ between latitudes of 7° and 13° N and longitudes 3° and 15° E. In such a vast area, there is a great deal of variation in climate, soils and vegetation. The length and severity of the dry season are the principal climatic factors that determine survival and growth of trees.

The study was conducted in three unthinned stands of *Pinus caribaea* aged 14 years planted at Afaka, Miango and Nimbia in the sub-humid savanna (rainfall > 1000 mm yr$^{-1}$). The sites present some contrasting soil and climatic conditions. A summary of the characteristics of the three sites and the stands is given in Table 1.

### 2.2 Soil sampling

In each stand and in adjacent savanna woodland a 0.25 ha sample plot was located. Composite mineral soil samples were taken at three profile segments, 0–10, 10–20 and 20–30 cm with a 5 cm-diameter soil auger. Ten composite soil samples (20 cores each) per segment, giving a total of 30 samples per site, were taken. In addition three soil pits were dug to a depth of 1 m in each

**Table 1:** Site and stand characteristics of the experimental areas.

| Characteristics | Afaka | Miango | Nimbia |
|---|---|---|---|
| Site | | | |
| Location | 10°33′N<br>7°15′E | 9°50′N<br>8°40′E | 9°26′N<br>8°30′E |
| Bio-climatic zone | Northern Guinea savanna | Jos Plateau | Derived savanna |
| Altitude (m) | 610 | 1200 | 600 |
| Mean annual rainfall (mm) | 1250 | 1560 | 1800 |
| Rainy days per year | 175 | 180 | 220 |
| Dry season | Nov.–March | Nov.–March | Nov.–March |
| Mean annual temperature (°C) | 25 | 22 | 25 |
| Soil type[a] | Plinthic Ferralsol derived from igneous and meta-morphic rocks | Chromic Cambisol derived from newer basalt | Eutric Cambisol derived from olivine basalt |
| Texture | Sandy loam | Clay loam | Clay loam |
| Stand | | | |
| Age (years) | 14 | 14 | 14 |
| Stocking (trees $ha^{-1}$) | 1063 | 1206 | 1050 |
| Stem diameter (cm) | 19.2 ± 0.3[b] | 22.2 ± 0.4 | 20.8 ± 0.3 |
| Mean height (m) | 15.1 ± 0.2 | 18.5 ± 0.3 | 16.2 ± 0.3 |
| Basal area ($m^2$ $ha^{-1}$) | 31.2 ± 1.2 | 46.9 ± 1.1 | 36.0 ± 0.9 |
| Tree biomass above-ground (t $ha^{-1}$) | 125.4 ± 3.9 | 232.1 ± 5.7 | 155.0 ± 3.8 |

a Classification according to the revised legend of the soil map of the world (FAO-UNESCO, 1988).
b standard error

plot from which soil bulk density measurements were made. Soil nutrient contents (kg $ha^{-1}$) were obtained from soil profile chemical analyses described in section 2.5 after adjustment for bulk density and the thickness of the soil segment.

### 2.3 Biomass and nutrient content

Allometric regression was used to derive equations relating the biomass of each tree component (bark, stemwood, branches and needles) to stem diameter at breast height (DBH). Previous data analysis (*Kadeba*, 1989) showed that the power function: Biomass = $e^a(DBH)^b$ where a and b are constants and e is the base of natural logarithm, provided the best fit for all tree components. Regression equations were derived using the above model from the data of twelve trees harvested from each site (*Kadeba*, 1989). The twelve trees which were proportionately distributed over the range of diameters found in each sample plot were harvested at ground level. The felled trees were then subdivided into bole, branches and foliage and their fresh weight determined in the field. The component parts were later sub-sampled for moisture determination and chemical analysis in the laboratory. To separate the bole of each tree into wood and bark, the technique of simple linear regression of bark weight on bole weight as desscribed by *Egunjobi* and *Bada* (1979) was employed. Nutrient quantities (kg $ha^{-1}$) of tree components were calculated from the product of average nutrient concentration and biomass of each component.

### 2.4 Litter sampling

In each sample plot, ten 50 cm-diameter plastic trays supported on wooden posts 90 cm above the ground and perforated at the base were randomly installed for litter collection. Needle litter was collected monthly for 2 years. The standing stock of litter on the forest floor was sampled in 20 random locations on three occasions at intervals of six months. The recycling time (mean residence time) of organic matter and nutrients in litterfall was calculated according to the method of *Olson* (1963).

### 2.5 Plant and soil analysis

Samples of tree components (needles, bark, stemwood and branches) were dried to constant weight at 70 °C and ground in a Wiley mill to pass 40 mesh screen.

Total nitrogen (N) was determined by the semi-micro Kjeldahl digestion method using selenium as a catalyst and N in the digest was determined colorimetrically with the Technicon auto-analyser II. Extracts for the analysis of phosphorus (P), potasium (K), calcium (Ca) and magnesium (Mg) were prepared by the wet ashing procedure using a mixture of nitric and perchloric acids. Calcium and Mg were determined with a Perkin Elmer atomic absorption spectrophotometer, P colorimetrically with an autoanalyser and K by flame photometry.

Prior to laboratory analysis all soil samples were air dried and passed through 2 mm sieve. Soil analytical procedures used were those previously described by *Kadeba* and *Aduayi* (1985). pH was measured with glass electrode at a soil to water ratio of 1:2.5. Organic carbon was determined by the Black-Walkey wet oxidation method. Total N was determined by semi-micro Kjeldhal method; P was extracted with 0.1 M sulphuric acid and measured colorimetrically by the ascorbic acid blue method. Exchangeable Ca, Mg and K were extracted with neutral 1 M ammonium acetate solution; Ca and Mg in the extract were determined by EDTA titration and K by flame photometry.

### 2.6 Nutrient fluxes

Annual nutrient storage in the above-ground biomass was calculated by dividing biomass nutrient content by stand age. Litter nutrient flux was estimated by dividing the nutrient content of the annual litterfall by the residence time; residence time is defined as F/L where F is the mean mass of litter on the forest floor and L is the mean annual litterfall (*Olson*, 1963). Litter nutrient flux is the fraction of the nutrient content in annual litterfall expected to mineralize in one year. Litter decomposition calculated this way may underestimate the decomposition rates because of the non-inclusion of the root death in litter input rate.

Estimates of nutrient requirements were derived from the sum of nutrients in the mean annual biomass increment, canopy leaching, and the nutrient required for the replacement of tree parts shed annually as littrefall (*Baker* and *Attwill*, 1985). Nutrient uptake was calculated by subtracting the amounts of nutrients supplied by retranslocation from nutrient requirement. Retranslocation was estimated by multiplying the difference in nutrient concentration between living canopy needles and dead needle tissue in litterfall by its corresponding litter mass.

## 3 Results and discussion

### 3.1 Soil nutrient content

The results of soil studies showed that the nutrient status of the soils was affected by the establishment and growth of *Pinus caribaea* (Table 2). Soil nutrients were significantly reduced in the 0–10 cm mineral soil layer under the pine plots when compared to adjacent savanna woodland. In an earlier publication (*Kadeba* and *Aduayi*, 1985), we found in a chronosequence of pine stands at Afaka a decrease in the levels of soil carbon, total N and exchangeable Ca, Mg und K until the plantation attained the age of about 10 years after which a gradual build up of nutrients began.

Changes in soil nutrient content expressed in kg $ha^{-1}$ and as percentage of nutrient reserves in the 0–30 cm depth are shown in Table 3. Decreases in soil nutrient contents were more pronounced at Afaka than at the more fertile sites of Miango and Nimbia. It has also been reported that reafforestation with *Pinus caribaea* of cleared natural forest had decreased soil N, P, Ca, Mg and K; within six years as much as 74 percent of the original site nutrients were lost (*Cornforth*, 1970).

### 3.2 Biomass and nutrient accumulation

One main reason for the differences in soil nutrient reserves between natural woodland and plantation is related to the accumulation of nutrients during the early stage of vigorous tree growth and little return of nutrients to the soil from the slowly decomposing litter. During this early stage of tree growth adequate tree nutrition depends on soil fertility.

The nutrient contents of the above-ground tree biomass and the within-tree distribution of nutrients for Afaka, Miango and Nimbia stands are given in Table 4. On the more fertile sites of Miango and Nimbia, the magnitude of biomass and nutrient accumulation was higher than at Afaka where site quality was lower. Besides soil fertility, the role of water as a determinant of forest growth in the dry savanna areas of Nigeria cannot be ignored. At Afaka soil moisture stress was frequently a cause of slow growth and tree mortality (*Kemp*, 1970).

**Table 2:** Comparison of means of soil chemical properties under natural savanna and 14-year-old pine stands.

| Site | Vegetation | pH ($H_2O$) | Org. C (%) | Total N (%) | C/N | P (ppm) | Exchangeable cations (meg/100 g) | | |
|---|---|---|---|---|---|---|---|---|---|
| | | | | | | | K | Ca | Mg |
| Afaka | Savanna | 6.0 | 1.18 | 0.072 | 16.4 | 6.1 | 0.25 | 2.37 | 1.59 |
| (0–10 cm) | Pine | 6.0 | 0.87 | 0.055 | 15.5 | 4.6 | 0.16 | 1.66 | 1.18 |
| | t-value | NS | ** | * | NS | * | * | * | * |
| Afaka | Savanna | 5.8 | 0.73 | 0.052 | 14.0 | 4.2 | 0.18 | 1.62 | 1.24 |
| (10–20 cm) | Pine | 5.7 | 0.54 | 0.040 | 13.5 | 3.8 | 0.12 | 1.16 | 0.73 |
| | t-value | NS | * | NS | NS | NS | * | NS | * |
| Miango | Savanna | 5.6 | 1.81 | 0.124 | 14.6 | 14.6 | 0.32 | 3.71 | 2.87 |
| (0–10 cm) | Pine | 5.7 | 1.30 | 0.102 | 13.7 | 10.5 | 0.23 | 2.95 | 2.31 |
| | t-value | NS | ** | * | NS | * | ** | * | * |
| Nimbia | Savanna | 6.3 | 2.25 | 0.153 | 14.7 | 17.2 | 0.41 | 4.82 | 2.97 |
| (0–10 cm) | Pine | 6.4 | 1.67 | 0.117 | 14.4 | 14.9 | 0.30 | 3.67 | 2.38 |
| | t-value | NS | ** | * | NS | * | ** | * | * |

NS = Not significant at 5% level: * significant at 5% level; ** significant at 1% level.

**Table 3:** Decrease in soil nutrient content (0–30 cm) when savanna woodland is replaced by *Pinus caribaea*.

| Site | Content (kg ha$^{-1}$) | | | | |
|---|---|---|---|---|---|
| | $N_{TOTAL}$ | $P_{extractable}$ | $K_{exch}$ | $Ca_{exch}$ | $Mg_{exch}$ |
| Afaka | 299 | 3.1 | 72 | 350 | 180 |
| Miango | 231 | 5.3 | 82 | 352 | 118 |
| Nimbia | 512 | 7.1 | 54 | 285 | 109 |
| | Decrease as a percent of initial value | | | | |
| Afaka | 13 | 15 | 25 | 23 | 27 |
| Miango | 7 | 12 | 23 | 15 | 11 |
| Nimbia | 11 | 13 | 12 | 10 | 10 |

Although differences occured in the magnitude and the rate of nutrient accumulation by the pine stands, there was similarity, however, in their pattern of nutrient allocation to the different tissues irrespective of site. In these stands, the accumulation of nutrients by trees followed the dynamics of biomass growth.

Data collected from plantations of the Caribbean pine ranging in age from 5 to 15 years (*Kadeba*, 1991) showed that the mean annual biomass increment increased with stand age. However, the data on mean periodic increment and net primary production showed that the rate of biomass and nutrient accumulation was highest between 6 to 10 years, during which a peak of 19.2 t ha$^{-1}$ year$^{-1}$ in net primary production was attained. These data also showed that the rates of nutrient accumulation increased rapidly in the early stages of stand development but started to decline after canopy closure. Once the canopy is closed at about age 10 years and foliage mass stabilized, nutrient demands are reduced because tree growth consists of mainly the accumulation of wood, a tissue that is nutrient-poor.

Compared with the nutrient contens of savanna woodland (*Nye*, 1961), the present data showed that the pine accumulated more nutrients in the above-ground biomass than the native vegetation. Biomass and nutrient estimates of this study were in the range of biomass and nutrient data reported for *Pinus caribaea* and other tropical pines (*Cuevas* et al., 1991 and *Waterloo*, 1994).

### 3.3 Litter dynamics

Ammount of organic matter returned to the forest floor and its accumulation on the forest floor are presented in Table 5. Litterfall (mainly needles) increased with plantation age and varied with site. Litterfall was generally low

**Table 4:** Biomass and nutrient content of tree components of *Pinus caribaea*. Data in parentheses are nutrient and biomass values as percentages of corresponding total above-ground value.

| Site | Tissue | Biomass (t ha$^{-1}$) | Nutrients (kg ha$^{-1}$) N | P | K | Ca | Mg |
|---|---|---|---|---|---|---|---|
| Afaka | Needles | 20.1 | 191.5 | 11.0 | 115.7 | 71.8 | 27.9 |
| | | (16) | (48) | (36) | (53) | (38) | (38) |
| | Branches | 15.0 | 42.0 | 4.4 | 22.5 | 27.0 | 9.0 |
| | | (12) | (11) | (14) | (11) | (14) | (12) |
| | Stembark | 16.3 | 45.7 | 4.2 | 19.6 | 18.0 | 9.8 |
| | | (13) | (11) | (14) | (9) | (9) | (13) |
| | Stemwood | 73.9 | 118.2 | 11.1 | 59.1 | 73.8 | 27.5 |
| | | (59) | (30) | (36) | (27) | (39) | (37) |
| | Above-ground | 125.3 | 397.4 | 30.7 | 216.9 | 190.6 | 74.2 |
| Miango | Needles | 38.0 | 323.7 | 19.0 | 201.9 | 118.1 | 41.9 |
| | | (17) | (52) | (43) | (57) | (40) | (36) |
| | Branches | 26.9 | 67.2 | 6.2 | 37.6 | 48.4 | 16.1 |
| | | (12) | (11) | (14) | (10) | (16) | (14) |
| | Stembark | 28.6 | 65.8 | 5.4 | 34.4 | 31.5 | 17.2 |
| | | (12) | (11) | (12) | (10) | (11) | (15) |
| | Stemwood | 137.9 | 165.5 | 13.8 | 82.8 | 96.4 | 41.4 |
| | | (59) | (27) | (31) | (23) | (33) | (35) |
| | Above-ground | 231.4 | 622.2 | 44.4 | 356.7 | 294.4 | 116.6 |
| Nimbia | Needles | 25.7 | 226.3 | 14.1 | 141.5 | 82.5 | 33.4 |
| | | (17) | (50) | (40) | (53) | (37) | (37) |
| | Branches | 18.4 | 44.2 | 4.6 | 27.6 | 38.8 | 14.7 |
| | | (12) | (10) | (13) | (10) | (16) | (16) |
| | Stembark | 20.6 | 51.4 | 4.1 | 26.7 | 24.7 | 14.4 |
| | | (13) | (11) | (11) | (10) | (11) | (16) |
| | Stemwood | 91.6 | 128.2 | 12.8 | 73.3 | 82.2 | 27.5 |
| | | (59) | (29) | (36) | (27) | (36) | (30) |
| | Above-ground | 156.3 | 450.1 | 35.6 | 269.1 | 228.2 | 90.0 |

**Table 5:** Litterfall, forest floor weight and recycling time for organic matter and nutrients in *Pinus caribaea* stands.

| Site | Stand age | Mean litterfall | Forest floor | Recycling time (year) | | | | | |
|---|---|---|---|---|---|---|---|---|---|
| | (year) | (kg $ha^{-1}$ $year^{-1}$) | (t $ha^{-1}$) | Org. matter | N | P | K | Ca | Mg |
| Afaka | 8 | 3068 | 10.63 | 3.5 | 4.1 | 3.5 | 2.0 | 4.0 | 2.9 |
| | 10 | 3665 | 12.13 | 3.3 | 4.0 | 3.4 | 2.2 | 4.3 | 3.0 |
| | 14 | 4515 | 15.66 | 3.4 | 3.6 | 3.3 | 1.9 | 3.7 | 3.0 |
| Miango | 14 | 6078 | 18.50 | 3.1 | 3.3 | 3.1 | 1.7 | 3.4 | 2.7 |
| Nimbia | 14 | 5054 | 17.17 | 3.4 | 3.5 | 3.1 | 1.7 | 3.5 | 3.2 |

($<2$ t $ha^{-1}$ $year^{-1}$) during the first few years after plantation establishment.

The mass of the standing crop of litter ranged between 10.6 and 17.2 t $ha^{-1}$. The huge accumulation of organic matter on the forest floor with associated nutrient immobilization suggests rate of litter decomposition that was lower than the rate of litter production. Litter decomposition and mineralization determine the rate at which nutrients in the litter become available for recycling. Estimates of recycling time for organic matter and the different elements are shown in Table 5. It would take 2 to 4 years to mineralize the nutrients present in an annual litterfall. The mobility of the different elements on the forest floor was in the order K $>$ Mg $>$ P $>$ N $>$ Ca. In Fiji, *Waterloo* (1994) reported similar order of nutrient mobility.

The present results contrast the situation under natural savanna where litter of mixed vegetation decomposed in 3 to 6 months (*Isichei*, 1982). Usually pine needles are more resistant to decomposition than the broad-leaved litters in which termites and other agents of biotic decomposition are more active (*Lundgren*, 1978 and *Nye*, 1961). This study and the one reported by *Egunjobi* and *Bada* (1979) showed that it is not always that litter decomposition proceeds at a fast rate in the tropics as decay rate depends on litter quality and moreover ecosystem types generally differ in their internal patterns, particularly in the relationship between the size of the nutrient pool and the rapidity of its turnover through production and decomposition.

## 3.4 Nutrient cycling

Nutrient and biomass distribution within the tree (Table 4) has particular significance to nutrient cycling because of the practice of removing portions of the tree in biomass harvest.

### *3.4.1 Nutrient removal*

Data on nutrient export in different intensities of biomass harvest expressed as percentage of soil nutrient reserve are presented in Table 6. From the data, it is apparent that the harvest of total above-ground biomass would remove relatively large amounts of N, P, K, Ca and Mg from the site. For P and K, their amounts in the tree biomass were relatively high and this could lead to depletion and deficiency following their loss from the site through biomass removal. On marginal sites such as Afaka, the fear of nutrient depletion is real because of the fragility of its soils which were characterised by the occurrence of low-activity clay, little reserves of weatherable minerals and low total soil P (*Ojanuga*, 1979). In absence of P fertilization, the Caribbean pine was often poorly adapted and exhibited symptoms of growth disorder when planted at Afaka (*Kadeba*, 1978).

But on the more fertile sites of Miango and Nimbia, the pine failed to respond to P and K addition in a fertilizer trial even though the amounts of extractable fractions of the two elements in the soil were low (*Kadeba*, 1978). This is an indication that sufficient quantities of P and K were available for tree growth despite their low soil storage. The conclusion may be drawn that on these fertile sites the removal of the two elements in biomass harvest is not likely to create their shortage ahead of other nutrients as the results merely revealed the weakness of the soil test as a predictor of availability of the two nutrients on these sites.

A factor that may have a decisive effect on nutrient removal in biomass harvest is the length of rotation. Rotation length can affect the nutrient cycle in two ways that may be of interest to forest managers. First, nutrient removed per unit of biomass is reduced by lengthening the rotation, even though more total nutrients are removed the longer the rotation. This has been confirmed in a study of an age series of *Pinus caribaea* at Afaka (*Kadeba*, 1991). Second, because maximum nutrient accumulation usually occurs during the early stages of stand development, the shorter the rotation the greater will be the nutrient removed on annual basis (*Jorgensen* and *Wells*, 1986). From the foregoing, it becomes apparent that the amount and number of times biomass can be removed before productivity is greatly reduced will vary depending upon site fertility. Where multiple short rotations are contemplated as in the pulpwood production in Nigeria, there is the need to minimize nutrient loss through careful choice of site, intensity of biomass harvest and by adding nutrients via fertilizer inputs.

### *3.4.2 Nutrient requirements*

Estimates of nutrient requirement, uptake, canopy leaching and other nutrient fluxes of the pine plantation ecosystems are given in Table 7. The annual nutrient requirements were considerably in excess of the net

**Table 6:** Nutrient export in different biomass harvest intensities expressed as percentage of "available" nutrient reserves[a] in 0–30 cm mineral soil.

| Site | Component harvested | Nutrients (%) | | | | |
|---|---|---|---|---|---|---|
| | | N | P | K | Ca | Mg |
| Afaka | Bole | 8 | 85 | 33 | 8 | 8 |
| | Bole and branches | 10 | 113 | 43 | 10 | 9 |
| | Above-ground biomass | 19 | 176 | 92 | 16 | 15 |
| | Soil reserves kg $ha^{-1}$ 3 $dm^{-1}$ | 2076 | 17.4 | 236 | 1191 | 496 |
| Miango | Bole | 6 | 45 | 30 | 6 | 6 |
| | Bole and branches | 8 | 60 | 40 | 8 | 8 |
| | Above-ground biomass | 17 | 104 | 92 | 13 | 13 |
| | Soil reserves kg $ha^{-1}$ 3 $dm^{-1}$ | 3595 | 42.6 | 387 | 2286 | 920 |
| Nimbia | Bole | 5 | 34 | 22 | 4 | 4 |
| | Bole and branches | 6 | 44 | 29 | 6 | 6 |
| | Above-ground biomass | 11 | 72 | 64 | 9 | 9 |
| | Soil reserves kg $ha^{-1}$ 3 $dm^{-1}$ | 3997 | 49.3 | 402 | 2576 | 1014 |

a N-total reserve

**Table 7:** Transfers of organic matter (t $ha^{-1}$ $year^{-1}$) and nitrogen, phosphorus and potassium (kg $ha^{-1}$ $year^{-1}$) in *Pinus acribaea* plantation ecosystems.

| Transfer | Afaka | | | | Miango | | | | Nimbia | | | |
|---|---|---|---|---|---|---|---|---|---|---|---|---|
| | Org. matter | N | P | K | Org. matter | N | P | K | Org. matter | N | P | K |
| Increase in live trees | 9.0 | 28.2 | 2.2 | 15.5 | 16.5 | 44.4 | 3.2 | 25.5 | 11.2 | 32.1 | 2.5 | 19.1 |
| Replacement of litterfall | 4.5 | 43.8 | 2.5 | 26.2 | 6.1 | 51.7 | 3.0 | 32.2 | 5.1 | 44.5 | 2.8 | 28.3 |
| Canopy leaching[a] | – | 0.3 | 0.2 | 11.0 | – | 0.3 | 0.2 | 11.0 | – | 0.3 | 0.2 | 11.0 |
| Litter decomposition | – | 6.6 | 0.4 | 5.3 | – | 11.8 | 1.0 | 8.7 | – | 8.0 | 0.7 | 7.1 |
| *Requirement* | 13.5 | 72.3 | 4.9 | 52.7 | 22.6 | 96.4 | 6.4 | 68.7 | 16.3 | 76.9 | 5.5 | 58.4 |
| Retranslocation | – | 19.4 | 1.1 | 16.3 | – | 14.5 | 0.6 | 17.6 | – | 16.7 | 0.6 | 16.2 |
| *Uptake* | – | 52.9 | 3.8 | 36.4 | – | 81.9 | 5.8 | 51.1 | – | 60.2 | 4.9 | 42.2 |
| Rainfall[a] | – | 4.5[b] | 0.4 | 7.6 | – | 4.5[b] | 0.4 | 7.6 | – | 4.5[b] | 0.4 | 7.6 |
| Soil reserves | – | 41.5 | 2.8 | 12.5 | – | 65.3 | 4.2 | 23.8 | – | 47.4 | 3.6 | 16.5 |

a Estimated from the data of *Waterloo* (1994)
b From *Jones* (1971)

annual accumulation in the biomass (i.e. nutrients in biomass increment of live trees). Substantial proportion of the annual uptake was allocated to the replacement of the needles shed as litterfall. Nutrients needed for the foliage production represented the highest proportion of the nutrient requirements of the ecosystems (Table 7).

The relative contributions of the different nutrient fluxes to the nutrition of the pine stands is given in Table 8. Retranslocation, litter decomposition, nutrient input from rainfall and canopy leaching together contributed substantially to the nutrient requirements of the stands, particularly to K requirement. These data showed that the gross nutrient demands of the trees could not be met from the supply of nutrients available for recycling within the tree and the litter components. The shortfalls had to be met by depletion of the mineral soil.

In the calculations made in section 2.6 the annual nutrient inputs from the soil reserves were unknown. Hence, the annual nutrient removal from the soil by the stands was estimated by subtracting from nutrient uptake the sum of nutrients returned to the forest floor via litter decomposition, canopy leaching and rainfall. At Miango, for example, these removals from soil reserves were 65.3, 4.2 and 23.8 kg $ha^{-1}$ $year^{-1}$ of N, P and K respectively. These data showed that most of the nutrients taken up by the biomass were supplied by the soil. The often reported nutrient deficits when natural tropical ecosystems are replaced by fast growing exotics as also found in this study may be related to the observation about the soil as the source of nutrients immobilised in tree biomass and litter layer as noted earlier. But where the below-ground inputs contribute as much or more to the organic matter decomposition cycle than the above-ground litterfall, the present conclusion may need to be qualified. In a recent study of the organic matter dynamics in *Pinus caribaea* plantation and a paired broadleaf secondary forest, *Cuevas* et al. (1991) found that the pine stand allocated more of its carbon to the above-ground growth and the litterfall contributed considerably more than below-ground inputs from fine root turnover to the organic matter decomposition cycle. The contrast was the case in the secondary forest.

**Table 8:** Relative contribution of different nutrient sources to the above-ground nutrient requirements of *Pinus caribaea* plantation ecosystems

| Nutrient sources | Contribution to requirement (%) | | | | | | | | |
|---|---|---|---|---|---|---|---|---|---|
| | Afaka | | | Miango | | | Nimbia | | |
| | N | P | K | N | P | K | N | P | K |
| Retranslocation | 27 | 23 | 31 | 15 | 9 | 26 | 23 | 11 | 27 |
| Litter decomposition | 9 | 8 | 10 | 12 | 16 | 13 | 10 | 13 | 12 |
| Rainfall | 6 | 8 | 14 | 5 | 6 | 10 | 5 | 7 | 13 |
| Canopy wash | < 1 | 4 | 21 | < 1 | 3 | 16 | < 1 | 4 | 19 |
| Soil reserves | 57 | 57 | 24 | 67 | 66 | 35 | 61 | 65 | 29 |

## 4 Conclusion

Nutrient cycling is apparently a central issue in plantation forestry as rates of nutrient transfer within the ecosystem vary at different stages of the rotation. Litter decomposition, nutrient retranslocation, atmospheric deposition, soil nutrient reserves, etc. make differential contributions to tree nutrition. This study showed that the impact of the establishment and growth of the Caribbean pine resulted in the decline of soil organic matter and nutrients relative to the baseline conditions under the natural savanna. The data obtained also showed that although the annual nutrient increment in biomass was modest, continued growth of the stands requires nutrient supply that is in excess of what is accumulated in the tree biomass.

Estimates of nutrient fluxes contributing to tree nutrition were calculated. The nutrient deficits recorded on these savanna sites as well as the deterioration of the rainforest soils when planted to this species (*Cornforth*, 1970, *Zech* and *Drechsel*, 1992), suggest an inability of the pine stand to immobilize large flushes of nutrients such as nutrient inputs from atmospheric deposition and organic matter mineralization; hence the stands were more dependent on the soil for nutrients. Some stand characteristics may have accounted for this inability. For example, *Lugo* (1992) reported lower root densities and biomass as well as shallower depth of root penetration, lower nutrient concentrations in roots and fewer microsites where roots grow in *Pinus caribaea* plantations than paired secondary forests of similar age. This impairement of root vitality could result in loss of some of the nutrient influxes via hydrological pathways.

In natural tropical forests, nutrition is maintained by a rapid cycling of nutrients between organic components and the substrate (*Nye*, 1961). The residence time (2 to 4 years) of organic matter and nutrients in the pine litter layer is in sharp contrast to the residence time of 2 to 5 months often reported for mixed natural vegetation and broad-leaved plantations. Nutrient immobilization in the slowly decomposing pine litter could produce nutrient imbalances or deficiencies that may cause poor growth in subsequent rotations.

Nutrient export in biomass harvest has the potential to drain site nutrient resources. The cycling of nutrients and the associated efficiency and conservation have, undoubtedly, contributed to the nutritional functioning of these pine stands. However, at age 14 years, it appears that the annual inputs of nutrients (perhaps with exception of K) from the soil remains the dominant source of tree nutrition in comparison with other nutrient fluxes.

## Acknowledgements

The work reported here was carried out as part of the research study on the establishment and growth of exotic tree species in the savanna region of Nigeria. Financial support provided by the *Forestry Research Institute of Nigeria* is gratefully acknowledged. The comments and suggestions of the two annonymous reviewers were helpful as these have added more strength and clarity to the text.

## References

*Baker, T. G.* and *P. M. Attwill* (1985): Above-ground nutrient distribution and cycling in Pinus radiata D. Don and Eucalyptus obliqua forests in southeastern Australia. For. Ecol. Manage. 13, 41–52.

*Bowen, G. D.* and *E. K. S. Nambiar* (1984): Nutrition of Plantation Forests. Academic Press, London, p. 316.

*Cornforth, I. S.* (1970): Reafforestation and nutrient reserves in the humid tropics. J. Appl. Ecol. 7, 609–615.

*Cuevas, E., S. Brown* and *A. E. Lugo* (1991): Above- and below-ground organic matter storage and production in a tropical pine plantation and a paired broadleaf secondary forest. Plant and Soil 135, 257–268.

*Egunjobi, J. K.* and *S. O. Bada* (1979): Biomass and nutrient distribution in stands of Pinus caribaea L. in dry forest zone of Nigeria. Biotropica 11, 130–135.

*Evans, J.* (1986): Plantation forestry in the tropics – trends and prospects. Int. Tree Crops J. 4 (1), 3–15.

*FAO-UNESCO.* (1988): Soil Map of the World. Revised legend. World Resources Report 60, Rome.

*Isichei, A. O.* (1982): Nitrogen in savanna grass and litter. In: MAB Nigerian savanna. Kainji Lake Research Institute, New Bussa. (Eds. W. W. Sanford, H. M. Yesufu and J. S. Ayeni), pp. 208–224.

*Jones, M. J.* (1971): Ammonium and nitrate nitrogen in rain water at Samaru, Nigeria. Tellus XXIII, 43.

*Jorgensen, M. J.* and *C. G. Wells* (1986): Tree nutrition and fast growing plantations in developing countries. Int. Tree Crops J. 3, 225–244.

*Kadeba, O.* (1978): Nutritional aspects of afforestation with exotic tree species in the savanna region of Nigeria. Commonw. For. Rev. 57, 191–199.

*Kadeba, O.* (1989): Biomass equations for even aged stands of Caribbean pine (Pinus caribaea) planted as an exotic in Nigeria. J. Trop. For. Sci. 1, 346–355.

*Kadeba, O.* (1991): Above-ground biomass production and nutrient accumulation in an age sequence of Pinus caribaea stands. For. Ecol. Manage. 41, 237–248.

*Kadeba, O.* and *E. A. Aduayi* (1985): Impact on soils of plantations of Pinus caribaea stands in natural tropical savanna. For. Ecol. Manage. 13, 27–39.

*Kemp, R. H.* (1970): Trials of exotic tree species in the savanna region of Nigeria. Res. Pap. No. 6 (Savanna series). Federal Dept. of For. Research, Ibadan., Nigeria, 16 p.

*Lugo, A. E.* (1992): Comparison of tropical tree plantations with secondary forests of similar age. Ecological Monographs 62, 1–41.

*Lundgren, B.* (1978): Soil conditions and nutrient cycling under natural and plantation forests in Tanzanian Highlands. Swedish Univ. Agric. Sci., Uppsala. Report For. Ecol. and For. Soils No. 31.

*Nye, P. H.* (1961): Organic matter and nutrient cycles under moist tropical forest. Plant and Soil 13, 333–346.

*Ojanuga, A. G.* (1979): Clay mineralogy of soils in the Nigerian tropical savanna regions. Soil Sci. Soc. Am. J. 43, 1237–1242.

*Olson, J. S.* (1963): Energy storage and the balance of producers and decomposers in ecological systems. Ecology 44, 323–346.

*Waterloo, M. J.* (1994): Ph. D. Thesis, Vrije Universiteit Amsterdam.

*Zech, W.* and *P. Drechsel* (1992): Multiple mineral deficiencies in forest plantations in Liberia. For. Ecol. Manage. 48, 121–143.

# Nutrient retranslocations during the early growth of two exotic plantation species

**Wan Rasidah Kadir**[1], **O. Van Cleemput**[2] and **A. R. Zaharah**[3]

[1]Forest Research Institute Malaysia, Kepong, 52109 Kuala Lumpur, Malaysia.
E-mail: rashidah@frim.gov.my
[2]Faculty of Agricultural and Applied Biological Sciences, University of Ghent, Coupure Links 653, 9000 Gent, Belgium
[3]Department of Soil Science, Universiti Pertanian Malaysia, 43400 Serdang, Selangor, Malaysia

## Abstract

Retranslocation of nutrients before leaf abscission is an important pathway of nutrient conservation. This article reports the extent of nutrient withdrawal by *Acacia mangium* Willd. and *Khaya ivorensis* A. Chev., before leaf fall. Measurements were carried out based on the nutrient content in intact green leaves and freshly fallen litter, collected at the plantation age of 24 months. Results showed that *A. mangium* consficated back about 78 % of P compared to only 26 % by *K. ivorensis*. *A. mangium* could probably need some of these P for the $N_2$ fixation process beside for surviving on a soil with poor P availability. The withdrawal of nitrogen seems less critical but the trend relates to the availability of nutrients in the soil. The fact that *A. mangium* took back about 22 % of N, which was far more higher than what has been taken back by *K. ivorensis*, is not fully understood as the former species has the capability to fix the atmospheric nitrogen. The withdrawal of K was about 13 % for both species and this value shows K is more limiting than N as far as *K. ivorensis* is concerned. Magnesium, on the other hand, was not retranslocated.

## 1 Introduction

The productivity of plants can often be improved through proper manuring. In forestry practices, very often that plants were not supplied with additional nutrients, thus they have to turn to other strategies such as retranslocation to conserve sufficient amount of nutrients to overcome nutrient limitations. Internal cycling of nutrients is important in the forest ecosystems especially for deciduous trees as it provides immediate nutrients for the development of new foliage (*Dawson* and *Funk*, 1981). In a study carried out on deciduous forest species in the North America, *Luxmore* et al. (1981) observed a very rapid translocation of N, P, K, Ca and Mg to leaves during spring and a retranslocation back to the stem around June for N and P. Other species such as *A. mangium*, eventhough not deciduous, shed a large number of leaves at a time. In a study on 6-year-old *A. mangium* plantation of BRIS soil, *Amir* and *Wan Rasidah* (1994) discovered significant positive relationships between foliar and soil nutrient content of P and K. Nitrogen, on the other hand, showed inverse relationship probably due to higher rate of N fixation on poor N soil.

In December 1991, an experimental plot was set up in a newly established forest plantation to study nutrient dynamics in a man-made forest at young age until canopy closure. Species planted were *Acacia mangium* and *Khaya ivorensis*. Nutrient analysis carried out at different stand ages showed a considerable decline in the nutrient content in the leaf litter compared to the green leaves particularly for phosphorus. Thus, this short study was conducted to estimate the percentage of nutrient retranslocation before leaf abscission. To confirm that nutrient loss from leaves occurs before leaf drop, analysis was also carried out on intact yellow leaves. However, calculation for retranslocation was only made with freshly fallen litter identified by their colour and form.

## 2 Materials and methods

### 2.1 Site and Species Description

The study area, located at the Kemasul Forest Reserve (3° 28′ N and 102° 23′ E), comprised of 34-month-old *Acacia mangium* and *Khaya ivorensis* trees. A survey of the area showed that the soil belongs to the Batu Anam series (Aquoxic tropudult), developed from iron poor shales with heavy clay texture. The annual rainfall for the period of 1992 to 1995 ranged between 1521 mm to 2257 mm while the mean temperature was constant at 26°C.

The two species used in this study are both plantation species and are considered as fast growing. *A. mangium* is a nitrogen fixing tree, grows well on a wide range of soil. The best growth, however, was observed on a deep, fertile soil with rainfall of more than 2000 mm year$^{-1}$ (*Zulkefli*, 1989). This species is suitable for use in light construction applications, for board making and in the production of tannin (*Chew* et al., 1992). The other species studied, *K. ivorensis*, is also known as African mahogany. This species originated from the coastal area of West African rainforests. Its timber properties is only slightly inferior to the genuine mahogany. Growth performance under the Malaysian conditions is satisfactory with the height reached 26.2 m after 26 years of growth (*Darus* et al., 1991).

### 2.2 Soil and Foliar Sampling and Analysis

Foliar samples were collected from trees selected randomly in a 0.2 ha plot at the age of 34 months old. Collection of sample was not done at earlier age due to the limited availability of freshly fallen *K. ivorensis* litter. Sampling were partitioned into three categories for *A. mangium* i.e. intact green leaves, intact yellow leaves and freshly fallen leaves. While for *K. ivorensis*, due to the limited availability of intact yellow leaves, only green leaves and freshly fallen leaves were sampled. These samples were dried, ground and extracted for the N content using Kjeldahl digestion and P, K, Mg and Ca contents using microwave digestion. Nitrogen and phosphorus were determined colorimetrically while K, Mg and Ca were quantified using inductively coupled plasma.

Nutrient withdrawal prior to leaf fall was measured based on the assumption that Ca is immobile the moment it reaches the leaves (*Vitousek* and *Sanford*, 1986). The following equation was used to estimate nutrient retranslocation.

$$\% \text{ retranslocation} = \frac{[\text{Nutrient/Ca}]_{\text{leaf litter}}}{[\text{Nutrient/Ca}]_{\text{green leaves}}} \times 100$$

Soil samples were taken at different growth stages, from 3-month-old to 24-month-old (canopy closure), in the same plot as for foliar sampling. Sampling was done randomly between the tree canopies and beneath the tree canopy of the two species, both at five locations and at four different depths. These soils were composited within the five locations and sent immediately to the laboratory for mineral-N analysis. The analysis was carried out on fresh soil sample, extracted with 1 N KCl solution, and results were expressed on dry weight basis based on the percentage of moisture content.

## 3 Results and discussion

The related physical and chemical properties of the soil at various depths are presented in Table 1. pH values showed that the soil is acidic and this agrees well with the low exchangeable bases content. The overall chemical data showed that the nutrient content in this soil was rather low, particularly extractable P.

The percentage of nutrient retranslocation by *A. mangium* and *K. ivorensis* before leaf abscission were presented in Table 2. Both species have a different capacity to retranslocate nutrients. Generally, nitrogen fixing species tend to retranslocate less N due to the availability of N through biological nitrogen fixation (*Cole*, 1981). Measurements from this study, however, revealed the opposite. Almost half of the N present in the green leaves of *A. mangium* disappear when the leaves turn into yellow before abscission. When the leaves fall, 22% N was estimated to be retranslocated back to the tree. On the other hand, the N content in the intact green leaves of *K. ivorensis* do not differ greatly compared with their corresponding litter. The trees retranslocated an average of 7% of N.

In this plantation, N availability reduces rather drastic especially during the first year after tree planting (Table 3). The trend is almost similar for the soil under and between the canopy except for a slight increase observed in the first 5 cm layer under the canopy at stand age of 24-month-old. At stand age of 6-month-old, there was an increase in mineral N concentration within 5 cm top soil layer between the canopy, showing an increase in N mineralisation capacity. This phenomenon was not obvious in the soil under the tree canopy as the mineralised N might have been taken up by the roots. The percentage of mineral nitrogen decrease was higher in the upper soils layer compared to those below the rooting depth (> 30 cm). This decrease could have been due to uptake as gaseous N emission is very unlikely in this low pH soil (Table 1). Mineral N concentration seems to reach the baseline at stand age of 18 months.

The demand for N by *A. mangium* is also obvious from the estimated fixation data which shows an increase fixation as the trees grow (Table 4). This trend is opposite to the changes in mineral N concentration. The percentage of nitrogen derived from atmosphere (Ndfa) was considered low. With a low soil pH and high moisture stress particu-

**Table 1:** The properties of soil in the plantation area

| Soil depth (cm) | pH | Clay (%) | C (%) | N (%) | C/N | Extr. P ($\mu g\ g^{-1}$) | Exch. cations ($cmol_c\ kg^{-1}$) | | | CEC ($cmol_c\ kg^{-1}$) |
|---|---|---|---|---|---|---|---|---|---|---|
| | | | | | | | K | Ca | Mg | |
| 0– 5 | 4.60 | 32 | 0.73 | 0.10 | 7.3 | 1.84 | .16 | n.d. | .07 | 4.75 |
| 5–15 | 4.45 | 32 | 0.63 | 0.09 | 7.0 | 1.70 | .16 | .21 | .15 | 3.35 |
| 15–30 | 4.47 | 44 | 0.34 | 0.07 | 4.9 | 1.25 | .10 | .07 | .06 | 2.28 |
| 30–45 | 4.42 | 44 | 0.22 | 0.06 | 3.7 | 1.20 | .08 | .03 | .02 | 3.62 |

**Table 2:** Selected nutrient contents (g $kg^{-1}$) in *Acacia mangium* and *Khaya ivorensis* leaves before and after senescence

| Parameter | Types of leaves | | | % estimated |
|---|---|---|---|---|
| | Intact green | Intact yellow | Freshly fallen | retranslocation |
| *Acacia mangium* | | | | |
| Nitrogen | 21.80 ± 0.69 | 13.61 ± 0.39 | 12.84 ± 0.20 | 22.3 |
| Phosphorus | 0.60 ± 0.05 | 0.18 ± 0.04 | 0.10 ± 0.00 | 78.0 |
| Potassium | 10.22 ± 0.87 | 6.82 ± 0.79 | 6.77 ± 0.82 | 12.6 |
| Magnesium | 1.03 ± 0.07 | 1.17 ± 0.15 | 0.92 ± 0.06 | −17.8 |
| Calcium | 8.48 ± 0.51 | 9.32 ± 0.57 | 6.43 ± 0.11 | |
| *Khaya ivorensis* | | | | |
| Nitrogen | 11.37 ± 1.15 | | 10.16 ± 2.03 | 7.3 |
| Phosphorus | 0.35 ± 0.03 | | 0.25 ± 0.03 | 25.9 |
| Potassium | 2.80 ± 0.28 | | 2.35 ± 0.95 | 13.0 |
| Magnesium | 2.20 ± 0.35 | | 2.70 ± 0.28 | −27.3 |
| Calcium | 2.24 ± 0.07 | | 2.16 ± 0.00 | |

(results are mean of six replicates for *Acacia mangium* and four replicates for *Khaya ivorensis*)

**Table 3:** Changes in the available N concentration ($NH_4^+ + NO_3^-$ in µg $g^{-1}$) in soil at different stand ages

| Soil depth | Stand age (month) | | | | |
|---|---|---|---|---|---|
| (cm) | 3 | 6 | 12 | 18 | 24 |
| | | | Between the canopy | | |
| 0– 5 | 72.32 | 79.92 | 27.39 | 11.78 | 7.56 |
| 5–15 | 55.23 | 53.39 | 18.07 | 9.34 | 6.92 |
| 15–30 | 39.29 | 29.03 | 17.17 | 7.20 | 6.16 |
| 30–45 | 40.39 | 22.15 | 19.81 | 8.10 | 6.07 |
| | | | Under the canopy | | |
| 0– 5 | 69.32 | 39.97 | 19.04 | 12.05 | 16.28 |
| 5–15 | 54.17 | 38.08 | 18.75 | 7.09 | 4.62 |
| 15–30 | 43.13 | 38.40 | 19.40 | 7.27 | 5.69 |
| 30–45 | 31.49 | 29.84 | 21.27 | 6.46 | 4.99 |

**Table 4:** Nitrogen fixation by *A. mangium* at different stand ages estimated using $^{15}N$ isotope dilution method

| Stand age (month) | Ndfa (% ± SE) | Nitrogen fixed (kg $ha^{-1}$) |
|---|---|---|
| 7–12 | 20.62 ± 6.87 | 3.41 |
| 13–18 | 44.85 ± 11.20 | 14.30 |
| 19–24 | 58.60 ± 4.71 | 35.70 |

Ndfa – nitrogen derived from atmosphere

larly in the first 12 months after planting, this low value is not suprising. Nitrogen fixation was estimated using $^{15}N$ isotope dilution method with *Khaya ivorensis* as reference tree (Kadir, 1995).

The continuous decrease in available N content and the increasing amount of nitrogen fixed by *A. mangium* indicate that nitrogen is critical for this species compared to *K. ivorensis*. Probably, efficient translocation of N provide instant source of N for new shoot development during the early growth. At a very low level of soil mineral N concentration, *K. ivorensis* retranslocated only a small portion of N.

The most pronounced withdrawal of all elements was found with phosphorus. This could be a strategy of these trees to survive in a soil with poor P availability. Values reported by Vitousek and Sanford (1986) were between 42 to 58% for N and 67 to 69% for P on an infertile Oxisol/Ultisol soils. In this study, the most distinct retranslocation was for *A. mangium*. Nearly 80% of P in *A. mangium* was retranslocated compared to only 26% by *K. ivorensis*. Many studies have confirmed the importance of P for biological nitrogen fixation. Sanginga et al. (1991) observed a significant increase in the nodule production by *Leucaena leucocephala* after P fertilization. In another study, P application increased the N concentration considerably in inoculated *Casuarina* but in N fertilized plants, the N concentration in the shoot decreased (Sanginga et al., 1989). *A. mangium* could have reserved some of these P for fixation besides plant growth.

The other two nutrients, K and Mg, did not differ much between the two species (Table 2). Both species returned back about 13% of K, which seemed to be more important than N for *K. ivorensis*. Magnesium, however, was not retranslocated.

## Acknowledgement

We thank the Forest Research Institute Malaysia for the financial assistance.

## References

*Amir, H. M. S.* and *Wan Rasidah Kadir* (1994): Growth differences, fertility status and foliar deficiency levels of six-year-old Acacia mangium on BRIS soil. J. Trop. Forest Sci. 6(3): 230–238.

*Chew, L. T., Nurulhuda, M. N., Saimin, B.* and *Yamada, N.* (1992): Chemical properties of Acacia mangium barks of different ages and locations. Proc. Second Chemistry Division Seminar, 29. Feb. 1992, Forest Research Institute Malaysia, p. 7–12.

*Cole, D. W.* (1981): Nitrogen uptake and translocation by forest ecosystems. In F. E. Clark and T. Rosswall: Terrestrial Nitrogen Cycles. Processes, Ecosystem Strategies and Management Impacts. Ecol. Bull. 33, 219–232.

*Darus, A.*, *Hashim, M. N.*, *Abdul-Rasip, A. G.* and *Lok, E. H.* (1991): Khaya ivorensis and Endospermum malaccense as potential species for future reforestation programme. In: S. Appanah, F. S. P. Ng and I. Roslan: Proc. Conf. Malaysian Forestry and Forest Products Research, Kuala Lumpur, 3–4 Oct. 1990. p. 60–65.

*Dawson, J. O.* and *Funk, D. T.* (1981): Seasonal change in foliar nitrogen concentration of Alnus Glutinosa. Forest Sci. 27(2): 239–243.

*Luxmore, R. J.*, *Grizzard, T.* and *Strand, R. H.* (1981): Nutrient translocation in the outer canopy and understory of an Eastern deciduous forest. Forest Sci. 27, 505–518.

*Sanginga, N.*, *Bowen, G. D.* and *Danso, S. K. A.* (1991): Intra-specific variation in growth and P accumulation of Leucaena leucocephala and Gliricidia sepium as influenced by soil phosphate status. Plant and Soil 133, 201–208.

*Sanginga, N.*, *Danso, S. K. A.* and *Bowen, G. D.* (1989): Nodulation and growth response of Allocasuarina and Casuarina species to phosphorus fertilization. Plant and Soil 118, 125–132.

*Vitousek, P. M.* and *Sanford, R. L.* (1986): Nutrient cycling in moist tropical forest. Ann. Rev. Ecol. Syst. 17: 137–167.

*Zulkefli, M.* (1989): Compensatory Forestry Plantation in Peninsular Malaysia: its scope and potential. Paper presented at the Persidangan Perhutanan Malaysia Kesepuluh, 24–29 July 1989, Pahang, Malaysia. 16 p.

# Effect of fertilizing on establishment and early growth of tree plantations on *Imperata cylindrica* grasslands

**Antti Otsamo**

Enso Forest Development Ltd, Reforestation and Tropical Forest Management Project (ATA-267), c/o Balai Teknologi Reboisasi Banjarbaru (BTR), P.O. Box 65 (Jl. Sei Ulin 28 B), 70714 Banjarbaru, Kal-Sel, Indonesia.

Present address: Department of Forest Ecology/Tropical Silviculture, P.O. Box 28, 00014 University of Helsinki, Finland.

## Abstract

Effects of fertilizer combinations on early growth of fast-growing plantation tree species were tested in two separate experiments on *Imperata cylindrica* dominated grasslands in South Kalimantan, Indonesia. Effects of commercial NPK and locally made NPK mixtures on the growth of *Acacia mangium, Gmelina arborea, Paraserianthes falcataria* and *Swietenia macrophylla* were tested by applying the fertilizer one year after planting. In the second experiment the effects of N, P, K, and their combinations with trace elements on the growth of *A. mangium, G. arborea* and *Eucalyptus deglupta* were tested by applying the fertilizer in two doses, immediately after planting and one year after planting. Two years after planting different NPK-mixtures did not have significant effect on the growth of *A. mangium* or *S. macrophylla*. On the contrary, all the NPK treatments roughly doubled the basal area of *P. falcataria*. For *G. arborea* the fertilizing effect was somewhat similar, but not statistically significant at $p < 0.05$. In the second experiment significant differences in growth of *A. mangium, G arborea* and *E. deglupta* between sites and fertilizer treatments were detected. Responses were strongest in the treatments including phosphorous. Generally *A. mangium* had steady performance compared to *G. arborea, P. falcataria* or *E. deglupta*, which were more sensitive to fertilization treatments and site effects. Fertilization is important on forest plantations on grasslands as it clearly speeds up the early growth that is important for gaining site control over the grass. Phosphorous must be included in the fertilization treatments on grasslands. NPK-fertilization is recommended as a standard solution if detailed information on the site is not available. Studies are needed on the long term effects of fertilizers, rotation-length fertilization requirements and the economic profitability of fertilizing, as well as the interactions between tree species, fertilizers and soil.

---

## 1 Introduction

Grasslands dominated by *Imperata cylindrica* (alang-alang) have developed after forest clearing and the subsequent fires, and they cover large areas of former forest land throughout moist tropical regions, particularly in Southeast Asia. In Indonesia alone, the total area of *Imperata* grasslands is estimated at 20 million ha (ITTO, 1990). Recent estimates indicate that so called mega-grasslands, often referred to as 'sheet *Imperata*', cover 8.6 million ha in Indonesia. An additional 18.9 million ha of land classified as bush or shrubland also contains a considerable fraction of *Imperata* grassland (Garrity et al., 1995).

In Indonesia forest plantations will be the major source of wood supply for forest industries in 10–20 years. Therefore, large areas have been allocated for forest plantations on degraded forest lands, which include vast areas of grass- and bushlands. However, there have been several failures in plantation projects. The survival rate in rehabilitation of forest land has been estimated at 57% (Pandley, 1992). Therefore, reliable knowledge on the factors affecting the plantation success is needed.

*Imperata* grasslands are difficult to reforest because of grass competition and allelopathy, fire susceptibility and soil degradation (Dela Cruz, 1986; Ohta, 1990). Promising results in grassland reforestation have been achieved by using fast-growing *Acacia*-species, *Gmelina arborea* and *Paraserianthes falcataria. Swietenia macrophylla* has been one of the promising well-known valuable timber tree species on grassland sites (Otsamo et al., 1997; Vuokko, 1996). Considering the relatively high establishment costs of forest plantations on *Imperata* grassland, the mean annual increment of a profitable timber or pulp plantation should be at least 25 $m^3ha^{-1}a^{-1}$ (Rissanen, 1995).

Grasslands have generally low nutrient levels compared to natural forest soils. Especially phosphorus levels have been noted to be low (Ohta, 1990). Furthermore, soil compaction is probably one of the factors making *Imperata* grasslands unfavorable sites for forest plantations. These constraints can at least partly been overcome by intensive site preparation and initial fertilization with NPK-fertilizer (Fig. 1).

The aim of the study was to test the effects of different fertilizer combinations on some promising fast growing

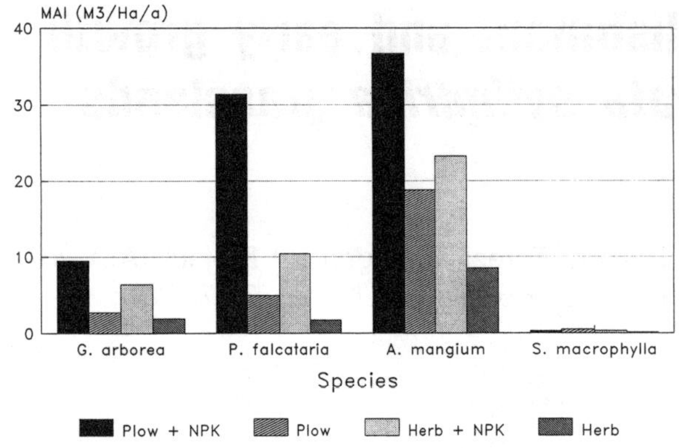


**Figure 1:** Mean annual increment estimates of *Gmelina arborea, Paraserianthes falcataria, Acacia mangium* and *Swietenia macrophylla* 36 months after planting on sites with four different site preparation methods on an *Imperata* grassland site in South Kalimantan, Indonesia (Otsamo et al. 1995a).

tree species used for the reforestation of *Imperata cylindrica* dominated grasslands. The emphasis was in monitoring the effects of fertilization on the establishment and early growth of plantations.

## 2 Material and methods

### 2.1 Study area

Riam Kiwa trial and pilot plantation area is located in South Kalimantan, Indonesia (3°30′S; 115°E). Topography is undulating and the altitude 100–200 m above the sea level. Mean annual rainfall in 1988–1994 was 2043 mm (1574–2364 mm). The long-time average is 2128 mm with a dry season from May to October, with only 20% of the annual rainfall occurring during this period.

Soils are deeply weathered, heavy textured and acid, suffering from various degrees of degradation. Depending on the classification system they belong to the red-yellow podzolic type (Stace et al., 1968), xanthic ferralsols (FAO-UNESCO, 1979) or udic ultisols (USDA, 1975), with low pH-values and low levels of nutrients, especially available phosphorous (Simpson, 1992). Parent material consists of igneous rocks. Compared to several other *Imperata*-dominated areas in Asia, soil at the study site is relatively deep and fertile and suitable for tree plantations (Jusop and Hanif, 1992). The good growth potential of the site was confirmed, as the growth of *Acacia auriculiformis* in Riam Kiwa was higher than in any other site, as seven sites in Indonesia, Malaysia, Thailand and Taiwan were compared in an international provenance trial (Awang et al., 1994). Details of the soil characteristics on experimental sites are presented in Table 1.

Vegetation on the trial sites is dominated by *Imperata cylindrica* grass (height 1–1.5 m; above ground dry biomass 10–20 t $ha^{-1}$), which has covered the area since the Second World War, probably even longer. Remnants of original woody vegetation only exist as narrow strips along the creeks.

### 2.2 Nursery practices

All seedlings used in the experiments were grown in plastic pot-tray containers (Enso-pot; 183 $cm^3$) in a mixture of local peat and rice husk (70:30), and tended according to standards used for seedling production of acacias (Ådjers and Srivastava, 1993). They were planted out at the age of three months.

### 2.3 Site preparation

The trial sites were prepared by plowing the whole area twice (5 and 3 months before planting) with a disc plow, reaching the depth of 25 cm, and harrowing once before planting, which is the most effective site preparation method tested in the area (Otsamo et al., 1995a). Site preparation was started during the dry season in July and the

**Table 1:** Soil characteristics on the trial sites of fertilizing trials on *Imperata* grassland sites in Riam Kiwa, South Kalimantan, Indonesia.

| Site | pH1 | pH2 | C-org (%) | N-tot (%) | P-avl (ppm) | CEC (meq) | K (meq) | Na (meq) | Ca (meq) | Mg (meq) | Al (meq) | H+ (meq) | Texture Clay % | Silt % | Sand % |
|---|---|---|---|---|---|---|---|---|---|---|---|---|---|---|---|
| Exp 1 | 4.0 | 4.2 | 0.7 | 0.03 | 0.3 | 21.0 | 0.03 | 0.23 | 0.2 | 0.18 | 1.15 | 1.52 | 56.1 | 15.3 | 28.7 |
| Exp 2.1 | 4.9 | 4.3 | 3.6 | 0.18 | 4.45 | 22.1 | 0.35 | 0.39 | 3.9 | 1.20 | 0.56 | 0.36 | 49.1 | 25.4 | 25.4 |
| Exp 2.2 | 4.4 | 4.0 | 2.7 | 0.15 | 3.02 | 20.1 | 0.38 | 0.36 | 2.0 | 1.06 | 3.42 | 0.60 | 41.2 | 28.3 | 30.5 |
| Exp 2.3 | 5.1 | 4.6 | 2.5 | 0.15 | 1.73 | 22.3 | 0.22 | 0.19 | 2.4 | 3.6 | 0.24 | 0.23 | 26.4 | 36.8 | 36.8 |

pH1 in $H_2O$
pH2 in KCl
org organic
tot total
avl available
CEC cation exchange capacity
meq meq/100 g of soil
NOTE: On the site of experiment 1 soil samples were collected as a composite sample at dephts of 1–80 cm, whereas on the three sites of experiment 2, only topsoil samples were collected at the depth of 1–20 cm. Each value concerning experiment 2 is an arithmetic mean of four separate samples.

planting was done in December, one month after the start of the rainy season.

### 2.4 Experiment 1

Experiment 1 consisted of separate trials for *Acacia mangium, Gmelina arborea, Paraserianthes falcataria* and *Swietenia macrophylla*. Each species was given the following fertilizer treatments (fertilizers as g tree$^{-1}$):

1) Control (no fertilizers)
2) NPK 1: Commercial NPK (150)
3) NPK 2: Mixed NPK (urea 50, TSP 50 and KCl 38)
4) NPK 3: Mixed NPK (urea 50, TSP 25, rock phosphate 45 and KCl 38)

Fertilizers used were NPK (15:15:15), urea (46% N), triple superphosphate TSP (20.7% P), KCl (50% K) and rock phosphate (15%P). The dosages were based on the field instructions of local forestry organizations.

Seedlings were planted in the plots of 25 (5 × 5) trees at the spacing of 2 × 2 m (2500 seedlings ha$^{-1}$). Randomized complete block (RCB) design with three replications was used. Fertilization was applied 12 months after planting. Fertilizers were mixed in the topsoil by a hoe at the radius of 50 cm around each tree. Manual weeding was done by slashing at six-month intervals.

Survival was assessed from the whole plot, whereas height, diameter at breast height and crown diameter were measured from nine inner trees in each plot 24 months after planting. The plot means data of the measured variables and the basal area were analyzed by analysis of variance. Pairwise comparisons were done by using Tukey's HSD-test.

### 2.5 Experiment 2

Experiment 2 consisted of separate trials for *Acacia mangium, Gmelina arborea* and *Eucalyptus deglupta*. Seedlings were planted in the plots of 25 (5 × 5) trees at the spacing of 3 × 3 m (1111 seedlings ha$^{-1}$). The trials were established at three sites: 1) yellow podzolic, 2) red podzolic, 3) alluvial. These sites were located within two-kilometer distance from each other. RCB design with two replications on each site was used. *Eucalyptus deglupta* was planted on sites 1 and 2 only. The fertilizer treatments were as follows (nutrients as kg ha$^{-1}$): 1) Control (0); 2) N (N50); 3) P (P50); 4) K (K50); 5) NP (N50+P50); 6)NK (N50+K50); 7) PK (P50+K50); 8) NPK (N50+P50+K50); 9) NPK+trace elements (N50+P50+K50+Ca107+Mg74+Cu5+Zn5+B5).

Fertilizers used were urea (46% N), triple superphosphate TSP (20.7% P), KCl (50% K), dolomite (13% Ca and 9% Mg), copper sulphate monohydrate (35.5% Cu), zinc sulphate (36% Zn), and borax (11.4% B).

One third of the total dose of fertilizers was applied immediately after planting, and two thirds 12 months after planting. Fertilizers were mixed in the topsoil by a hoe at the radius of 50 cm around the tree. Manual weeding was done by slashing at six-month intervals.

Survival, height, diameter at breast height and crown diameter were measured 24 months after establishment. Survival was assessed from the whole plot, whereas other variables were measured from nine inner trees only. For indicating the differences between treatments, tree volume estimates for all species were calculated by using stem form factor 0.45, determined for *Acacia mangium* (Sadono and Setyarso, 1992). Therefore, volume estimates were not included in statistical analysis and should not be considered absolute values nor used for comparison of different species. The effects of fertilization and site were analyzed with two-way analysis of variance for each species. Pairwise comparisons were done by Tukey's HSD-test.

## 3 Results

### 3.1 Experiment 1

Survival for all the species was 100%. Fertilization treatments did not have any statistically significant effects on the growth of *Acacia mangium* and *Swietenia macrophylla*. However, a significant difference between the treatments for *Paraserianthes falcataria* was detected, as the commercial mixture of NPK more than doubled the basal area growth compared to control. Mixed NPK-treatments also had significant effects, which were not quite as strong as with the commercial fertilizer (Table 2). For *Gmelina arborea* the fertilization effects had similar trend with *Paraserianthes falcataria*, but the differences were not as distinct and barely not statistically significant at $p < 0.05$ (Table 2).

### 3.2 Experiment 2

Statistically significant differences in growth of *Acacia mangium* between fertilization treatments were detected (Table 3, Fig. 2). The lowest basal area (control) was 49% of the highest recorded value (NPK+trace elements). There were also significant differences between the sites ($F = 8.528$; $p < 0.001$ for basal area; $F = 16.410$; $p < 0.001$ for stand volume) but no interaction between sites and fertilization. The mean (all treatments pooled) basal area (S.E.) was 7.1 (0.7), 9.0 (0.5) and 5.7 (0.5) m$^2$ha$^{-1}$, and stand volume (S.E.) 25.4 (2.9), 32.0 (1.9) and 19.2 (2.1) m$^3$ha$^{-1}$ at sites 1, 2, and 3 respectively.

Statistically significant differences in growth of *Gmelina arborea* between fertilization treatments were detected (Table 4). The lowest basal area (K) was 24% of the highest recorded value (NPK). There were also significant differences between the sites ($F = 23.666$; $p < 0.001$ for basal area) but no interaction between sites and fertilization. The mean (all treatments pooled) basal area (S.E.) was 5.3 (0.7), 3.4 (0.6) and 0.6 (0.1 m$^2$ha$^{-1}$ at sites 1, 2, and 3 respectively.

**Table 2:** Effect of three different NPK treatments on height, diameter at breast height (DBH), basal area and crown diameter of *Acacia mangium, Gmelina arborea, Paraserianthes falcataria*, and *Swietenia macrophylla* 24 months after planting on an *Imperata* grassland site in South Kalimantan, Indonesia. Standard error of mean in parentheses.

| Species | Treatment | Height (m) | DBH (cm) | Basal area ($m^2ha^{-1}$) | Crown diameter (m) |
|---|---|---|---|---|---|
| *Acacia mangium* | Control | 7.5(0.5) | 6.4(0.5) | 9.0(1.3) | 2.9(0.1) |
| | NPK 1 | 8.0(0.2) | 6.4(0.5) | 9.2(1.4) | 3.0(0.5) |
| | NPK2 | 8.3(0.4) | 7.1(1.0) | 10.7(2.8) | 2.8(0.1) |
| | NPK 3 | 7.9(0.2) | 6.5(0.3) | 9.0(1.2) | 2.9(0.3) |
| | F | 1.462 | 0.837 | 0.816 | 0.147 |
| | p | 0.316 | 0.521 | 0.530 | 0.928 |
| *Gmelina arborea* | Control | 3.4(0.3) | 5.5(0.6) | 6.6(1.3) | 3.2(0.2) |
| | NPK 1 | 4.4(0.3) | 7.5(0.5) | 11.7(1.5) | 3.2(0.1) |
| | NPK 2 | 4.3(0.3) | 7.4(0.5) | 11.6(1.5) | 3.1(0.0) |
| | NPK 3 | 4.2(0.1) | 7.0(0.2) | 10.6(0.8) | 3.1(0.1) |
| | F | 4.571 | 3.837 | 3.018 | 0.100 |
| | p | 0.054 | 0.076 | 0.116 | 0.957 |
| *Paraserianthes falcataria* | Control | 3.4(0.2) | 4.7(0.4) | 4.8(0.9) | 2.8(0.3) |
| | NPK 1 | 6.1(0.2) | 7.6(0.1) | 11.7(0.3) | 3.8(0.1) |
| | NPK 2 | 4.6(0.6) | 6.2(0.9) | 8.5(2.2) | 3.8(0.4) |
| | NPK 3 | 4.9(0.2) | 6.6(0.4) | 9.5(1.4) | 4.2(0.2) |
| | F | 12.874 | 9.230 | 9.128 | 2.342 |
| | p | 0.005 | 0.012 | 0.012 | 0.173 |
| | HSD | 1.5 | 2.0 | 4.7 | - |
| *Swietenia macrophylla* | Control | 2.2(0.2) | 3.1(0.2) | 2.1(0.2) | 0.9(0.1) |
| | NPK 1 | 2.4(0.2) | 3.3(0.2) | 2.4(0.3) | 1.0(0.1) |
| | NPK 2 | 2.7(0.3) | 4.2(0.4) | 3.5(0.6) | 1.1(0.2) |
| | NPK 3 | 2.1(0.3) | 3.4(0.4) | 2.4(0.4) | 0.8(0.1) |
| | F | 1.057 | 3.308 | 2.904 | 1.317 |
| | p | 0.434 | 0.099 | 0.123 | 0.353 |

**Table 3:** Effect of different fertilizer combinations on survival, height, diameter at breast height (DBH) basal area, and crown diameter of *Acacia mangium* 24 months after planting of *Imperata* grassland sites in South Kalimantan, Indonesia. Pooled results from three sites. Standard error of mean in parentheses.

| Fertilizer | Survival (%) | Height (m) | DBH (cm) | Basal area ($m^2ha^{-1}$) | Crown diameter (m) |
|---|---|---|---|---|---|
| Control | 100(-) | 6.7(0.2) | 7.0(0.7) | 4.7(0.9) | 3.0(0.2) |
| N | 100(-) | 7.1(0.3) | 7.6(0.6) | 5.4(0.8) | 3.2(0.2) |
| P | 100(-) | 7.7(0.5) | 9.4(0.8) | 8.3(1.2) | 3.4(0.2) |
| K | 96(2) | 6.8(0.3) | 7.5(0.6) | 4.9(0.7) | 3.2(0.2) |
| NP | 100(-) | 8.1(0.2) | 9.4(0.5) | 8.0(0.8) | 3.5(0.1) |
| NK | 100(-) | 7.0(0.5) | 8.3(0.8) | 6.5(1.1) | 3.3(0.2) |
| PK | 100(-) | 7.8(0.2) | 9.9(0.4) | 8.9(0.7) | 3.6(0.2) |
| NPK | 100(-) | 7.8(0.1) | 10.1(0.5) | 9.3(0.9) | 3.6(0.2) |
| NPK+trace | 100(-) | 8.0(0.1) | 10.3(0.3) | 9.6(0.6) | 3.7(0.1) |
| F | 2.457 | 3.262 | 7.029 | 8.528 | 4.317 |
| p | 0.027 | 0.010 | $< 0.001$ | $< 0.001$ | 0.002 |
| HSD | 4 | 1.4 | 2.3 | 3.3 | 0.5 |

Statistically significant differences in growth of *Eucalyptus deglupta* between fertilization treatments were detected (Table 5). The lowest basal area (K, NK) was 10% of the highest recorded value (NPK). There were also significant differences between the sites on height ($F = 8.703$; $p < 0.001$) and DBH ($F = 4.610$; $p = 0.001$), but not on basal area. No interaction between sites and fertilization was detected. The mean (all treatments pooled) basal area (S.E.) was 0.7 (0.2) and 1.1 (0.2 $m^2ha^{-1}$ at sites 1 and 2 respectively.

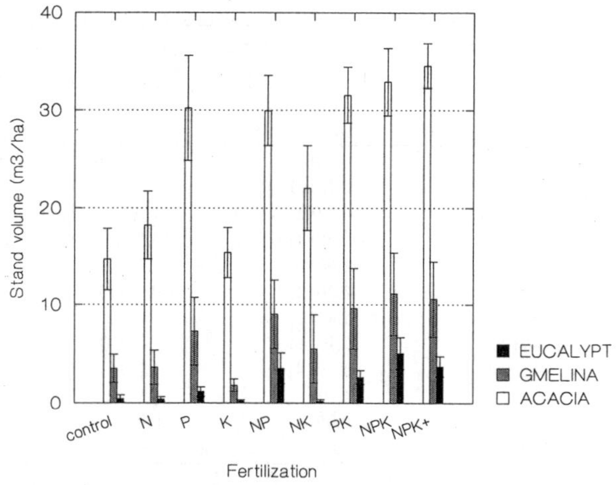


**Figure 2:** Effect of different fertilizer combinations on the stand volume estimates of *Acacia mangium, Gmelina arborea* and *Eucalyptus deglupta* 24 months after planting on *Imperata* grassland sites in South Kalimantan, Indonesia. Pooled results from three sites. Standard error of mean indicated by thin vertical bars.

## 4 Discussion

The results of the study indicated that *Acacia mangium* had a reasonably good early performance on all sites even though the site had a significant effect on the growth. Responses to fertilizers were less distinct compared to other species tested (except *Swietenia macrophylla*). *Gmelina arborea* and *Paraserianthes falcataria* were more sensitive to fertilization treatments and site effects, which indicates need for careful site selection and supplementary fertilization if these species are used in plantations on grasslands. Performance of *Eucalyptus deglupta* was poor, which is in accordance with the earlier records on eucalypts on the area (Otsamo et al. 1995b; Otsamo et al., 1997; Vuokko, 1996). However, the response of *E. deglupta* to fertilizing was stronger than with any other tested species. Site differences for the species could not be effectively analyzed, as it was not planted on the alluvial site.

The low response of *Swietenia macrophylla* to fertilizer treatments confirms the earlier records at the same area (Otsamo et al., 1995a). The lack of response of *Swietenia macrophylla* may be due to the young age of the stands and characteristically slower growth compared to other tested species.

For all the species responses were strongest in the treatments where phosphorous was included. This result is in accordance with those of Ohta (1990), which indicate that especially phosphorous levels on grasslands are low. On the contrary to the conclusions of Simpson (1992), fertilizing with potassium did not have any positive effect on any of the tested species.

The results of this study were obtained on a comparatively fertile *Imperata* grassland site. They are strongly supported by the results from fairly similar edaphic and climatic conditions in Pulau Laut, South Kalimantan, where fertilizing with phosphorus or NPK clearly increased the early growth of *Acacia mangium*, even though the fertilizing effect seemed to level down considerably before the age of three years (Turvey, 1995). Corresponding fertilizing effects were also observed in *Acacia mangium* trials on an *Imperata* site with clearly lower soil nutrient levels and slightly higher annual rainfall in West Kalimantan (Honkanen, 1995). These independent studies clearly indicate the importance of fertilization for reliable establishment of plantations on *Imperata* grassland sites.

Random variation within treatments was quite distinct, which may be an indication of considerable microvariation within sites. The differences in growth between the three different sites indicate the difficulty in accurate growth prediction on large-scale forest plantations on grasslands and further emphasize the need for careful site evaluation and pilot plantations before starting extensive planting operations.

The results of the study present the early effects of fertilization. This information is especially important, since

**Table 4:** Effect on different fertilizer combinations on survival, height, diameter at breast height (DBH), basal area, and crown diameter of *Gmelina arborea* 24 months after planting on *Imperata* grassland sites in South Kalimantan, Indonesia. Pooled results from three sites. Standard error of mean in parentheses.

| Fertilizer | Survival (%) | Height (m) | DBH (cm) | Basal area ($m^2ha^{-1}$) | Crown diamter (m) |
|---|---|---|---|---|---|
| Control | 87(8) | 3.0(0.6) | 4.1(0.9) | 1.9(0.7) | 2.8(0.4) |
| N | 84(6) | 2.8(0.5) | 3.7(0.9) | 1.7(0.7) | 2.5(0.4) |
| P | 85(11) | 3.6(0.8) | 5.4(1.1) | 3.1(1.2) | 3.3(0.6) |
| K | 93(4) | 2.7(0.3) | 3.2(0.4) | 1.1(0.3) | 2.6(0.2) |
| NP | 98(2) | 3.9(0.6) | 6.1(1.1) | 4.1(1.3) | 3.4(0.3) |
| NK | 87(9) | 3.0(0.7) | 4.4(1.3) | 2.6(1.3) | 2.9(0.5) |
| PK | 93(7) | 3.7(0.8) | 5.8(1.5) | 4.1(1.5) | 3.3(0.5) |
| NPK | 93(5) | 3.8(0.8) | 6.2(1.5) | 4.6(1.6) | 3.0(0.5) |
| NPK+trace | 95(6) | 4.0(0.7) | 6.5(1.1) | 4.5(1.3) | 3.3(0.3) |
| F | 1.020 | 1.452 | 2.590 | 2.397 | 1.270 |
| p | 0.436 | 0.221 | 0.031 | 0.043 | 0.300 |
| HSD | - | - | 2.1 | 2.6 | - |

**Table 5:** Effect on different fertilizer combinations on survival, height, diameter at breast height (DBH), and crown diameter of *Eucalyptus deglupta* 24 months after planting on *Imperata* grassland sites in South Kalimantan, Indonesia. Pooled results from two sites. Standard error of mean in parentheses.

| Fertilizer | Survival (%) | Height (m) | DBH (cm) | Basal area ($m^2ha^{-1}$) | Crown diameter (m) |
|---|---|---|---|---|---|
| Control | 67(18) | 2.0(0.6) | 1.7(0.6) | 0.3(0.2) | 1.1(0.4) |
| N | 64(16) | 2.1(0.4) | 1.8(0.5) | 0.3(0.1) | 1.1(0.3) |
| P | 84(7) | 2.9(0.4) | 2.7(0.5) | 0.7(0.2) | 1.7(0.4) |
| K | 75(15) | 1.9(0.3) | 1.3(0.3) | 0.2(0.1) | 0.9(0.2) |
| NP | 89(8) | 3.9(0.5) | 4.0(0.7) | 1.6(0.6) | 2.1(0.2) |
| NK | 56(18) | 1.6(0.5) | 1.5(0.3) | 0.2(0.1) | 0.9(0.3) |
| PK | 92(5) | 3.7(0.4) | 3.8(0.5) | 1.4(0.3) | 2.1(0.2) |
| NPK | 86(5) | 4.8(0.3) | 4.8(0.7) | 2.1(0.6) | 2.4(0.3) |
| NPK+trace | 86(8) | 4.4(0.4) | 4.3(0.4) | 1.6(0.4) | 2.2(0.1) |
| F | 1.519 | 8.703 | 6.671 | 4.100 | 4.583 |
| p | 0.199 | < 0.001 | 0.001 | 0.007 | 0.004 |
| HSD | - | 2.1 | 2.6 | 1.8 | 1.5 |

maximum growth right after planting is considered essential in forest plantations on *Imperata* grasslands, so that grass competition can be minimized and later eliminated with closed canopy. The results of the study indicate the importance of early fertilization in ensuring plantation establishment and vigorous early growth. Long-lasting effects on actual wood production during the whole rotation cannot be predicted from these results. However, according to the review of Schönau and Herbert (1989) on eucalypts, the effects of early fertilization are usually long-lasting and will last the whole rotation period. This would be logical, since fertilizer input on poor sites like *Imperata* grasslands increases the total nutrient quantity that remains in the nutrient cycle of the stand until harvesting. The optimal quantities and application schedules are probably highly site- and species-specific and need well-organized field research combined with disciplined construction of theoretical models on nutrient cycling.

The study concentrated on presenting the effects of fertilizing from the silvicultural point of view. However, if the results are applied on large-scale wood production programs, detailed cost-benefit analyses have to be included in order to assess the economic profitability of fertilizing. According to Rissanen (1995) even a small increase in yield is worth a considerable investment. The results of this study indicate that fertilization is an effective means to increase yield, which supports the idea of fertilizing on forest plantations on grasslands being a profitable operation. However, the concept of profitability and the need for high yield may be considerably different e.g. in industrial plantations, in protective plantations established by public funding, or in small-holder based agroforestry systems, which are all being examined as alternatives for utilizing *Imperata* grasslands.

## 5 Conclusions

Fertilization plays an important role in forest plantations on grasslands since it clearly speeds up the early growth which is important for gaining site control over the grass. Despite differences between sites and tree species, the response to fertilizers was somewhat similar on all sites. Therefore, these results can be applied on other grassland sites as well, even though the magnitude of response and the detailed fertilization recommendations must be carefully determined for each planting site and tree species.

It is evident that phosphorous must be included in the fertilization treatments on grasslands. NPK-fertilization is recommended as a standard practice for grassland reforestation if more detailed information on the site is not availabe. Studies are needed on the long-term effects of fertilizers, rotation-length fertilization requirements and the economic profitability of fertilizing, as well as the interactions between tree species, fertilizers and soil.

## Acknowledgements

The work was done under the auspices of the Reforestation and Tropical Forest Management Project financed by the Finnish International Development Agency (FINNIDA) and the Ministry of Forestry, Indonesia and implemented as a joint effort of the Reforestation Technology Center of Banjarbaru (BTR) and Enso Forest Development Ltd., Finland. The author wishes to thank Dr. J. A. Simpson of Queensland forestry services for assistance in planning the experiment 2, the field staff of BTR for their efforts during the field work and Mr. M. S. Hairunsyah from the Agricultural Research Laboratory in Banjarbaru for the soil analyses.

## References

*Ådjers, G.* and *Srivastava, P. B. L.* (1993): Nursery practices. In: Awang, K. and Taylor, D. (eds.). *Acacia mangium*, growing and utilization. MPTS Monograph Series No. 3. Winrock International and FAO, Bangkok, Thailand, pp. 75–99.

*Awang, K., Venkateswarlu, P., Nor Aini, A. S., Ådjers, G., Bhumibhamon, S., Kietvuttinon, B., Pan, F. J., Pitpreecha, K.* and *Simsiri, A.* (1994): Three year performance of international provenance trials *Acacia auriculiformis*. Forest Ecology and Management 70 (1994): 147–158.

*Dela Cruz, R. E.* (1986): Constraints and strategies for the regeneration of *Imperata* grasslands. In: Forest regeneration in Southeast Asia. Proceedings of the symposium in Bogor, Indonesia, 9–11 May, 1984. Biotrop special publication No. 25. Biotrop. Bogor, Indonesia, pp. 23–34.

*FAO-UNESCO* (1979): Soil map of the World, volume X. Southeast Asia. UNESCO Place ade Fontenoy, Paris.

*Garrity, D., Sukardi, M., de la Cruz, R., Majid, N. M., Pathak, P. S. van So, N.* and *van Noordwijk, M.* (1995): The *Imperata* grasslands of tropical Asia – distribution and typology. Paper presented at the International Workshop on Agroforestry Innovations for *Imperata* Grassland Rehabilitation, January 23–27, 1995 in Banjarmasin, South Kalimantan, Indonesia. 23 p.

*Honkanen, A.* (1995): Species and provenance selection. Results of trials in Sape and Bonti areas. Sanggau Timber Estate, Research Paper No. 5. Enso Forest Development Ltd. 37+9 p.

*ITTO (International Tropical Timber Organization)* (1990): Rehabilitation of logged-over forests in Asia/Pacific region. Draft project report. ITTO/Japan Overseas Forestry Consultants Association (JOFCA). Yokohama. 95 p.

*Jusop, S.* and *Hanif, A. H. M.* (1992): Characterization and classification of soils at MPTS trial sites in Thailand, Sri Lanka, Indonesia, Philippines and Malaysia. Universiti Pertanian Malaysia and Winrock International. 100 p.

*Ohta, S.* (1990): Influence of deforestation on the soils of the Pantabangan area, Central Luzon, the Philippines. Soil Sci. Plant Nutr. 36(4): 561–573.

*Otsamo, A., Ådjers, G., Hadi, T. J., Kuusipalo, J., Tuomela, K.* and *Vuokko, R.* (1995a): Effect of site preparation and initial fertilization on the establishment and growth of four plantation tree species used in reforestation of *Imperata cylindrica* (L.) Beauv. dominated grasslands. Forest Ecology and Management 73: 271–277.

*Otsamo, A., Ådjers, G., Hadi, T. J., Kuusipalo, J.* and *Vuokko, R.* (1995b): Performance and yield of 14 eucalypt species on *Imperata cylindrica* (L.) Beauv. grassland 3 years after planting. New Forests 10: 257–265.

*Otsamo, A., Ådjers, G., Hadi, T. J., Kuusipalo, J.* and *Vuokko, R.* (1997): Evaluation of reforestation potential of 83 tree species planted on *Imperata cylindrica* dominated grassland – a case study from South Kalimantan, Indonesia. New Forests (in press).

*Pandley, D.* (1992): Assessment of tropical forest plantation resource. Department of Forest Survey, Swedish University of Agricultural Science. Draft. 100 p.

*Rissanen, H.* (1995): Reforestation as an investment. In: Tampubolon, A., Otsamo, A., Kuusipalo, J. and Jaskari, H. (eds.) From grassland to forest: Profitable and sustainable reforestation of alang-alang grasslands in Indonesia. Proceedings of a seminar held in Jakarta, 11–12 January 1994. Reforestation and Tropical Forest Management Project (ATA-267), Ministry of Forestry, Indonesia, Enso Forest Development Ltd. and Finish International Development Agency (FINNIDA), pp. 37–53.

*Sadono, R.* and *Setyarso, A.* (1992): Study on stem-form factor and local tree volume table of *Acacia mangium* Willd. in South Kalimantan and South Sumatera. Gadjah Mada University, Yogyakarta-Indonesia, Bulletin Fakultas Kehutanan No. 21: 65–73.

*Schönau, A. P. G.* and *Herbert, M. A.* (1989): Fertilizing eucalypts at plantation establishment. Forest Ecology and Management 29: 221–244.

*Simpson, J. A.* (1992): An assessment of the soils and nutrient status of trial plantations in the Indonesian-Finland Project at Riam Kiwa, South Kalimantan, Indonesia. Technical Report 25. Reforestation and Tropical Forest Management Project (ATA-267). Ministry of Forestry, Indonesia, Enso Forest Development Ltd. and Finish International Development Agency (FINNIDA). 36+71 p.

*Stace, H. C. T., Hubbles, G. D., Brewer, R., Northcote, K. H., Sleeman, J. R., Mulchay, M. J.* and *Hallsworth, E. G.* (1968): A handbook of Australian soils. Rellim. Tech Publications, Glenside, Australia. 435 p.

*Turvey, N.* (1995): Afforestation of *Imperata* grasslands in Indonesia. Results of industrial tree plantation research trials at Teluk Sirih on Pulau Laut, Kalimantan Selatan. ACIAR Technical Reports No. 33. Canberra, Australia. 43 p.

*USDA* (1975): Soil survey staff. Soil Taxonomy. A basic system of soil classification for making and intepreting soil surveys. Agri, Handbook No. 436, 754 pp.

*Vuokko, R.* (1996): Species selection for the afforestation of *Imperata cylindrica* (L.) Beauv. grasslands in Indonesia. Ph.D.-thesis. Department of Crop Science, Horticulture and Forestry, Faculty of Agriculture, University College Dublin, Ireland. 220 p.

# Effect of tillage and woody hedgerows in alley cropping on the productivity of an Alfisol

**Biauw T. Kang**[1 2], **Felix K. Salako**[1] and **Nilantha R. Hulugalle**[3]

[1]International Institute of Tropical Agriculture (IITA), PMB 5320 Ibadan, Nigeria.
[2]Present address; Department of Crop and Soil Sciences, University of Georgia, Athens, Georgia, GA 30602, USA.
[3]Australian Cotton Research Institute, Myall Vale Mail Run, Narrabri, NSW 2390, Australia.

## Abstract

Multipurpose woody species can play an important role in the improvement and maintenance of soil productivity in traditional and planted fallow systems. This study examined the longterm effects of woody species grown in hedgerows in alley cropping and of soil tillage, on soil properties, runoff and erosion, and crop performance. Observations conducted from 1988 to 1993 on a trial established in 1982 on a sloping land is reported in this paper. Six land use systems were compared: *Leucaena leucocephala* and *Gliricidia sepium* hedgerows planted at 2 and 4 m interhedgerow spacings and tilled, and no-till and tilled controls without hedgerows.

The presence of hedgerows and no-till treatments maintained higher soil organic carbon (Org. C) and exchangeable cations contents. Highest Org. C and exchangeable cations were observed in plots with 4 m *Leucaena* hedgerows. Treatments with hedgerows also showed lower soil bulk density than the controls. The no-till treatment showed the highest soil bulk density. No relationships were observed between soil bulk density, aggregate stability and infiltration rate. Mean maize yields for the five years showed the following order: 4 m *Leucaena* > 4 m *Gliricidia* > 2 m *Leucaena* > 2 m *Gliricidia* > no-till control > tilled control plots. Intercropped cassava root yields showed the highest stability in the three cropping seasons in the tilled control and *Gliricidia* alley cropped plots. Lower cassava root yields were attributed to soil compaction in no-till treatment and to shading with 2 m *Leucaena* alley cropping

## Introduction

Trees and shrubs are important components of the traditional farming systems in the tropics because of their many uses, and their environmental and socioeconomic benefits. In the bush-fallow slash-and-burn cultivation system that still dominates traditional farming in many areas of the humid and subhumid tropics, farmers have for many generations exploited the potentials of multipurpose woody species (MVS) for the regeneration of soil productivity reduced during the cropping period. With increasing population pressure and other competing land use demands, farmers in many areas of the humid and subhumid tropics can no longer afford to practice ecologically balanced production systems with long fallow periods to restore soil productivity.

In some areas of humid west Africa, farmers respond to this problem by the selective retention of MWS which they consider useful and more effective in enriching the fallow in a short time (*Obi* and *Tuley*, 1973; *Okigbo*, 1976; *Getahun* et al., 1982; *Kang* and *vanden Beldt*, 1990). The ameliorating effects of trees on soil vary, depending on tree species, soil type, age of fallow or plantation and husbandry practices (*Chalinor*, 1968; *Juo* and *Lal*, 1977; *Lundgren*, 1977; *Chijioke*, 1980; *Kang* et al., 1994). In alley cropping, where food crops are grown in alleys formed by managed hedgerows of MWS, the woody hedgerows still retain the soil ameliorating function as usually observed in the bush fallow (*Kang* et al., 1990).

In rainfed upland cropping in the humid tropics, accelerated soil degradation is frequently associated with deforestation and high rates of soil erosion. Rainfall impact results in both soil compaction and the formation of surface crust which reduce water infiltration rates and thereby increase runoff and erosion. Furthermore, traffic during routine farm operations results in additional soil compaction. To sustain crop production, land management systems for the humid tropics, therefore, should minimize soil exposure and traffic. Two such land management systems are no-tillage (*Lal*, 1990; *Roose*, 1981) and alley cropping (*Kang* et al., 1984). In alley cropping the hedgerows of trees and shrubs are regularly pruned during the cropping period to minimize shading of the associated food crop(s), and the prunings are used as mulch or incorporated as green manure. The presence of mulch as soil cover minimizes rates of soil structural breakdown, soil erosion, and excessive soil temperature, and can also increase earthworm activity, soil water retention, available water holding capacity, and profile water content (*Hauser* and *Kang*, 1993; *Hulugalle* and *Kang*, 1990a; *Kang* and

*Ghuman*, 1991; *Kang* et al., 1990; *Kang*, 1993; *Lal*, 1989a, 1989b, 1989d, 1989e, 1990; *Yamoah* et al., 1986; *Young*, 1990). Crop yields are also observed to be higher and can be sustained for longer periods with either no-tillage or alley cropping (*Kang* et al., 1990; *Lal*, 1989a, 1990). Considerable information is, therefore, available on the effects of either alley cropping or no-tillage on soil properties and on crop yields. Comparative studies of the two systems and the effect of hedgerow species in alley cropping on the soil productivity of an Alfisol have been undertaken since 1982 in a long term trial on sloping land (*Lal*, 1989a). Results of the 1982–1987 observations showed that decline in soil fertility parameters were less with *Leucaena* based systems than the other treatments, particularly plow-till control treatment (*Lal*, 1989a).

The trial was modified in 1988, to look at the effect of lower and more affordable N rate and intercropping which is more widely practiced in the area. The results of observations undertaken from 1988 to 1993 are updated in this paper.

## Materials and methods

### Experimental site

The experiment was conducted at the International Institute of Tropical Agriculture (IITA) Ibadan (7° 30′ N; 3° 54′ E), Nigeria from March 1988 to May 1993. IITA is located in the forest-savanna transition zone of southwestern Nigeria. The mean annual rainfall is 1280 mm with bimodal distribution. The soil at the site (with 7% slope) is an Alfisol, classified as a clayey kaolinitic isohyperthermic Oxic paleustalf with a sandy loam topsoil which changes to clay loam and clay in the sub-surface horizons (*Moormann* et al., 1975).

### Plot layout and treatments

The experiment consisted of 6 land use systems as treatments: hedgerows of *Leucaena leucocephala* (var.K28) and *Gliricidia sepium* (local cultivar) were planted on the contour at intrahedgerow spacing of 0.25 m and interhedgerow spacings of 2 m and 4 m and tilled, and no-till and tilled control (no hedgerows) treatments, which were imposed on the six unreplicated plots of 70 x 10 m each (*Lal*, 1989a). Plots were separated by ditches and earthen bunds. Runoff and soil loss measuring tanks were connected to each plot at the down slope for soil erosion assessment.

Preplanting land preparation in the alley cropped treatments until 1987 involved pruning of hedgerows at a height of 0.75 m above the soil surface, uniform spreading of prunings in the alleys and incorporation therein by rotovation (*Lal*, 1989a). Post 1987 land preparation was limited to hand hoeing followed by uniform spreading of prunings in alleys as a surface mulch (*Kang* and *Ghuman*, 1991). Additional prunings (3–4 times) of the hedgerows annually was required during the crop growing season to minimize shading of the associated crop in the alley cropped treatments. The later prunings were applied on the soil surface as mulch. Land preparation in the tilled (control) treatment was disc ploughing to a depth of 20 cm followed by rotovation from 1982 to 1987, and hand hoeing to a depth of 0–15 cm thereafter. Land preparation with no-till involved spraying 0.5 kg a.i/ha of paraquat one week prior to seeding. The annual cropping systems used in the trial were a maize-cowpea sequential cropping from 1982 to 1989 and a maize + cassava intercropping thereafter. Plant populations were 5.3 plants/m$^2$ for maize and 8.8 plants/m$^2$ for cowpea in maize-cowpea sequential cropping. In the maize + cassava intercrop, plant populations were 4 plants/m$^2$ for maize and 1 plants/m$^2$ for cassava. Crops were planted on flat and crop residue were kept in the field. Fertilizer application rates from 1982 to 1987 were 120 kg ha$^{-1}$ of N as urea (1/3 applied at sowing and 2/3 applied four weeks later), 26 kg/ha of P (as single superphosphate) and 30 kg/ha of K as muriate of potash (*Lal*, 1989a). Subsequent to 1987, the fertilizer rate was reduced to the locally recommended level of 45 kg N, 20 kg P, and 37 kg K/ha (as 15:15:15 fertilizer), broadcast-applied at the time of sowing maize. Maize was side dressed with 30 kg N/ha applied as calcium ammonium nitrate at 4 weeks after planting maize.

### Data collection

*Soil chemical properties:* Composite surface (0–15 cm) soil samples were collected from the 4 plot segments (i.e., lower quarter, the lower middle quarter, the upper middle quarter and the upper quarter) in each of the 6 treatment plots at the beginning of the main cropping season, prior to the addition of prunings and fertilizer. Soil samples were air dried, sieved, and analyzed for pH-$H_20$, Org. C, exchangeable K, Ca, Mg, and extractable Bray P-1 (IITA, 1982).

*Soil physical properties:* Measurements carried out in November 1992 on the 4 plot segments included soil bulk density (using 5 cm diameter core) at 0–5 and 5–10 cm depth, gravimetric soil water retention using pressure plate apparatus at potentials of −10 kPa and −1500 kPa for undisturbed core samples taken at 0–10 cm depth, aggregate stability for surface (0–10 cm) soil using wet-sieving method at 30 rpm for 30 minutes, and water infiltration rate using double ring infiltrometer (50 cm diameter outer ring and 30 cm diameter inner ring, both 30 cm deep). Runoff, soil loss, and nutrient concentration in runoff water (*Kang* and *Ghuman*, 1991) were measured from 1990 to 1992.

*Hedgerow biomas and crop yields:* Hedgerow pruning (leaf + green stems) and wood yields were estimated from 3 m long strips in each of the 4 segments of each treatment. Subsamples were taken and dried at 65°C to determine dry pruning and wood yields. Maize and cassava tuber yields were collected from 4 x 10 m plots from the 4 segments of each of the 6 treatment plots. Maize grain yield (t/ha) is expressed at 12% moisture content, and cassava yield as fresh root yield (t/ha).

## Experimental results

### Hedgerow biomass yield

The *Gliricidia* and *Leucaena* hedgerows still produced large amounts of wood and pruning biomass in spite of intensive (3 to 4) annual prunings since 1983 (Table 1).

In 1990 and 1991, wood and pruning biomass yields were higher with 2 m than 4 m intrahedgerow spacing. *Leucaena* hedgerow also produced more wood and pruning biomass than *Gliricidia*. These figures were comparable to wood and pruning biomass yields observed in 1988 (*Kang* and *Ghuman*, 1991).

### *Soil chemical characteristics*

Results of surface soil chemical analysis conducted during 1988, 1989, 1990, and 1992 showed insignificant annual variations. It appears that the different treatments have reached an equilibrium level following longterm cropping. For this reason, only the mean data for the four years are presented in Table 2. A comparison of the six teatments shows noticeable differences as follows:

- The no-tillage control treatment shows higher pH-$H_2O$ value than the other treatments.
- The tilled control treatment shows the lowest Org. C level. Plots alley cropped with *leucaena* showed higher Org. C content than the other treatments.
- The no-till control plot showed the lowest extractable P level. When compared to the tilled control tratment, the alley cropped plots show lower P levels.
- The tilled control plot shows lower exchangeable K, Ca, and Mg levels than the other treatments. The 4 m *leucaena* hedgerow treatment shows higher exchangeable cations levels than the other treatments.

### *Soil physical characteristics*

The no-till control treatment shows the highest soil bulk density (Table 3). At 0–5 cm depth, the soil bulk density was substantially lower in the alley cropped plots, particularly in the 2 m *Leucaena* hedgerow treatment. At 5–10 cm depth, the treatment with 4 m leucaena showed slightly lower bulk density. Water retention was better in the alley cropped and no-till plots than in the tilled control plot. Water infiltration was relatively high on the tilled control plot with loose topsoil. However, aggregates of the 4 m *leucaena* plot were more stable than those of the other plots. Although, the data suggest that soil physical conditions under the 4 m *Leucaena* treatment were comparatively good, however, variable soil physical effects would be obtained with different alley cropping practices.

The average runoff and soil loss data (Table 4) indicate, that soil erosion was generally low after 9 years of plot establishment. The highest rates of runoff and soil loss were observed in the tilled control treatment. Average soil losses from 1990–1992 (Table 4) were in the following order: tilled control > no-till control > 4 m *Gliricidia* > 2 m *Gliricidia* > 4 m *Leucaena*. Measurements of nutrient losses with runoff also showed low losses and no significant differences between treatments, except for *Leucaena* plots which showed higher nutrient loss (data not shown).

**Table 1.** Dry pruning biomass yield of hedgerows of *Leucaena leucocephala* and *Gliricidia sepium* as affected by interhedgerow spacing.

| Treatments | Pruning biomass (t/ha/yr) Prunings* | | | Wood | | |
|---|---|---|---|---|---|---|
| | 1990 | 1991 | Mean | 1990 | 1991 | Mean |
| 2 m *Gliricidia* | 6.88 | 5.16 | 6.02 | 3.41 | 3.21 | 3.31 |
| 4 m *Gliricidia* | 4.77 | 2.95 | 3.86 | 2.00 | 1.92 | 1.96 |
| 2 m *Leucaena* | 10.40 | 8.40 | 9.40 | 6.70 | 5.24 | 5.92 |
| 4 m *Leucaena* | 6.72 | 6.96 | 6.84 | 4.30 | 3.90 | 4.10 |

* Foliage + green branches

**Table 2:** Mean for 1988, 1989, 1990 and 1992 chemical characteristics of surface (0–15 cm) soil collected at the beginning of the cropping seasons.

| Treatments | pH-$H_2O$ | Org.C (g/kg) | Extract. Bray P-1 (mg/kg) | Exchangeable cations (cmol/kg) K | Ca | Mg |
|---|---|---|---|---|---|---|
| Control (no hedgerow) | | | | | | |
| Tilled | 5.1±0.22* | 5.9±1.0 | 8.52±3.6 | 0.21±0.03 | 1.06±0.26 | 0.20±0.02 |
| No-tillage | 5.3±0.13 | 8.9±1.8 | 6.12±2.3 | 0.28±0.02 | 1.61±0.87 | 0.28±0.05 |
| Alley cropped | | | | | | |
| 2 m *Gliricidia* | 5.0±0.15 | 8.0±1.9 | 8.31±3.3 | 0.33±0.05 | 1.10±0.17 | 0.25±0.05 |
| 4 m *Gliricidia* | 5.1±0.08 | 8.5±1.1 | 6.92±3.2 | 0.35±0.04 | 1.33±0.38 | 0.26±0.03 |
| 2 m *Leucaena* | 4.9±0.15 | 10.3±1.5 | 8.18±2.2 | 0.37±0.03 | 1.32.±0.46 | 0.25±0.03 |
| 4 m *Leucaena* | 5.0±0.15 | 11.1±0.9 | 6.69±2.4 | 0.47±0.03 | 1.62±0.37 | 0.32±0.04 |

* Mean ± standard deviation

**Table 3:** Physical properties of surface soil measured under cassava crop in 1992 as affected by tillage and alley cropping.

| Treatments | Bulk density ($Mg/m^3$) Soil depth (cm) 0–5 | 5–10 | Moisture content (g/kg) at 0–10 cm depth Matric potential (kPa) −10 | −1500 | Infiltration rate (cm/h) Interval 1 min. | 2 h | MWD[a] (mm) Soil depth (cm) 0–10 |
|---|---|---|---|---|---|---|---|
| Control (no hedgerow) | | | | | | | |
| Tilled | 1.36±0.06 | 1.40±0.08 | 121±10 | 23.5±10 | 179±81 | 32±14 | 0.84±0.48 |
| No-tillage | 1.40±0.14 | 1.50±0.08 | 131±10 | 39.0±10 | 167±42 | 16± 7 | 0.79±0.38 |
| Alley cropped | | | | | | | |
| 2 m *Gliricidia* | 1.31±0.03 | 1.43±0.11 | 131±30 | 37.3±10 | 122±27 | 27±18 | 0.41±0.60 |
| 4 m *Gliricidia* | 1.29±0.17 | 1.43±0.08 | 139±10 | 34.3± 3 | 134±31 | 2± 2 | 0.70±0.22 |
| 2 m *Leucaena* | 1.26±0.17 | 1.44±0.11 | 133±10 | 37.8±10 | 89±46 | 6± 8 | 0.96±0.36 |
| 4 m *Leucaena* | 1.28±0.14 | 1.38±0.08 | 144±50 | 52.3±20 | 63±40 | 9±14 | 1.42±0.94 |

* MWD = Mean weight diameter

**Table 4:** Runoff (mm) and soil loss (t/ha) as affected by tillage and alley cropping.

| Treatment | Runoff (mm) 1990 | 1991 | 1992 | Mean | Soil loss (t/ha) 1990 | 1991 | 1992 | Mean |
|---|---|---|---|---|---|---|---|---|
| Control (no hedgerow) | | | | | | | | |
| Tilled | 84.2 | 9.9 | 3.2 | 32.4 | 1.7 | 10.5 | 0.79 | 4.3 |
| No-tillage | 22.0 | 7.7 | 3.2 | 11.0 | 2.5 | 0.65 | 0.27 | 1.1 |
| Alley cropped | | | | | | | | |
| 2 m *Gliricidia* | 26.8 | 6.6 | 5.4 | 12.9 | 1.0 | 0.68 | 0.29 | 0.66 |
| 4 m *Gliricidia* | 14.6 | 7.7 | 2.1 | 8.1 | 1.4 | 0.75 | 0.20 | 0.78 |
| 2 m *Leucaena* | nd | nd | 2.1 | nd | nd | nd | 0.28 | nd |
| 4 m *Leucaena* | 13.4 | 16.5 | 4.3 | 11.4 | 0.7 | 0.44 | 0.35 | 0.50 |

nd = not determined because of faulty equipment.

### Crop performance

Grain yields of sole and intercropped maize (Table 5) were lower in the tilled and no-tillage treatments than in the alley cropped treatments. Grain yield of intercropped maize declined from 1990 to 1992 in all the treatments. The data in Table 5 indicate that the average yields are in the following order: 4 m *Leucaena* > 2 m *Leucaena* > 2 m *Gliricidia* > 4 m *Gliricidia* > no-tillage control > tilled control.

The root yield of intercropped cassava (Table 6) indicates, that yield was more stable over years on the tilled control and *Gliricidia* plots. The lowest cassava root yield was observed with the 2 m *Leucaena* treatment.

## Discussion and conclusion

Observations made between 1990 to 1992 (Table 4) confirm earlier reported results from the trial (*Kang* and *Ghuman*, 1991; *Lal*, 1989a, 1989b, 1989c) and other observations (*Paningbatan*, 1990; *Sajjapongse* and *Syers*, 1995), that alley cropping reduces water runoff and soil loss. Trial results (Table 4) also show that even with tillage, the alley cropped plots had lower runoff and soil loss than the no-till treatment. Nevertheles, runoff and soil loss were low even for the tilled control plot and this indicates that all plots had reached an equilibrium level for soil erosion. Following repeated tillings, the alleys have developed into mini-bench terraces, which helped to decrease runoff and erosion. In addition, manual tillage in the tilled control treatment may also reduce soil erosion due to surface roughness. The higher nutrient loss with runoff in the *Leucaena* as compared to other treatments, is related to the high nutrient content of the decomposing *Leucaena* mulch. Incorporation of the prunings in the soil could have reduced nutrient loss with runoff.

The lower surface soil bulk density and the higher soil moisture retention and aggregate stability in the alley cropped plots (Table 3), agree with earlier reported results (*Hulugalle* and *Kang*, 1990a; *Kang* and *Ghuman*, 1991). The high soil bulk density observed in the no-till plot may have resulted from consolidation of the fine particles and inadequate amount of mulch added to sustain this land use system (*Hulugalle* and *Kang*, 1990b). The better soil physical conditions in the alley cropped plots may be due to higher soil organic matter content (Table 2), which is conducive to higher earthworm acivity and physical soil improvement (*Hauser* and *Kang*, 1993). Higher pruning biomass and crop residue with *Leucaena* alley cropping contributed to the higher organic carbon and exchangeable cations in this treatment. Lower runof and erosion also contributed to higher soil organic carbon and exchangeable cation contents of the no-till and alley cropped treatment. The lower extractable P level in the alley

**Table 5:** Grain yield of maize (t/ha) as affected by tillage and different hedgerow species and spacing in alley cropping.

| Treatments | 1988 | 1989 | 1990* | 1991* | 1992* | Mean |
|---|---|---|---|---|---|---|
| Control (no hedgerows) | | | | | | |
| Tilled | 2.27±0.21 | 0.89±0.02 | 1.95±0.31 | 1.69±0.18 | 1.81±0.16 | 1.72 |
| No-tillage | 2.40±0.29 | 2.29±0.27 | 2.95±0.29 | 1.47±0.19 | 1.67±0.23 | 2.15 |
| Alley cropped | | | | | | |
| 2 m *Gliricidia* | 3.24±0.28 | 2.21±0.16 | 3.12±0.27 | 2.45±0.21 | 1.99±0.11 | 2.60 |
| 4 m *Gliricidia* | 2.83±0.34 | 2.02±0.19 | 3.13±0.22 | 2.56±0.31 | 2.38±0.29 | 2.58 |
| 2 m *Leucaena* | 3.45±0.12 | 2.96±0.80 | 3.44±0.24 | 2.05±0.19 | 2.03±0.19 | 2.78 |
| 4 m *Leucaena* | 3.13±0.17 | 2.73±0.75 | 3.24±0.17 | 2.81±0.28 | 2.71±0.14 | 2.93 |

* Intercropped with cassava

**Table 6:** Fresh root yield of intercropped cassava (t/ha) as affected by different hedgerows in alley cropping and by tillage treatments.

| Treatments | 1991 | 1992 | 1993 | Mean |
|---|---|---|---|---|
| Control (no hedgerows) | | | | |
| Tilled | 10.3±1.7 | 10.7±2.1 | 10.8±1.4 | 10.6 |
| No-tillage | 12.7±1.4 | 7.5±0.9 | 7.3±1.0 | 9.2 |
| Alley cropped | | | | |
| 2 m *Gliricidia* | 12.8±1.0 | 9.6±1.5 | 12.9±1.2 | 11.8 |
| 4 m *Gliricidia* | 13.5±1.8 | 10.9±1.4 | 11.0±1.3 | 11.8 |
| 2 m *Leucaena* | 8.6±1.1 | 6.0±2.0 | 5.9±0.6 | 6.8 |
| 4 m *Leucaena* | 16.4±2.4 | 12.3±2.6 | 8.7±1.0 | 12.5 |

cropped plots (Table 2), may in part be due to P uptake by the hedgerows and P removal when wood is harvested with the prunings. Despite the advantages of alley cropping, incorporation of prunings with high N content tends to lower soil pH partly as a result of depletion of exchangeable Ca in the soil, Similar Ca and pH relationship was also reported by *Lal* (1989c) and *Lulugalle* (1992).

The higher longterm maize yield with alley cropping confirmed other results (*Atta-Krah*, 1990; *Kang* and *Ghuman*, 1991; *Kang*, 1993). The decline in grain yield of intercropped maize from 1990 to 1992, may be due to increased nutrient removal by the intercropped cassava. With *Gliricidia*, interhedgerow spacing has no effect on intercropped maize and cassava yields (Tables 5 and 6). However, with *Leucaena* that has more vigorous growth, closer hedgerow spacing of 2 m lowers maize (Table 5) and particularly cassava root yield (Table 6). This is mainly due to light competition (*Lawson* and *Kang*, 1990). These result indicate, that alley cropping intercropped maize and cassava with *Leucaena* at 2 m interhedgerow spacing is not recommended. The lower maize and cassava yields (Tables 5 and 6) in the no-till treatment can be attributed to the high soil bulk density in this treatment (Table 3).

From the results of this longterm study it can be concluded, that the rate of soil productivity decline was lower with alley cropping than the other two control treatments with continuous cropping. Rehabilitation of soil productivity with fallowing would most likely be faster in the alley cropped plots as compared to the no-till and tilled control treatments.

## References

*Atta-Krah, A. N.* (1990): Alley farming with Leucaena: Effect of short grazed fallows on soil fertility and crop yields. Exp. Agric. 26:1–10.

*Chalinor, D.* (1968): Alterations of surface soil characteristics by four tree species. Ecology 49:286–290.

*Chijioke, E. O.* (1980): Impact on soil of fast-growing species in lowland humid tropics. FAO Forestry Paper No. 21. FAO, Rome, Italy.

*Getahun, A., Wilson, G. F.* and *Kang, B. T.* (1982): The role of trees in farming systems in the humid tropics. In: L. H. MacDonald (ed.): Agroforestry in the African humid tropics. The United Nations Univ., Tokyo, Japan. pp. 28–35.

*Hauser, S.* and *Kang, B. T.* (1993): Nutrient dynamics, maize yield and soil organic matter in alley cropping. In: Mulongoy, K. and R. Merckx. (eds.) Soil organic matter dynamics and sustainability of tropical agriculture, John Wiley and Sons, Chichester, UK. pp. 215–222.

*Hulugalle, N. R.* (1992): Amelioration of a highly degraded tropical Alfisol. I. Changes in soil physical and chemical properties 1989–91. Land. Degr. and Rehab. 3: 141–152.

*Hulugalle, N. R.* and *Kang, B. T.* (1990a): Effect of hedgerow species in alley cropping systems on surface soil physical properties of an Oxic paleustalf in southwestern Nigeria. Journal of Agricultural Science, Cambridge 114 301–307.

*Hulugalle, N. R.* and *Kang, B. T.* (1990b): Effect of land management system on surface soil physical properties of an Alfisol. In: A. Singh et al. (eds.) International symposium on water erosion, sedimentation and resource conservation. Central Soil and Water Conservation Research and Training Institute, Dehradun, Uttar Pradesh, India.pp. 390–402.

*IITA (International Institute of Tropical Agriculture)* (1982): Automated and semi-automated methods for soil and plant analysis. Manual series No. 7, IITA, Ibadan, Nigeria. 32 pp.

*Juo, A. S. R.* and *Lal, R.* (1977): Effects of fallow and continuous cultivation on soil chemical and physical properties of an Alfisol in western Nigeria. Plant and Soil 47: 567–584.

*Kang, B. T.* (1993): Alley cropping: Past achievements and future directions. Agroforestry Systems 23: 141–155.

*Kang, B. T., Akinnifesi, F. K.* and *Lapido, D. O.* (1994): Performance of selected woody agroforestry species grown on an Alfisol and an Ultisol in the humid lowland of west Africa and their effects on soil properties. Journal of Tropical Forest Science 7: 303–312.

*Kang, B. T.* and *Ghuman, B. S.* (1991): Alley cropping as a sustainable system. In: Moldenauer, W. C. et al. (eds.) Development of conservation farming on hill slopes. Soil and Water Conservation Society, Ankeny, Iowa, USA. pp. 172–184.

*Kang, B. T., Reynolds, L.* and *Atta-Krah, A. N.* (1990): Alley farming. Advances In Agronomy 43: 315–357.

*Kang, B. T.* and *vanden Beldt, R. J.* (1990): Agroforestry systems for sustained crop production in the tropics, with special reference to west Africa. In: Moore, E. (ed.) Agroforestry Land Use Systems. Nitrogen Fixing Tree Association, Waimanalo, Hawaii, USA. pp. 13–33.

*Kang, B. T., Wilson, G. F.* and *Lawson, T. L.* (1984). Alley cropping. A stable alternative to shifting cultivation. IITA, Ibadan, Nigeria, p. 23.

*Lal, R.* (1989a): Agroforestry systems and soil surface management of a tropical alfisol: I. Soil moisture and crop yields. Agroforestry Systems 8: 7–29.

*Lal, R.* (1989b): Agroforestry systems and soil surface management of a tropical alfisol: II. Water runoff, soil erosion and nutrient loss. Agroforestry Systems 8: 97–111.

*Lal, R.* (1989c): Agroforestry systems and soil surface management of a tropical alfisol: III. Changes in soil chemical properties. Agroforestry Systems 8: 113–132.

*Lal, R.* (1989d): Agroforestry systems and soil surface management of a tropical alfisol: IV. Effects on soil physical and mechanical properties. Agroforestry Systems 8: 197–215.

*Lal, R.* (1989e): Agroforestry systems and soil surface management of a tropical alfisol: V. Water infiltrability, transmissivity and soil water sorptivity. Agroforestry Systems 8: 217–238.

*Lal, R.* (1990): Soil erosion in the tropics: Principles and management. MC Graw-Hill, New York, USA p. 588.

*Lawson, T. L.* and *Kang, B. T.* (1990): Yield of maize and cowpea in an alley cropping system in relation to available light. Agriculture and Forest Meteorology 52: 347–357.

*Lundgren, B.* (1977): Soil conditions and nutrient cycling under natural and plantation forests in Tanzanian highlands. Report on Forest Ecology and Forest Soils, Vol. 31. Department of Forest Soils, Swedish University of Agriculture Science, Uppsala, Sweden.

*Moormann, R. F., Lal, R.* and *Juo, A. S. R.* (1975): The soils of IITA. Technical Bulletin No. 3. IITA, Ibadan, Nigeria. p. 48.

*Sajjapongse, A.* and *Syers, J. K.* (1995): Tangible outcomes and impacts from the ASIALAND management of sloping lands network. In: International Workshop on Conservation Farming for Sloping Uplands in Southeast Asia: Challenges, Opportunities and Prospects. Proc. No 4. IBSRAM, Bangkok, Thailand. pp. 3–14.

*Obi, J. K.* and *Tuley, P.* (1973): The bush fallow and ley farming in the oil palm belt of southeastern Nigeria. Agricultural Research Station Umudike and Land Resources Division. Misc. Rep. No 161, ODM, UK p. 24.

*Okigbo, B. N.* (1976): Role of legumes in small holdings of the humid tropics of Afrika. In: J. M. Vincent, A. S. Whitney and J. Bose (eds.). Exploiting the Legume Rhizobium Symbiosis in Tropical Agriculture. College of Agriculture, Misc. Publ. 145, Univ. of Hawaii, USA pp. 367–377.

*Paningbatan, E. P.* (1990): Alley cropping for managing soil erosion in sloping lands. Transaction 14th. International Congress of Soil Science, Kyoto, Japan. Vol. VII. pp. 367–377.

*Roose, E.* (1981): Dynamique actuelle de sols ferralitiques et ferrugineux tropicaux d'Afrique occidentale. ORSTROM, Paris, France.

*Yamoah, C. F., Agboola, A. A., Wilson, G. F.* and *Mulongoy, K.* (1986): Soil properties as affected by the use of leguminous shrubs for alley cropping with maize. Agriculture, Ecosystems and Environment 18: 167–177.

*Young, A.* (1990): Agroforestry for soil conservation. CAB International, Wallingfod, UK p. 276.

# Longterm studies of soil fertility in cacao-shade trees-agroforestry systems: Results of 15 years of organic matter and nutrients research in Costa Rica

**Hans W. Fassbender**

Technical Forest Faculty Hildesheim/Holzminden, Büsgenweg 1a, 37077 Göttingen, Germany

## Abstract

At the "Experimento Central" of CATIE (Centro Agronomico Tropical de Enseñanza, Turrialba, Costa Rica) long term studies have been carried out to determine productivity and the soil fertility (organic matter and nutrients) in agroforestry systems. In this paper results are presented and discussed as examples of sustainable agroforestry, using the systems of cacao (*Theobroma cacao*) under Laurel (*Cordia alliodora*) or Poro (*Erythrina poeppigiana*) as a case of study.

The following productivity and soil fertility parameters were found:

- Average cacao bean harvest during 13 years is very stable, reaching average values of 718 and 745 kg $ha^{-1}$ $a^{-1}$ under *C. alliodora* and *E. Poeppigiana*, respectively.
- Growth values for *Cordia alliodora* are very high, reaching 30 cm average diameter and 20 m height and a timber volume of 140 $m^3$ $ha^{-1}$ at age 15 years.
- Accumulation of organic matter, measured for the different species (leaves, branches, trunks, roots, fruits and litter) at an age of 5 years (1982) and of 10 years (1987) are as follows: 50.3 and 110.6 t $ha^{-1}$ for *T. cacao/C. alliodora* and 39,1 and 87.3 for *T. cacao/E. poeppigiana*, respectively.
- Net primary production between ages 6 and 10 of the systems are in average 28.2 and 36.3 t $ha^{-1}$ $a^{-1}$ for *T. cacao/C. alliodora* and *T. cacao/ E. poeppigiana*.
- The natural leaf fall and residues of prunings, studied at age 6 to 10 and also at age 14 to 15 years indicate a total production of plant residues of 11.4 and 22.9 t $ha^{-1}$ $a^{-1}$ for the given agroforestry systems. The latter value under *E. poeppigiana* is especially related to pruning regime of the leguminose tree.
- The soil organic matter increased in the first ten experimental (1977–1987) years by 15.3 and 41.6 t $ha^{-1}$ $a^{-1}$. The soil N-reserves also increased, specially under *E. poeppigiana*.
- The losses of nutrients by leaching are negible.

Therefore it is concluded that the agroforestry systems analysed maintain their soil organic matter status and promote efficient nutrient cycling. They are also sustainable and appropiate for soil fertility conservation.

## Introduction

High expectations have been raised in recent years about the potential of agroforestry as a major land management alternative for sustainable production and for the maintenance of soil fertility in the tropics (*Beer* et al., 1987; *Beer* et al., 1990; *Fassbender* et al., 1988, 1991; *Fassbender*, 1993; *Nair*, 1987; *Sanchez*, 1987; *Young*, 1987). Most recently the discussion on appropiate land use systems in the tropics has been dominated by the term of "sustainability". The CGIAR-TAC-report describes sustainable agriculture as on which "involves the successful management of resources for agriculture to satisfy changing human needs while maintaining or enhancing the quality of environment and conserving natural resources" (*Javier* and *Rendborg*, 1988).

During research activities of the agroforestry cooperation project GTZ/CATIE (German Agency for Development, Gesellschaft für Technische Zusammenarbeit / Centro Agronómico Tropical de Investigación y Enseñanza, Turrialba, Costa Rica) long term studies has been carried out in the Experimento Central of CATIE to provide basis for the evaluation of systems productivity and soil fertility conservation with models of organic matter, water and nutrient cycles for the following agroforestry systems:

- Cacao (*Theobroma cacao*) with shade of Laurel (*Cordia alliodora*) or Poro (*Erythrina poeppigiana*);
- Coffee (*Coffea arabica*) under the same shade tress and
- Stargrass (*Cynodon plectostachyus*) associated with the same shade trees.

In this paper, the tree and crop production values along with the distribution and fluxes of organic material and nutrients in the systems of *Theobroma cacao* with *Cordia alliodora* and *Erythrina peoppigiana* are presented, synthesized into models and discussed as examples of sustainable agroforestry systems for the lowland humid tropics.

## The "Experimento Central" of CATIE

Details of the "Experimento Central" at CATIE, Turrialba, Costa Rica, in which the systems were studied, have already been described (*Alpizar* et al., 1986; *Beer* et al., 1987, 1990; *Fassbender*, 1993; *Fassbender* et al., 1988, 1991; *Heuveldop* et al., 1988; *Imbach* et al., 1989).

The Turrialba region is located at 9° 53 min N and 88° 38 min W and has an altitude of 600–650 m above sea level. The annual average temperature (23 years) is 22.3 °C; the minimum and maximum temperatures reach 17.7 °C and 27.0 °C respectively. The average annual rainfall (38 years) is 2648 mm with an average of 246 days of rain per year. Average monthly rainfall of under 100 mm was recorded only in the month of March (70 mm). The monthly average relative humidity is 87.6 percent.

The soil of the experimental site is classified as typic humitropept, fine, halloysitic and isohyperthermic. The texture is clay loam. The soil has level topography, poor drainage, and is of fluvial-lacustrine origin.

In August 1977 cacao (*Theobroma cacao*) hybrids (Catongo x Pound) were planted at 3 x 3 meter spacing (1111 trees $ha^{-1}$) with shade trees of laurel (*Cordia alliodora*) and poro (*Erythrina poeppigiana*) interplanted at 6 x 6 meter spacing (278 trees $ha^{-1}$) in two split plots measuring 18 x 36 meters in random blocks.

The agronomical treatment of the experiments is representative of the farms in the Turrialba region. An initial fertilizer application (13.9 kg N $ha^{-1}$, 18.2 kg P $ha^{-1}$, 11.5 kg K $ha^{-1}$) was given, the long term the annual rate of application averaged 87.5 kg N, 34.6 kg P and 34.9 K per hectare.

Tables 1 and 2 contains the results from the soil analysis of the plots of *T. cacao* under the *E. poeppigiana* at the beginning of the experiment (1977). The pH values are homogeneous and acid. The content of organic mater is high and decreases normally with soil depth. These characteristics are similar for total nitrogen and phosphorus. The values of the exchangeable bases are uniformly high, especially of potassium. The anotated standard deviations generate high coefficients of variation of all of the findings, ranging between 10 and 50 percent.

The changing values of organic matter, nitrogen and exchangeable bases (K, Ca and Mg) after nine years of the experiment are sumarized in Table 2.

## Cocoa production

*Theobroma cacao* pods were harvested every 15 days beginning with the second experimental year (June 1979) until 1991. Separate fresh weights of beans and husks, produced by each of 32 cacao trees per plot, were determined from 1979–1983 in order to determinate the procentual distribution of both fruit parts. Over two years (1981 and 1982) samples were taken from each harvest to determine oven dry weights.

In both systems *T. cacao* production started at age 2 years and gradually increased to age 6 years (1983), after which it stabilized (Table 3). On average, the cocoa beans made up 41% of the dry pod weights. The long term average production valus over 15 years in both systems (1979–1991) are 718 and 745 kg $ha^{-1}$ $a^{-1}$ of cacao beans under *C. alliodora* and *E. poeppigiana* respectively. The values are statistical comparable, there are no differences in the production of cacao of both systems. The respective value for the non-utilizable husk is also comparable for both systems, around 1000 kg $ha^{-1}$ $a^{-1}$.

**Table 1:** Physical characteristics of the soil of the *T. cacao/E. poeppigiana* system.

| Depth (cm) | Apparent density g $cm^{-3}$ | Sand % | Silt % | Clay % | Moisture content at 1.5 kPa | Moisture content at 0.33 kPa | Available water % |
|---|---|---|---|---|---|---|---|
| 0–15 | 1.15 | 29 | 32 | 38 | 21.0 | 40.7 | 19.7 |
| 15–30 | 1.29 | 32 | 31 | 37 | 31.1 | 44.0 | 12.9 |
| 30–45 | 1.24 | 26 | 31 | 43 | 25.7 | 42.1 | 16.4 |

**Table 2:** Chemical characteristics of the soils of the *T. cacao/E. poeppigiana* system (standard deviations in parenthesis).

| Depht (cm) | pH 1977 | pH 1986 | Humus % 1977 | Humus % 1986 | % Total nitrogen 1977 | % Total nitrogen 1986 | ppm Total P 1977 |
|---|---|---|---|---|---|---|---|
| 0–15 | 3.80(0.08) | 4.26(0.05) | 4.76(0.58) | 5.50(0.49) | 0.22(0.02) | 0.25(0.01) | 583(49) |
| 15–30 | 3.88(0.09) | 4.22(0.03) | 3.87(0.32) | 4.23(0.56) | 0.19(0.01) | 0.19(0.02) | 616(68) |
| 30–45 | 4.12(0.22) | 4.22(0.07) | 2.23(0.29) | 3.40(0.52) | 0.11(0.01) | 0.16(0.02) | 559(57) |

| | exch. K meq. 100 g soil 1977 | exch. K meq. 100 g soil 1986 | exch. Ca meq. 100 g soil 1977 | exch. Ca meq. 100 g soil 1986 | exch. Mg meq. 100 g soil 1977 | exch. Mg meq. 100 g soil 1986 |
|---|---|---|---|---|---|---|
| 0–15 | 0.57(0.20) | 0.28(0.04) | 2.40(0.23) | 3.29(0.63) | 0.95(0.13) | 1.34(0.26) |
| 15–30 | 0.27(0.04) | 0.17(0.05) | 2.65(0.21) | 2.62(0.38) | 0.98(0.05) | 1.09(0.22) |
| 30–45 | 0.17(0.06) | 0.10(0.04) | 3.27(0.30) | 1.88(0.85) | 1.03(0.05) | 0.78(0.28) |

**Table 3:** Annual *Theobroma cacao* production under the shade of *Cordia alliodora* or *Erythrina poeppigiana* (Dry weights cocoa beans; kg $ha^{-1}$ $a^{-1}$).

| | *Cordia alliodora* | *Erythrina poeppigiana* |
|---|---|---|
| 1979 | 105 | 72 |
| 1980 | 452 | 474 |
| 1981 | 284 | 579 |
| 1982 | 691 | 758 |
| 1983 | 962 | 1040 |
| 1984 | 977 | 992 |
| 1985 | 909 | 1134 |
| 1986 | 1370 | 995 |
| 1987 | 960 | 1123 |
| 1988 | 851 | 831 |
| 1989 | 783 | 527 |
| 1990 | 837 | 711 |
| 1991 | 491 | 662 |
| Average 15 years of the systems 13 years cacao production | 718 | 745 |

**Table 4:** Growth of *Cordia alliodora* (Diameter, d; hight h and mean annual increase, MAI) associated with *Theobroma cacao* (n = 16; standard deviation in parenthesis)

| Age | Diameter | | Height | | Stem |
|---|---|---|---|---|---|
| | d | MAI | h | MAI | volume* |
| (yr) | (cm) | (cm $a^{-1}$) | (m) | (m $a^{-1}$) | ($m^3$ $ha^{-1}$) |
| 1.2 | 4.1(1.6) | 3.4 | – | – | – |
| 1.5 | 5.8(1.7) | 5.7 | 5.0(1.2) | 3.3 | |
| 2.2 | 8.7(1.7) | 4.1 | 6.3(1.1) | 1.9 | |
| 3.7 | 14.0(2.1) | 3.5 | 8.7(1.0) | 1.6 | |
| 4.5 | 16.1(2.1) | 2.6 | 10.1(1.1) | 1.7 | 22 |
| 5.3 | 17.8(2.2) | 2.1 | 10.6(1.4) | 0.6 | 32 |
| 5.8 | 17.7(2.7) | 0.0 | 10.9(1.3) | 0.6 | 32 |
| 6.3 | 18.7(2.4) | 2.0 | 11.5(1.7) | 1.2 | 39 |
| 7.3 | 20.1(2.9) | 1.4 | 12.6(2.5) | 1.1 | 47 |
| 8.3 | 21.7(3.2) | 1.6 | 13.7(2.8) | 1.1 | 61 |
| 9.3 | 22.8(3.5) | 1.1 | 13.7(2.7) | 0.0 | 66 |
| 10.5 | 24.1(3.8) | 1.1 | 15.0(2.8) | 1.1 | 78 |
| 12.4 | 26.6(4.1) | 1.3 | 15.9(2.8) | 0.5 | 100 |
| 13.3 | 27.2(4.1) | 0.7 | 16.5(3.1) | 0.6 | 110 |
| 14.8 | 28.6(4.3) | 0.9 | 17.5(3.0) | 0.8 | 127 |
| 15.6 | 29.6(4.5) | 1.2 | 18.0(3.1) | 0.5 | 140 |

* Estimated from volume table using the average values of d and h, Somarriba and Beer (1987).

## Growth rates and timber production of laurel *(Cordia alliodora)*

Diameter at breast height (DBH; cm) and total height (h; m) of eight *C. alliodora* per plot have been measured at different intervals (once or twice per year) since 1978, and the average annual growth rates (MAId and MAIh) for each year were calculated. Stem volume (V) increments were determined with a volume table using d and h values (*Somarriba* and *Beer*, 1978). Therefore the timber production was evaluated for a period of 15 years.

The average data of diameter and heigth of the *Cordia alliodora* from Table 4 are the basis for the calculated values for timber production. The average diameter of almost 30 cm at age 15.6 years is typical for *C. alliodora* in the Atlantic zone of Costa Rica, but the average height of 18 m is less than would be expected. As a consequence the BHD-h relationship at this study site is only just within the measured distribution for this species in this zone (*Somarriba* and *Beer*, 1987).

At age 10.5 years, estimated *C. alliodora* stem volumes totalled 77.6 $m^3$ $ha^{-1}$, equivalent to a growth rate of 7.4 $m^3$ $ha^{-1}$ $a^{-1}$, those at age 15 years corresponded to 140 $m^3$ $ha^{-1}$ and 9.0 $m^3$ $ha^{-1}$ $a^{-1}$ (Table 4). The inventory data at age 10 (see latter), gives also a value of 9 $m^3$ $ha^{-1}$ $a^{-1}$, which corroborates the estimated growth rate even thought it is based on a volume table (*Sommarriba* and *Beer*, 1987), which may not be applicable to the abnormal BDH-h relationship at this site.

## Measurements of the biomass

Standing biomass accumulation of the three species, (compartments: leaves, branches and trunks) was measured in February 1982 at age 5 and again in May and June 1987, at age 10 (*Alpizar* et al., 1986; *Beer* et al., 1990).

The accumulation of organic matter in the biomass of the studied agroforesty system has been summarized for the determination at age 5 and 10 years in Figure 1 and also in total values in Table 5. In the last five years, the increase in the standing *T. cacao* biomass (excluding roots) under *C. alliodora* was apparently greater than the increase in shade tree biomass (Figure 1) but the latter value was affected by the seasonal shade tree leaf fall. The increase for *T. cacao* was due to a large change in branch biomass (20 t.ha-1), the compartment which showed the greatest increase between age 5 and 10 in this system. The increase in *T. cacao* biomass under *E. poeppigiana* was significantly less than under *C. alliodora* (18.9 vs 26.1 t.ha-1) and there was significantly less *T. cacao* biomass under *E. poeppigiana* at age 10 years. Nevertheless, in percentage terms (1982–1987 increments as % of 1982 values, Figure 1) the corresponding increases in total, stem or branch biomass of *T. cacao* are similar in both systems.

## Litter production

To determine natural litterfall (leaves and branches), twelve litter traps (1 $m^2$) were randomly placed in only one plot of each system. Weekly measurements were made continuously from November 1981 until October 1987 at age 6 to 10 years of the systems (*Heuveldop* et al., 1988). With some small changes in the collecting traps the litterfall was studied between june 1992 and june 1993, at age 15 years of the systems.

Total pruning residues per tree were determined from the same *T. cacao* and *E. poeppigiana* which were measured during the 1987 standing biomass inventory and also

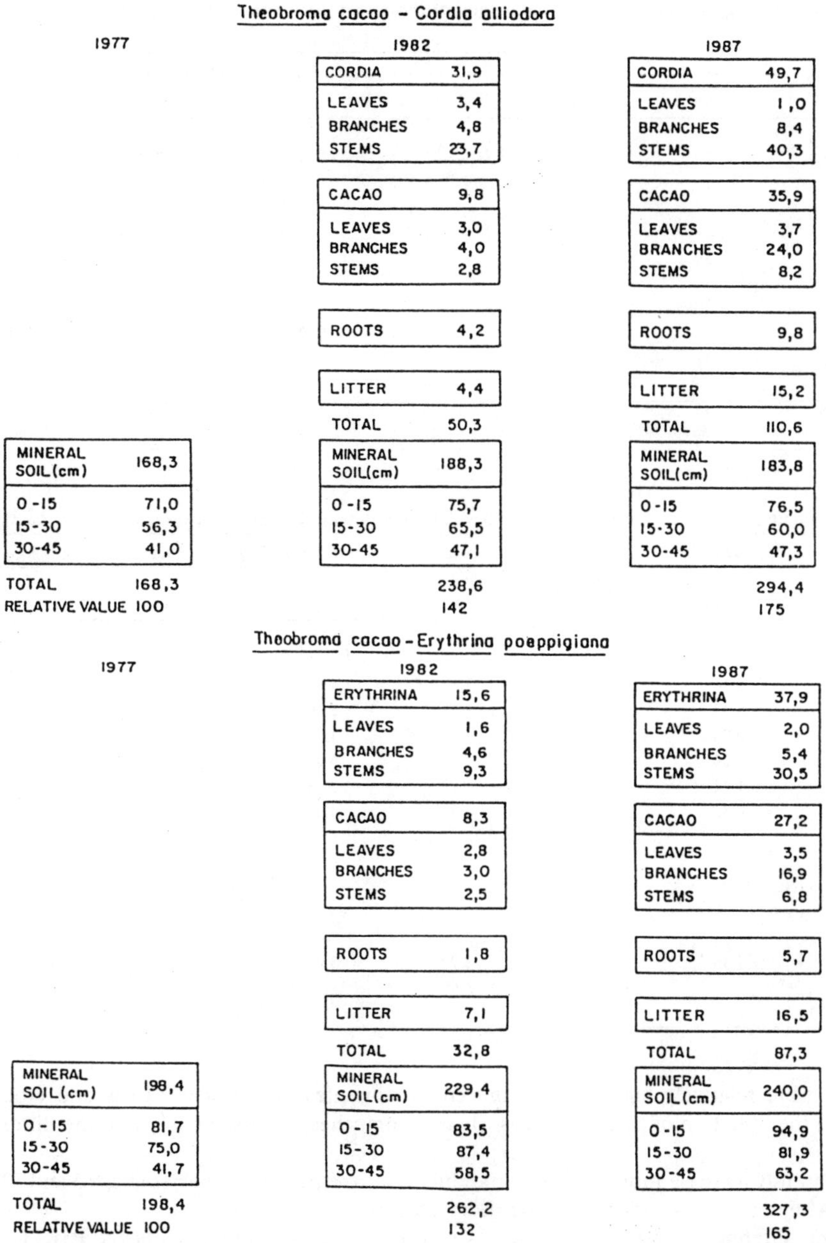


**Figure 1:** Organic material reserves in the agroforestry systems of *Theobroma cacao*-*Cordia alliodora* and *Theobroma cacao*-*Erythrina poeppigiana* at ages 5 and 10 years (t ha$^{-1}$).

between March 1985 and May 1987 (3 prunings in 26 months) and for *E. poeppigiana* between June 1986 and November 1987 (3 prunings in 17 months).

The values in Table 6 demostrate the variability of the natural litterfall in both studied systems at age 6 to 10 years. The pruning litterfall is presented as the average of the different years of studies. The total amount of litter production in these age (6 to 10 years) averaged 11.40 and 22.86 t ha$^{-1}$ a$^{-1}$ for *T. cacao*/*C. alliodora* and *T. cacao*/*E. poeppigiana* respectively. The results in natural and pruning residues at age 14 to 15 years reached a total of 10.69 and 19.34 t ha$^{-1}$ a$^{-1}$ respectively and are between of the variability founded for age 6 to 10 years.

According to *Heuveldop* et al. (1988) the decomposition rate is exponential, after a year the plant residues are decomposed and incorporated to humus and mineral nutrients in the soil.

There is a general tendency for soil organic material concentrations to increase with time (Figure 1, Table 1), the highest values being attained in the *Erythrina* system. However, these temporal increases are not statistically significant, (5% level), in part because the soil organic mate-

**Table 5:** Organic matter (O. M., t $ha^{-1}$) and nutrient accumulation (kg $ha^{-1}$) in the systems of *Theobroma cacao* with *Cordia alliodora* and *Erythrina poeppigiana*.

| | *T. cacao/C. alliodora* | | | | *T. cacao/E. poeppigiana* | | | |
|---|---|---|---|---|---|---|---|---|
| | O. M. | N | P | K | O. M. | N | P | K |
| Age 5 years | | | | | | | | |
| Tree | 31.9 | 233 | 29 | 243 | 21.9 | 156 | 16 | 122 |
| Crop | 9.8 | 99 | 12 | 99 | 8.3 | 103 | 10 | 49 |
| Roots | 4.2 | 39 | 4 | 21 | 1.8 | 22 | 2 | 18 |
| Litter layer | 4.4 | 76 | 5 | 10 | 7.1 | 112 | 9 | 12 |
| Total | 50.3 | 447 | 50 | 373 | 39.1 | 393 | 37 | 201 |
| Age 10 years | | | | | | | | |
| Tree | 49.7 | 245 | 16 | 103 | 37.9 | 206 | 20 | 283 |
| Crop | 35.9 | 223 | 42 | 248 | 27.2 | 249 | 29 | 195 |
| Roots | 9.8 | 99 | 6 | 57 | 5.7 | 63 | 2 | 27 |
| Litter layer | 15.2 | 198 | 15 | 11 | 16.5 | 204 | 19 | 15 |
| Total | 110.6 | 765 | 79 | 419 | 87.3 | 722 | 70 | 520 |

**Table 6:** Natural litterfall and pruning residue inputs in the systems of *Theobroma cacao* with Cordia alliodora and *Erythrina poeppigiana* (t $ha^{-1}$ $a^{-1}$).

| | *T. cacao/C. alliodora* | | | | *T. cacao/E. poeppigiana* | | | |
|---|---|---|---|---|---|---|---|---|
| | leaves | leaves | branches | TOTAL | leaves | leaves | branches | TOTAL |
| Age 6–10 years | | | | | | | | |
| Natural | | | | | | | | |
| 6 years | 2.96 | 2.26 | 0.34 | 5.56 | 3.16 | 4.83 | 0.71 | 8.70 |
| 7 year | 4.31 | 2.43 | 0.37 | 7.11 | 3.73 | 5.36 | 0.72 | 9.81 |
| 8 year | 5.46 | 3.03 | 0.70 | 9.19 | 4.78 | 4.77 | 0.88 | 10.43 |
| 9 year | 4.35 | 3.95 | 1.04 | 9.34 | 4.13 | 4.98 | 0.75 | 9.86 |
| 10 year | 4.94 | 2.71 | 1.72 | 9.37 | 3.83 | 3.08 | 0.66 | 7.57 |
| Average | 4.40 | 2.88 | 0.83 | 8.11 | 3.93 | 4.62 | 0.74 | 9.29 |
| Pruning | | | | | | | | |
| by species | 3.29 | – | – | 3.29 | 3.80 | – | 9.77 | 13.57 |
| SYSTEM TOTAL | | 11.40 | | | | 22.86 | | |
| Age 14–15 years | | | | | | | | |
| Natural | 1.73 | 4.22 | 1.58 | 7.53 | 2.22 | 3.24 | 0.87 | 6.33 |
| Pruning | 3.13 | – | | 3.13 | 3.69 | 9.32 | | 13.01 |
| SYSTEM TOTAL | | 10.69 | | | | 19.34 | | |

rial concentrations (0–15 cm; 4%) were relatively high (*Fassbender* and *Bornemisza*, 1986) at the initiation of this experiment.

Soil organic material reserves, apparently increase more (1977–1987) under *E. peoeppigiana* (41.6 t $ha^{-1}$) than under *C. alliodora* (15.5 t $ha^{-1}$) but again the differences are not significant. However in 1987, there were some significant inter-system differences when comparing the same soil horizons (0–15; 30–45 cm), presumably attributable to the greater litter inputs (pruning residues) under *E. poeppigiana* (Figure 1).

### Models of organic matter

For the models of organic matter presented in this publication, the following values were considered:

– above ground biomas at age 5 (*Alpizar* et al., 1986) and 10 (*Beer* et al., 1990), using non-destructive methods for the compartimentes of the shade trees and crop trees (leaves, branches, stems, pods) and the litter layer;

– fine root biomass, irrespective of species at age 5 and 10 (literature above)

– reserves of organic matter and nutrients (N, P, K, Ca and Mg) in the mineral soil and vegetation, based in chemical analysis, at ages 0, 5 and 10 years;

– cacao production (beans and husks) during 5 years (1983–1987) (*Heuveldop* et al., 1988 and *Beer* et al., 1990;

– growth rates of *C. alliodora* in the same period measured as stem diameter at breast height, total hight of the trees and calculated timber (*Heuveldop* et al., 1988, *Beer* et al., 1990);

– natural leaf fall measured for 5 years (1983–1987) and pruning residues of *T. cacao* and *E. poeppigiana* measured over 2 years (1985–1987) (*Beer* et al., 1990);

– nutrient input with rainfall and output with percolating water measured during one year (*Imbach* et al., 1989);
– rate of fertilization with N, P and K applied to the experiment (*Alpizar* et al., 1986).

Figure 2 shows the five-yearly models (age 6 to 10) for organic matter in the agroforestry systems *T. cacao/C. alliodora* and *T. cacao/E. poeppigiana* in Turrialba, Costa Rica. The reserves are the average of the biomass measurements, litter and humus at ages 6 and 10; the transfer values are the averages of the determinations during the five year period. Values of *T. cacao* pods are shown in Figure 2 in brackets to indicate that the cacao pod biomass was not always present but is transitory. The data refer to the removal of biomass as an annual mean over five years (1983–1987).

Besides the information directly available in Figure 2, from the model other important criteria can be derived for the characterisation of the agroforestry systems as:
– Net primary production
– Productivity and export indices
– Recycling indices

Net primary productivity (NPP), between ages 6 to 10 years, was calculated by adding the following average annual values in t ha$^{-1}$ a$^{-1}$ for the principal parameters:
– Biomass accumulation (both standing and roots); derived from the difference between the 5 and 10 year inventories.
– Root turn-over; annual regeneration of all fine roots (< 5 mm) was assumed, using the average value between age or between age 5 and 10.
– *T. cacao* production (husks and beans) for the period 1983–1987.
– Plant residue production; litterfall and pruning residues (leaves and branches) as the value is an average of the annual measurements during this period (1983–1987).

During the second quinquennial period, the values for standing biomass accumulation, fine root turn-over and litterfall, for *T. cacao/C. alliodora*, are similar to the highest published values for tropical humid lowland forests (*Fassbender*, 1993) which is to be expected given the natural fertility and management (drainage, fertilization) of these plots. In the same period, the NPP of 35 t ha$^{-1}$ a$^{-1}$ for the *E. poeppigiana/T. cacao* system is equal to the maximum of the range of values for the NPP of tropical rain forests (10–35 t ha$^{-1}$ a$^{-1}$, *Fassbender*, 1993, *Jordan*, 1983, 1985).

The export index can be used as another parameter for the characterization of the productivity and sustainability of the systems, it can be calculated as the percent of bio-

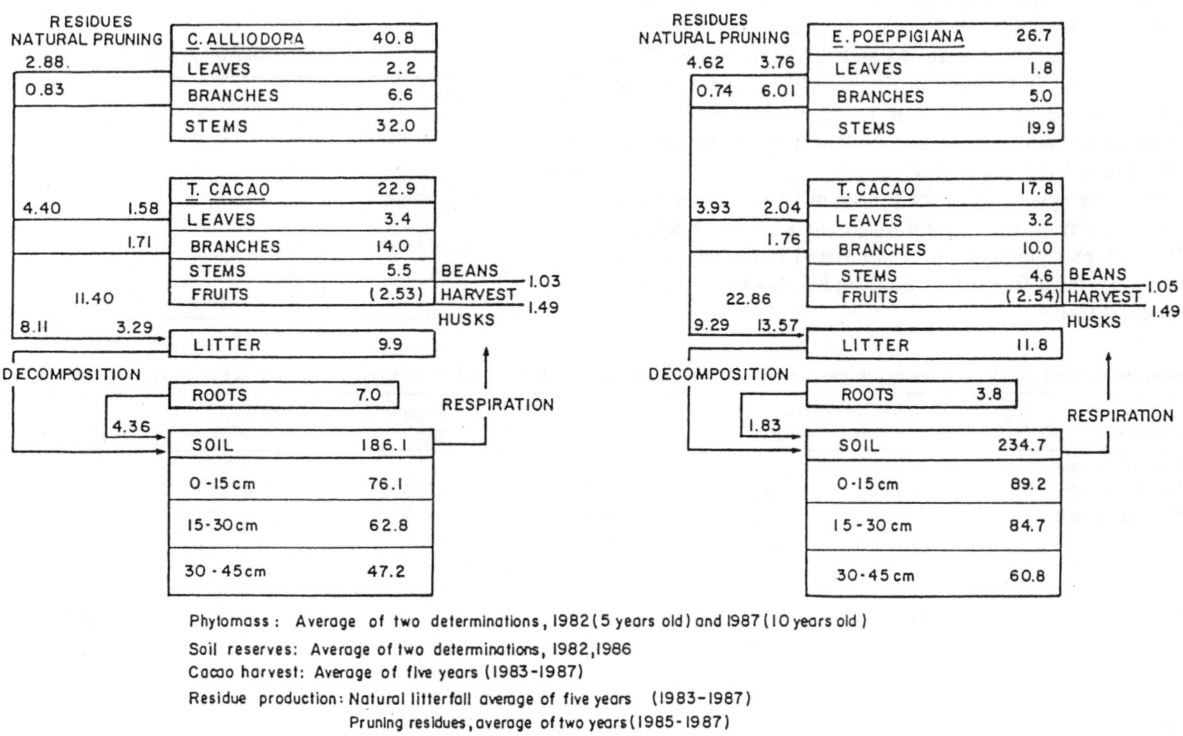


**Figure 2:** Quinquennial average models for organic matter in the systems *T. cacao-C. alliodora* and *T. cacao-E. poeppigiana* (age 6–10 years, reserves t ha$^{-1}$, transfers t ha$^{-1}$ a$^{-1}$).

mass extracted with yields of the system. The annual harvest export of cacao pods as a proportion of the total biomass (Figure 2) is only 3.1 and 4.2% for the *Cordia alliodora* and *Erytrina poeppigiana* systems respectively, or when expressed as a proportion of the NPP (Table 5) 9 and 7% respectively.

The recycling index represents the procentual value of litter production (natural and pruning) of the biomass. The annual production of above-ground plant residues as percentages of standing biomass (Fig. 2) are as follows:

| | *C. alliodora* | *T. cacao* | *E. poeppigiana* | *T. cacao* |
|---|---|---|---|---|
| by species | 9.1 | 33.6 | 56.7 | 43.4 |
| by system | 17.9 | | 51.4 | |

## Nutrient pools in the agroforestry systems

In order to study other aspects of soil fertility similar evaluations and models can be compiled for nutrients in agroforestry systems (*Fassbender*, 1993; *Fassbender* et al., 1988). However, nutrient models are more complex because of the different forms (total, extractable, available) of the nutrients in the mineral soil, management practices (fertilization, liming), and some important ecological processes (N fixation by legumes, input with rainfall and output with leaching water).

The accumulation of total N in biomass shows fairly well defined tendencies for each system (Table 5). At age 5, *T. cacao/C. alliodora* accumulated 447 kg $ha^{-1}$ of N compared to 393 kg $ha^{-1}$ for *T. cacao/E. poeppigiana*. At age 10 the respective values were 765 and 722 kg $ha^{-1}$ of N. Thus the difference in N uptake during the 10 year experimental period was 43 kg $ha^{-1}$ of N in favour of *T. cacao/C. alliodora*, as a result of extra accumulation in the stems and branches of the shade tree.

The very small values for P accumulation indicate the low mobility of this element in agroecosystems. Available P, which may gradually be accumulated in the vegetation, is a very small fraction of total P (*Fassbender* and *Bornemisza*, 1987).

The average values of K uptake at age 10 years were 419 and 520 kg $ha^{-1}$ for *T. cacao* in association with *C. alliodora* and *E. poeppigiana*, respectively. Without fertilization this would imply a notable depletion of the original exchangeable soil K (1977: 577 and 713 kg $ha^{-1}$, respectively).

## Nutrient internal transfers, inputs and outputs

The rates of nutrient transfer for the *T. cacao/E. poeppigiana* system were normally higher than those of the *T. cacao/C. allidora* system, largely because of pruning of the leguminous tree (Table 8). The total value of deposting N to the soil with the plant residues (natural litter and pruning) shows a tremendous difference in favour of the *T. cacao/E. poeppigiana* system (447.1 kg $ha^{-1}$ $a^{-1}$) as compared with *T. cacao/C. alliodora* (169.1 kg $ha^{-1}$ $a^{-1}$). Assuming that the annual rate of decomposition of plant residues is almost comparable with those of deposition (*Heuveldop* et al., 1988), the turn over of N is very important. The export of N from the systems with cacao yields reached 23.1 and 28.4 kg N $ha^{-1}$ $a^{-1}$ for the studied systems respectively and are only a small fraction of the accumulated or circulating in the agroforestry systems. The input of N with rainfall, reaching 5.0 kg $ha^{-1}$ $a^{-1}$ is comparable

**Table 7:** Net primary production between ages 6 and 10 for the systems of *Theobroma cacao* with *Cordia alliodora* or *Erythrina poeppigiana* (t $ha^{-1}$ $a^{-1}$).

| Product | System | |
|---|---|---|
| | *T. cacao/C. alliodora* | *T. cacao/E. Poeppigiana* |
| Phytomass increase | | |
| Crop | 5.22 | 3.78 |
| Tree | 3.56 | 4.46 |
| Roots | 1.12 | 0.78 |
| Agricultural production | | |
| Cacao harvest | 2.52 | 2.54 |
| Production of residues | 11.40 | 22.86 |
| Fine root renewal | 4.36 | 1.83 |
| Total | 28.18 | 36.25 |

**Table 8:** Nutrient inputs and outputs in the systems of *Theobroma cacao* with *C. alliodora* or *E. poeppigiana* at age 6–10 years (kg $ha^{-1}$ $a^{-1}$).

| | *T. cacao/C. alliodora* | | | *T. cacao/E. poeppigiana* | | |
|---|---|---|---|---|---|---|
| Transfer | N | P | K | N | P | K |
| Internal transfer | | | | | | |
| Natural litterfal | 129.0 | 19.8 | 37.7 | 198.3 | 17.6 | 25.8 |
| Pruning residues | 40.1 | 4.1 | 35.4 | 248.8 | 22.1 | 151.5 |
| Total | 169.1 | 23.9 | 73.1 | 447.1 | 39.7 | 177.3 |
| Outputs | | | | | | |
| Cacao yield | 23.1 | 4.7 | 34.3 | 28.4 | 4.7 | 29.5 |
| Leaching | 6.0 | 0.6 | 2.2 | 6.0 | 0.7 | 1.8 |
| Total | 29.1 | 5.3 | 36.5 | 34.4 | 5.4 | 31.3 |
| Inputs | | | | | | |
| Rainfall | 5.0 | 0.2 | 2.5 | 5.0 | 0.2 | 2.5 |
| Fertilization | 87.5 | 34.4 | 32.4 | 87.5 | 34.4 | 32.4 |
| Total | 92.5 | 34.6 | 34.9 | 92.5 | 34.6 | 39.9 |

with other values founded in the literature for other sites in tropical conditions (*Fassbender*, 1993, *Jordan*, 1985).

The values recorded for P for the inputs and outputs from the systems are more or less comparable. There is one exception, those of the input with pruning materials, reaching 22.1 and 4.1 kg P $ha^{-1}$ $a^{-1}$ for *T. cacao/E. poeppigiana* and *T. cacao/C. alliodora* respectively.

The transfer rates of K with plant residues is once again larger under *Erythrina* (151.5 kg K $ha^{-1}$ $a^{-1}$) as under *Cordia* (35.4 kg K $ha^{-1}$ $a^{-1}$). Therefore, the nutrient cycles for *T. cacao/E. poeppigiana* turnover are faster for all nutrients in comparison with *T. cacao/C. alliodora*.

## Nutrient balances in the agroforestry systems

Figure 3 summarizes all details of the nitrogen balance for the system *T. cacao/E. poeppigiana* for the 10 first years of the study. The N accumulated in the biomass, resulting from N uptake from the soil, during the 10 years studies reached 722 $kg^{-1}$ $ha^{-1}$ (crop 249, tree 206, litter 204 and roots 63). Adding the value of N export with the cacao yields (206 kg $ha^{-1}$ 10 $a^{-1}$), the total N uptake from the soil represents 986 kg N $ha^{-1}$ 10 $a^{-1}$. This value is comparable to the rate of fertilization (962 kg N $ha^{-1}$ 10 $a^{-1}$). The rate of input with rain fall (50 kg N $ha^{-1}$ 10 $a^{-1}$) is almost equal of the rate of leaching with percolating water (60 kg N $ha^{-1}$ 10 $a^{-1}$). An important observation of the study by that the N-reserve in the soil increased in the 10 years of studies for 1409 kg N $ha^{-1}$ 10 $a^{-1}$ (1987; 10964, 1977: 9555 kg $ha^{-1}$). This accumulation may be a result of the N fixation by the leguminous tree. The values for the chemical analysis of nitrogen shown very high variation coefficients (Table 2), therefore the N increase in the soil in not statistical significant and the postulated increase is only hypothetical. In all the cases no decline of the N-reserves was shown demostrating maintenace of soil fertility.

## Conclusions, performances of the systems and both shade trees

Under the experimental conditions of this study, which included normal fertilization ratused by farmers, the agroforestry systems of *E. poeppigiana* or *C. alliodora* with *T. cacao* have reached large both agricultural production and timber production as well as high net primary production during the studies, thus showing that the production of these systems is sustainable for at least 15 years.

The apparent increase in soil organic material and nitrogen, especially under *E. poeppigiana*, also gives evidence for the ecological sustainability of these systems over longer periods. Although litter inputs and soil organic material levels are lower under *C. alliodora*, they are at the very least maintained. When the *C. alliodora* timber production of 9.0 $m^3$ $ha^{-1}$ $a^{-1}$ is added to the cocoa production of 0.7 t $ha^{-1}$ $a^{-1}$, then there is a clear economic advantage in using this species rather than the leguminous tree *E. poeppigiana* (management cost for *C. alliodora* are estimated to be less than the costs of using *E. poeppigiana*). However, if nutrient export in *C. alliodora* timber, together with crop damage during tree felling is taken into account, then there may be no long-term economic advantage in using *C. alliodora*. Organic material and hence nutrient cycling is greater under *E. poeppigiana* and on sites where management intensity (especially fertilization) and/or site fertility is less, then the relative advantages of using *E. poeppigiana* would increase.

Finally it may be concluded that the analyzed agroforestry systems maintain soil organic matter and promote nutrient cycling and therefore are also sustainable and appropiate for the soil conservation.

## Acknowledgements

The author wish to acknowledge the participation of G. Enriquez, L. Alpizar, J. Heuveldop, A. Imbach, A. Bonnemann and special J. Beer

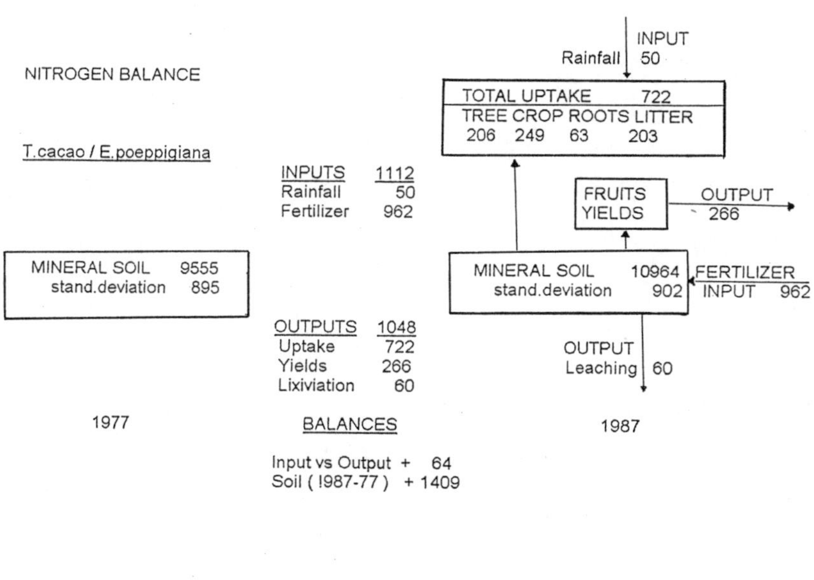


**Figure 3:** Ten years nitrogen balance for the system *T. cacao-E. poeppigiana* (reserves kg $ha^{-1}$; transfers kg $ha^{-1}$ 10 $a^{-1}$).

in earlier evaluations and publications of the carried reseach at "Experimento Central". Financial and technical support were provided by the Deutsche Gesellschaft für Technische Zusammenarbeit (GTZ) and CATIE.

## Literature

*Alpizar L., Fassbender H. W., Heuveldop J., Folster H.* and *Enriquez G.* (1986). Modelling agroforestry systems of cacao *(Theobroma cacao)* with *Cordia alliodora* and *Erythrina poeppigiana* in Costa Rica. I Inventory of organic matter and nutrients. Agroforestry Systems 4: 175–189.

*Beer, J., Fassbender H. W.* and *Heuveldop J.* (1987). Advances in agroforestry research. Proceedings of a seminar CATIE, Turrialba, Costa Rica, 379 p.

*Beer, J., Bonnemann A., Chavez W., Fassbender H. W., Imbach A. C.* and *Martel I.* (1990) Modelling agroforestry systems of cacao *(Theobroma cacao)* with *Cordia alliodora* and *Erythrina poeppigiana* in Costa Rica. V Productivity indices, organic matter models and sustainability over 10 years. Agrofor. Syst., 12: 229–249.

*Fassbender H. W.* and *Bornemisza E.* (1986). Quimica de suelos. San Jose/ Costa Rica. IICA. 420 p.

*Fassbender H. W., Alpizar L., Heuveldop J., Folster H.* and *Enriquez G.* (1988). Modelling agroforestry systems of cacao *(Theobroma cacao)* with laurel *(Cordia alliodora)* and poro *(Erythrina poeppigiana)* in Costa Rica. III. Cycles of organic matter and nutrients. Agroforestry Systems 6: 49–62.

*Fassbender H. W., Beer J., Heuveldop J., Imbach A., Enriquez G.* and *Bonnemann A.* (1991). Ten years balances of organic matter and nutrients in agroforestry systems at CATIE, Costa Rica. In Jarvis PG (1991) Agroforestry, principles and practice. Elsevier.

*Fassbender H. W.,* (1993). Modelos edafologicos de sistemas agroforestales., 2nd edition. Turrialba, Costa Rica, GTZ/CATIE 420 p.

*Heuveldop J., Fassbender H. W., Alpizar L., Enriquez G.* and *Folster H.* (1988). Modelling agroforestry systems of cacao *(Theobroma cacao)* with laurel *(Cordia alliodora)* and poro *(Erythrina poeppigiana)* in Costa Rica. II Cacao and wood production, litter production and decomposition. Agrof. Syst. 6.: 37–48.

*Imbach A. C., Fassbender H. W., Borel R., Beer J.* and *Bonnemann A.* (1989). Modelling agroforestry systems of cacao *(Theobroma cacao)* with laurel *(Cordia allidora)* and poro *(Erythrina poeppigiana)* in Costa Rica. IV. Water balances, nutrient inputs and leaching. Agrof. Systems 8: 267–287.

*Javier E., Renborg* (1989). The change dynamics of global agriculture. Proceedings of a seminar held at Feldafing, West Germany, 22.–28.9.1988, 378 p.

*Jordan C. F.* (1983). Productivity of tropical rain forest ecosystems and the implications for their use as future wood and energy sources. In: Golley FB, ed, Tropical rain forest ecosystems: structure and function, pp. II 7–136. Amsterdam, Elsevier. (Ecosystems of the World 14 A).

*Jordan C. F.* (1985). Nutrient cycling in tropical forest ecosystems. New York: Wiley 190 p.

*Nair P. K. R.* (1987). Soil productivity under agroforestry. In Ghlz H. L. (editor), Agroforestry: reality and possibilities. Nijhoff, Holland, 227 p.

*Proctor J.* (1984). Tropical forest litterfall II: The data set. In: Chadwick A. C., Sutton S. L., eds., Tropical rain forest: The Leeds symposium, pp. 83–113. Leeds/England. Leeds Philosophical and Literary Society.

*Sanchez P.* (1987). Soil productivity and sustainability of agroforestry systems. In: Steppler H. A., Nair P. K. R., eds., Agroforestry, a decade of development, pp. 205–223. Nairobi/Kenya. ICRAF.

*Somarriba E.* and *Beer J.* (1987). Dimensions, volumes and growth of Cordia alliodora in agroforestry systems. For Ecol Man 18: 113–126.

*Young A.* (1987). Soil productivity, soil conservation and land evaluation. Agroforestry Systems 5: 277–292.

# Soil environment and tree species diversity in tropical rain forest, West Sumatra, Indonesia

**Daisuke Kubota, Tsugiyuki Masunaga, Hermansah*, Azwar Rasyidin*, Mitsuru Hotta**, Yoshiaki Shinmura** and **Toshiyuki Wakatsuki**

Faculty of Life and Environmental Science, Shimane University, Matsue 690, Japan
*Faculty of Agriculture, Andalas University, Padang, Indonesia
**Faculty of Science, Kagoshima University, Kagoshima 890, Japan

## Abstract

Soil characteristics were examined in relation to tree species diversity of a tropical rain forest on a 100 m transect plot covering from ridge to valley with 0.10 ha and a long-term observation plot, 1.0 ha, in Mt. Gadut area near Padang, West Sumatra, Indonesia. The area experienced an annual rainfall of more than 6000 mm with no real dry season. In the 0.10 ha transect plot, detailed soil characterization was performed. From each sub-plot having the measurements of 5 x 5 m, 5 samples were taken from each 0–5 and 5–15 cm depth. In this transect plot, the subplots positioned on the ridge showed a greater diversity of tree species and higher biomas than those positioned in the valley. Soils found on the ridge showed less fertility and greater variation of fertility characteristics than those found in the valley. The examination of vertical soil profile distribution of ridge, middle and valley positions in this transect also revealed that exchangeable Calcium, total carbon and total nitrogen of ridge soil tended to accumulate of these nutrients at the upper horizons and of dissipation at lower horizons more than those of valley soil profile. Detailed characterization of 115 subplots of 1.0 ha permanent plot showed that horizontal variations of soil fertility characteristics were also important in relating the tree species diversity of tropical rain forest. These results suggest that both vertical and horizontal variations of soil characteristics are imperative in supporting the tree species diversity.

## Introduction

The number of tree species greater than 10 cm in diameter at breast height (DBH) found in one hectare of tropical rain forest can be over 200, as compared to 10–20 for the most temperate forests (*Gentry* 1988; *Huston* 1994; *Richards* 1996; *Whitmore* 1984; *Whitmore* and *Sidiyasa* 1986, *Whitmore* et al., 1987; *Whittaker*, 1956). The enormous diversity of tropical rain forest trees is related to both the evolutionary and ecological (*Huston*, 1994) processes. The objective of this report is to characterize soil edaphic characteristics, which are associated with high tree species diversity in tropical forests. *Ashton* (1982) reported the positive relationship between species diversity and soil fertility, such as magnesium content in southeast Asian tropics. On the other hand *Holdridge* et al. (1971) and *Huston* (1980) showed that soil productivity was negatively correlated with species diversity in Costa Rican tropical forests. *Korning* et al. (1994) suggested that a high local species diversity in western Amazonia might be caused by local variations in soil characters. *Newbery* et al. (1984, 1986) reported the influence of soil factors on tropical rain forests in Cameroon and Sarawak. *Baillie* et al. (1987) in *Sarawak* and *Gartlan* et al. (1986) in Cameroon also studied the site characteristics and distribution of tree species. However the information between soil environmental characteristics, especially edaphic natures, and tree species diversity in tropical rain forest, much remain to be investigated.

In this paper, the authors present detailed information on spatial, both vertical and horizontal, variations of soil edaphic characteristics in relation to tree species diversity in long term forest ecology observation plots (*Hotta*, 1984, 1986, 1989).

## Study sites

Pinang Pinang Atas transect plot (denoted as PAT plot) is located at an altitude between 460–550 m on a foothill of Mt. Gadut (altitude 1855 m), 18 km east of Padang, West Sumatra as shown in Fig. 1. The study area experiences an annual rainfall of more than 6000 mm and no real dry season (*Hotta*, 1984). This toposequence plot extends from one valley to another valley over the ridge extending from Mt. Gadut. The plot has been divided into 37 subplots with a 10 x 10 m grid as 370 x 10 m, these measurements were established by *Kohyama* et al. in September 1984. All trees over 5 cm in DBH were numbered and recorded. Only a part of the northwestern slope of PAT plot i.e., 10 subplots from No. 7–16, 0.1 ha, and partly No. 17 which were denoted as NW slope as shown in Fig. 2 was investigated for this paper. The reason being is that these parts of the PAT plot were the most typical transect covering from ridges and valleys with the least amount of human and stream disturbance among the other parts of the PAT plot (*Hotta*, 1984). Total number of trees, the number of species and the basal area with more than 5 cm in DBH, were 152, 95 and 13.5 $m^2$, respectively.

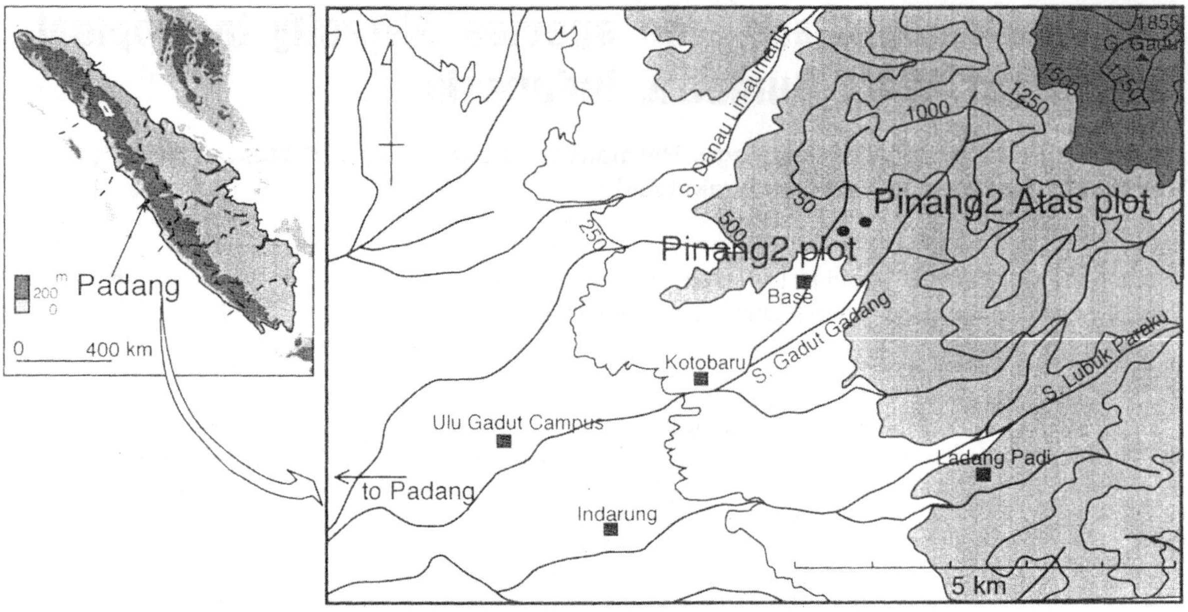


**Figure 1:** Maps showing resarch site, at Ulu Gadut area, Padang, West Sumatra.

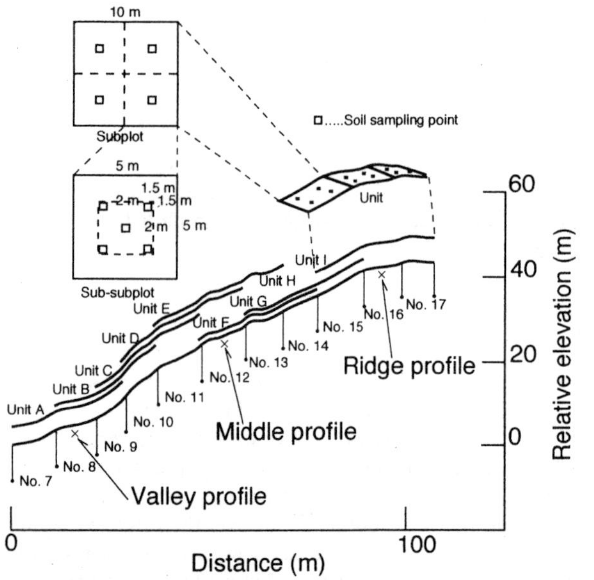


**Figure 2:** A cross sectional profile of northwest slope of Pinang Pinang Atas transect plot. For statistical analysis purposes the subplots were grouped into nine units as follows: unit A = subplot No. 7–9, unit B = subplot No. 8–10, unit C = subplot No. 9–11; ... unit I = subplot No. 15–17. For these grouping the data set of subplot No. 17 which was similary obtained to other subplots was used.

The Pinang Pinang plot (denoted as PIN2) was established in 1981 on a ridge top of one of the foothills of Mt. Gadut (*Hotta*, 1984). This plot (1.0 ha) has been divided into 115 subplots of about 10 x 10 m in size. All trees over 8 cm in DBH were numbered and recorded. There were 795 trees which had more than 8 cm in DBH and there were 304 different species, including some unidentified trees (*Hotta*, 1994). *Nagano* et al. (1978) estimated the above-ground biomass and the net production rate of big woody organs over 10 cm in diameter to be 411 metric to $ha^{-1}$, and 19.8 ton $ha^{-1}$ $year^{-1}$. The dominant species based on basal area are Swintonia schwenckii, Lithocarpus spp., Palaquium spp., Syzygium spp., Canarium denticulatum and Nephehelium juglandifolium (*Yoneda* et al., 1994). Routine soil survey of these plots were conducted by *Wakatsuki* (1986). A detailed soil map as shown in Fig. 3. Soils in the PIN2 plot were developed from mainly andesite with the inclusion of some limestone. Although the total area was only 1.0 ha, soils were extremely heteroge-

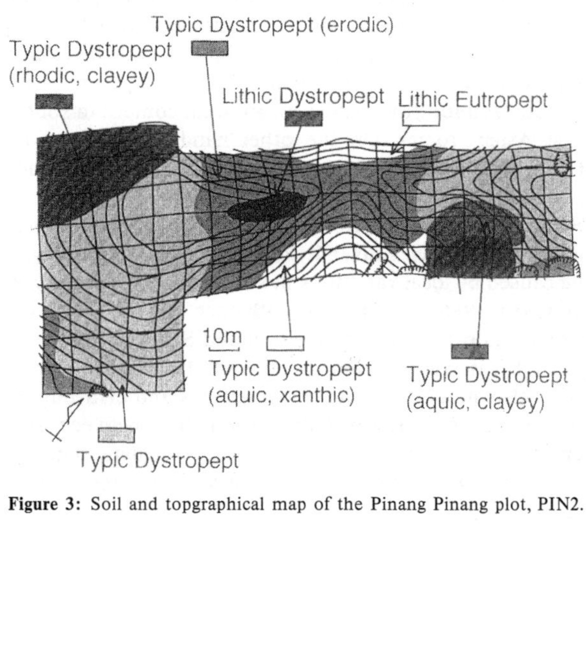


**Figure 3:** Soil and topgraphical map of the Pinang Pinang plot, PIN2.

neous. Three different soil subgroups, Typic Dystropept, Lithic Dystropept and Lithic Eutropept were recognized, which were further subdivided into seven families as shown in the figure 3.

## Field methods

In order to characterize vertical variations of soil characteristics, we collected soil samples at 0–5 cm, 5–10, 10–25, 25–40, 40–55, 55–70, 70–85, 85–100, 100–115, 115–130, 130–145 and 145–160 cm depth from three profiles dug at ridge, middle and valley positions in NW slope. The three pits were selected at 5 m outside from subplot No. 16, No. 13 and No. 8 as shown in Fig. 2. In order to characterize horizontal soil variations, we collected four composite soil samples from each subplot, 10 x 10 m, which was further sub-divided into four grids with the measurements of 5 x 5 m. Five soil samples from each 0–5 and 5–15 cm depths were taken from each 5 x 5 m grid as shown in Fig. 2. A composite sample was made from these 5 samples from the 5 x 5 m grid for laboratory analysis. Eighty composite soil samples were collected from subplots No. 7–16, at both 0–5 and 5–15 cm depths along the NW slope of PAT plot (10 x 100 m, Sub-plots No. 7–17) as shown in Fig. 2. Soil samples were also collected from PIN2 plot. Soil samples were collected at 0–15 cm depth from five points of each subplot and 115 composite samples were made for analyses (*Masunaga* et al. 1997).

## Laboratory methods

All soil samples were air dried, ground and passed through a 2-mm mesh sieve. Exchangeable Ca, K, Mg, and Na, and extractable Al, B, Cu, Fe, Mn, Mo, P, S, Si, Sr, and Zn, wre extracted with 0.1 N hydrochloric acid (soil to solvent ratio at 1:10). The extracts were determined by inductively coupled plasma atomic emission spectrometry (ICP) using Shimazu ICPS-2000. The dry combustion method as described by *Allison* et al. (1965) was used to measure total carbon and total nitrogen.

## Statistics

In order to statistically analyze the coefficient of variation, the subplots were grouped into nine units. Each unit has three consecutive subplots, i.e., Unit A = No. 7–9, Unit B = No. 8–10, Unit C = No. 9–11, Unit D = No. 10–12, Unit E = No. 11–13, Unit F = No. 12–14, Unit G = No. 13–15, Unit H = No. 14–16, Unit I = No. 15–17. Subplot No. 17's data set was included to calculate the coefficient of variation, which data set was obtained in similar ways to the other subplots of No. 7–16. The SYSTAT software package (SYSTAT Inc., 1992) was used for all statistical analyses.

For the correlation of spatial distribution between tree species diversity and various soil environmental factors, we calculated the ω index of *Iwao* (1977) (*Khoyama* et al., 1994).

The ω index was defined as

$$\omega_{(+)} = \frac{\gamma - \gamma_{(ind)}}{1 - \gamma_{(ind)}} = \sqrt{\frac{m^*_{XY} m^*_{YX} - m_X m_Y}{(m^*_X + 1)(m^*_Y + 1) - m_X m_Y}}, \quad \text{if } \gamma \geq \gamma_{(ind)}, \text{ or}$$

$$\bar{\omega}_{(-)} = \frac{\gamma - \gamma_{(ind)}}{\gamma_{(ind)}} = \sqrt{\frac{m^*_{XY}}{m_X} \frac{m^*_{YX}}{m_Y}} - 1, \quad \text{if } \gamma \leq \gamma_{(ind)}.$$

Here, $\gamma$ and $\gamma_{(ind)}$ were defined as follows,

$$\gamma = \sqrt{\left(\frac{m^*_{YX}}{m_X} \frac{m^*_{XY}}{m_Y}\right) \Big/ \left(\frac{(m^*_X + 1)}{m_X} \frac{(m^*_Y + 1)}{m_Y}\right)},$$

$$\gamma_{(ind)} = \sqrt{\frac{m_X}{(m^*_X + 1)} \frac{m_Y}{(m^*_X + 1)}} = \frac{1}{\sqrt{\frac{(m^*_Y + 1)}{m_X} \frac{(m^*_Y + 1)}{m_Y}}},$$

$$m^*_{XY} = \sum_{j=1}^{Q} x_{Xj} x_{Yj} \Big/ \sum_{j=1}^{Q} x_{Xj},$$

$$m^*_{YX} = \sum_{j=1}^{Q} x_{Xj} x_{Yj} \Big/ \sum_{j=1}^{Q} x_{Yj},$$

$$m^*_X = \sum_{j=1}^{Q} x_{Xj} (x_{Xj} - 1) \Big/ \sum_{j=1}^{Q} x_{Xj},$$

$$m^*_Y = \sum_{j=1}^{Q} x_{Yj} (x_{Yj} - 1) \Big/ \sum_{j=1}^{Q} x_{Yj},$$

$x_{Xi}$ = population of X at i-th grid
$x_{Yi}$ = population of Y at i-th grid
Q = total number of grid
$m_X$ = mean population of X per given grid
$m_Y$ = mean population of Y per given grid.

The ω index gives the value from −1 to 1, zero for mutually independent distribution, positive values up to +1 for a positive correlation of distribution, and negative values up to −1 for a negative correlation of distribution.

## Results and discussion

### Tree species composition in PAT plot

The number of tree species, the number of trees and basal area per unit in the NW slope are shown in Fig. 4. Apart from a slight increase in the unit B and a slight decrease in the unit E, tree population per unit area was almost the same throughout the toposequence. The maximum, average and minimum of population per each unit were 54, 45.6, and 38 respectively. On the other hand, as the elevation increased so did the number of tree species. Located on the ridge of unit I was the highest number of species, which was 44. Unit A, located in the valley, showed a species number of 26. The basal area showed similar but a more pronounced trend to that of the number of tree species. There was a steep rise from unit F to unit I, suggesting the effect of topographical conversion

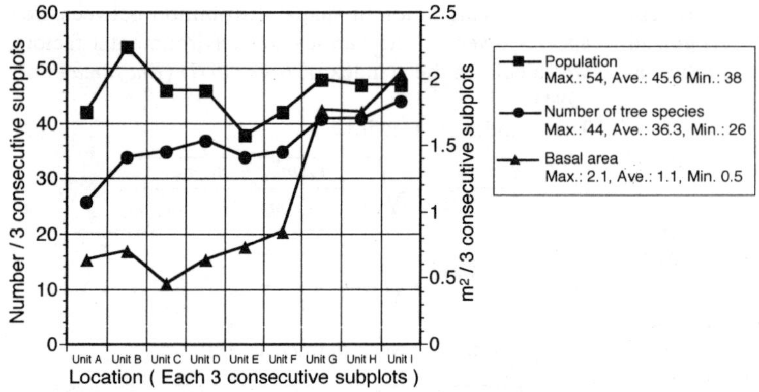


**Figure 4:** Number of tree species, population and basal area per 3 consecutive subplots (10 x 30 m) along the northwest slope of PAT, Pinang Pinang Atas, plot.

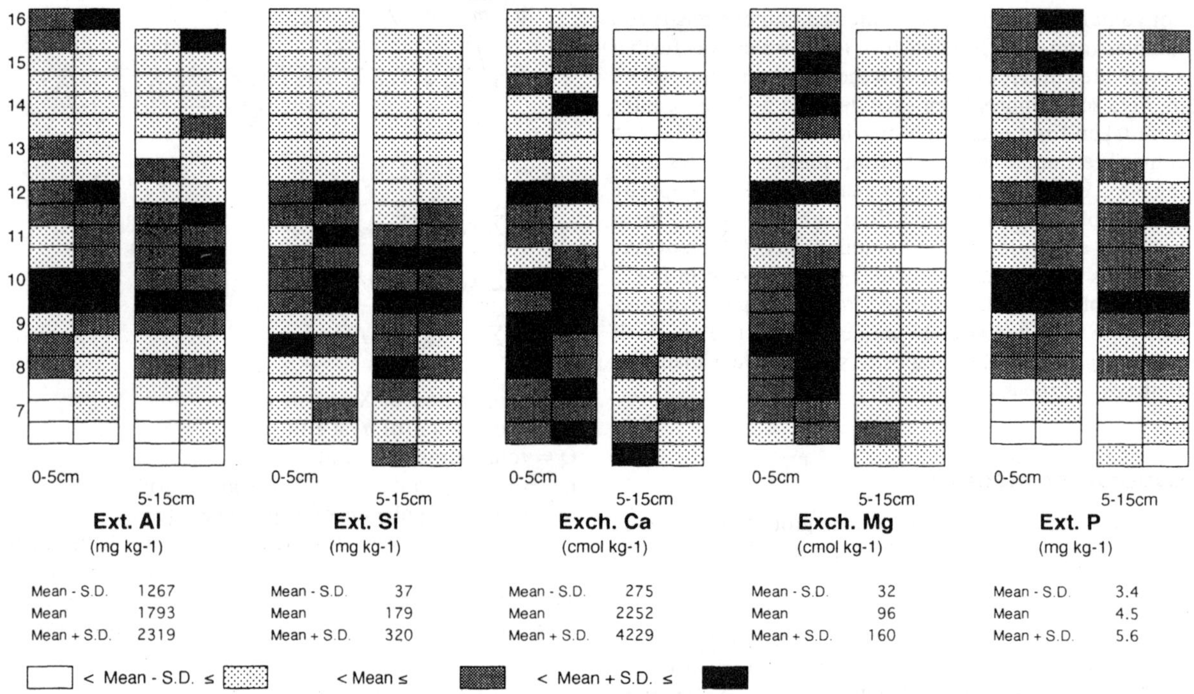


**Figure 5:** The distribution of concentration of Ext. Al, Si, and P, and Exch. Ca, and Mg in PAT.

from colluvial slope to convex ridge (Fig. 2). These figures show that the ridge site has higher diversity of tree species and biomas than the valley and middle slope.

The tree species with the greatest basal area in each of the units were Sterculiaceae Sterculia sp. (one trunk), Sapindaceae Nephelium mutabile (one trunk), and Meliaceae Dysoxylum sp. (tree trunks) which had 23.9, 26.8 and 56.1% respectively of the basal area per each unit.

### Soil characteristics of northwest slope of PAT plot

Fig. 5 shows the selected analytical results of surface soils, both 0–5 and 5–15 cm depth, in each sub of the subplots in 5 x 5 m grids. Table 1 shows the mean concentration of exchangeable Na, K, Ca and Mg, extractable. Al, Fe, Mn, P and Si, T-N and T-C of surface soils, both 0–5 and 5–15 cm depth, in each three consecutive subplots. The top 0–5 cm data shows that the exchangeable Ca, Mg and K and extractable Mn had the highest concentration at valley sites. Extractable Al, Fe, P and Si had their highest concentration at the middle to lower slope. Whereas T-N, T-C and C/N ratio had their highest values at the ridge site. Topographical factor seem to have strongly influenced the soil development in NW slope of PAT plot. The trend of CN ratio, i.e., on the ridge was higher than in the valley, showed lower activity of microorganism in ridge site than in valley site. Since the difference of C/N ratio was small between the ridge and the valley, the difference of the microbial activity may not be as big. The trends

**Table 1:** The mean concentrations of Exch. Na, Exch. K, Exch. Ca and Exch. Mg, Ext. Al, Ext. Fe, Ext. Mn, Ext. P and Ext. Si in two soil depths 0–5 and 5–15 cm, respectively of each three consecutive subplots.

| | | cmol(+)/Kg Dry soil | | | | mg/Kg Dry soil | | | | | % | | |
|---|---|---|---|---|---|---|---|---|---|---|---|---|---|
| | (0–5 cm) | Exch. Na | Exch. K | Exch. Ca | Exch. Mg | Ext. Al | Ext. Fe | Ext. Mn | Ext. P | Ext. Si | N | C | C/N |
| Valley | Unit A | **0.006** | **0.38** | 11.77 | 1.08 | 1447 | 31 | 283 | 4.02 | 187 | 0.21 | 2.3 | 11.2 |
| ↑ | Unit B | **0.006** | 0.36 | **12.47** | **1.28** | 2012 | 24 | **409** | 5.07 | 288 | 0.24 | 2.8 | 11.3 |
| | Unit C | **0.006** | 0.31 | 10.29 | 1.13 | 2151 | 131 | 401 | 5.25 | **334** | 0.23 | 2.7 | 11.7 |
| | Unit D | **0.006** | 0.30 | 9.53 | 1.08 | **2232** | 240 | 377 | **5.42** | 331 | 0.31 | 3.8 | 12.1 |
| | Unit E | 0.005 | 0.23 | 7.17 | 0.76 | 1842 | **253** | 261 | 4.72 | 200 | 0.39 | 4.8 | 12.4 |
| | Unit F | 0.005 | 0.23 | 7.07 | 0.80 | 1689 | 160 | 206 | 4.65 | 128 | **0.44** | **5.4** | 12.3 |
| ↓ | Unit G | 0.004 | 0.23 | 5.77 | 0.72 | 1534 | 61 | 149 | 4.50 | 65 | 0.39 | 5.0 | **12.7** |
| Ridge | Unit H | 0.005 | 0.24 | 5.10 | 0.77 | 1685 | 80 | 112 | 4.89 | 61 | 0.36 | 4.5 | 12.5 |
| | Average | 0.005 | 0.28 | 8.65 | 0.95 | 1824 | 122 | 275 | 4.81 | 199 | 0.32 | 3.9 | 12.0 |
| | S.D. | 0.001 | 0.06 | 2.76 | 0.21 | 286 | 89 | 115 | 0.45 | 111 | 0.09 | 1.2 | 0.6 |
| | Max./Min. | 1.4 | 1.7 | 2.4 | 1.8 | 1.5 | 10.6 | 3.7 | 1.3 | 5.5 | 2.1 | 2.3 | 1.1 |
| | | cmol(+)/Kg Dry soil | | | | mg/Kg Dry soil | | | | | % | | |
| | (5–15 cm) | Exch. Na | Exch. K | Exch. Ca | Exch. Mg | Ext. Al | Ext. Fe | Ext. Mn | Ext. P | Ext. Si | N | C | C/N |
| Valley | Unit A | 0.005 | **0.26** | **5.91** | **0.60** | 1520 | 79 | 186 | 3.82 | 189 | 0.09 | 1.0 | 10.5 |
| ↑ | Unit B | 0.005 | 0.20 | 3.92 | 0.56 | 2046 | 160 | **250** | 4.76 | 266 | 0.10 | 1.0 | 10.5 |
| | Unit C | **0.006** | 0.18 | 3.11 | 0.48 | 2261 | 279 | 247 | 5.15 | **322** | 0.09 | 0.9 | 10.1 |
| | Unit D | 0.005 | 0.17 | 2.16 | 0.38 | **2403** | **403** | 208 | **5.35** | 307 | 0.09 | 1.0 | 11.0 |
| | Unit E | 0.004 | 0.13 | 1.53 | 0.27 | 2048 | 355 | 124 | 4.63 | 207 | 0.09 | 1.1 | 11.6 |
| | Unit F | 0.004 | 0.11 | 1.12 | 0.23 | 1787 | 250 | 79 | 4.12 | 122 | 0.10 | 1.3 | 12.4 |
| ↓ | Unit G | 0.004 | 0.10 | 0.85 | 0.20 | 1538 | 129 | 43 | 3.68 | 74 | 0.10 | 1.3 | 12.6 |
| Ridge | Unit H | 0.004 | 0.10 | 0.81 | 0.21 | 1571 | 123 | 26 | 3.77 | 61 | 0.10 | **1.4** | **14.0** |
| | Average | 0.005 | 0.16 | 2.43 | 0.37 | 1897 | 222 | 145 | 4.41 | 193 | 0.10 | 1.1 | 11.6 |
| | S.D. | 0.001 | 0.06 | 1.79 | 0.16 | 343 | 118 | 90 | 0.65 | 101 | 0.01 | 0.2 | 1.3 |
| | Max./Min. | 1.4 | 2.6 | 7.3 | 2.9 | 1.6 | 5.1 | 9.8 | 1.5 | 5.2 | 1.2 | 1.6 | 1.4 |

observed in the top 0–5 cm were also observed in the 5–15 cm depth. The differences between the valley and ridge sites i.e. maximum/minimum, were, however, more pronounced in the cases of exchangeable K, Ca, and Mg and extractable Mn. Except for carbon and nitrogen status, especially of 0–5 cm, nutritional status was higher in valley soils than in ridge soils. On the other hand, there was less tree species diversity in the valley that of the ridge area. It seems that the greater the species diversity is the less the soil fertility is. These results were somewhat similar to those of Costa Rican forests where tree species diversity was negatively correlated with soil nutrient availability (Huston 1994).

The diversity of niches within a local ecosystem may be important to maintain great tree species diversity. The diversity of niches in soil may more closely related to the variation in various soil characteristics than the concentrations themselves. Table 2 shows the coefficient of variations of selected analytical results of surface soils, both 0–5 and 5–15 cm depth, in each unit. The top 0–5 cm data shows exchangeable Ca, Mg, and K, and extractable Mg and Si had the highest coefficient of variations at ridge sites. Whereas extractable Al, Fe and P showed the highest coefficient of variations at the middle to valley positions. These trends observed in coefficient of variations were somewhat opposite characteristics observed in fertility parameter as shown in the Table 2.

Fig. 6 shows the results of profile distributions of exchangeable Ca, extractable Fe, total carbon and total nitrogen contents at the ridge, middle and valley positions in the toposequence. Their profiles were located close to No. 8, 12 and 16 respectively as shown in Fig. 2. The profile distribution patterns of exchangeable Ca, T-N, T-C and extractable Fe, showed sharp decrease towards lower horizons and accumulation at top 0–5 and 5–10 cm horizons. This trend was more clear in exchangeable Ca, Mg, total carbon and total nitrogen. The degree of variations, i.e. the enrichment of the upper horizon and/or dissipation to the lower horizons, were different among the three profiles. The ridge profile which had the highest profile variations. On the other hand, the valley bottom profile which had the least species diversity showed the lowest profile variations. These results suggest that vertical profile variations of soils are also important in supporting great tree species diversity.

### Horizontal variations of soil characteristics in relation of tree species diversity in PIN2 plot

Horizontal distribution patterns of various soil characteristics at 5–15 cm depth in relation to tree species diversity in PIN2 is presented in Fig. 7. The tree species diversity was calculated using moving means of species number per 9 consecutive subplots (30 x 30 m). According to the

**Table 2:** The coefficient of variations of Exch. Na, Exch. K., Exch. Ca and Exch. Mg, Ext. Al, Ext. Fe, Ext. Mn, Ext. P and Ext. Si in two soil depths 0–5 and 5–15 cm, respectively of each three consecutive subplots.

| | | | | | | Coefficient of variation (%) | | | | | | | |
|---|---|---|---|---|---|---|---|---|---|---|---|---|---|
| | (0–5 cm) | Exch. Na | Exch. K | Exch. Ca | Exch. Mg | Ext. Al | Ext. Fe | Ext. Mn | Ext. P | Ext. Si | N | C | C/N |
| Valley | Unit A | 26.4 | 36.2 | 30.2 | 26.5 | **30.9** | 100.0 | 47.2 | 21.5 | 57.0 | 59.0 | 65.0 | 12.9 |
| ↓ | Unit B | 27.0 | 35.6 | 33.3 | 27.4 | 30.7 | 92.0 | 35.9 | **22.2** | 69.7 | 47.8 | 49.3 | **17.7** |
| | Unit C | **32.2** | 44.8 | 48.6 | 41.4 | 23.7 | **124.3** | 38.5 | 18.1 | 54.2 | 56.1 | 56.5 | 16.5 |
| | Unit D | 26.4 | 37.4 | 55.7 | 47.0 | 19.3 | 83.4 | 50.5 | 18.1 | 50.3 | **68.7** | **76.9** | 16.7 |
| | Unit E | 22.3 | 38.8 | 61.3 | 52.6 | 18.6 | 73.0 | 63.0 | 17.6 | 59.9 | 55.0 | 62.6 | 12.5 |
| | Unit F | 19.6 | **51.0** | **64.7** | **59.4** | 20.8 | 114.8 | **81.8** | 17.3 | **81.3** | 39.1 | 45.8 | 10.6 |
| ↓ | Unit G | 11.8 | 50.3 | 53.5 | 52.6 | 13.2 | 48.9 | 57.3 | 15.5 | 33.3 | 29.5 | 31.5 | 11.4 |
| Ridge | Unit H | 19.7 | 43.7 | **64.7** | 48.4 | 21.7 | 57.6 | 43.9 | 20.8 | 23.5 | 33.0 | 31.6 | 11.2 |
| | Average | 23.2 | 42.2 | 51.5 | 44.4 | 22.4 | 86.7 | 52.3 | 18.9 | 53.6 | 48.5 | 52.4 | 13.7 |
| | S.D. | 6.2 | 6.2 | 13.4 | 12.0 | 6.0 | 26.4 | 15.0 | 2.3 | 18.5 | 13.7 | 16.0 | 2.8 |
| | Max./Min. | 2.7 | 1.4 | 2.1 | 2.2 | 2.3 | 2.2 | 2.3 | 1.4 | 3.5 | 2.3 | 2.4 | 1.7 |
| | | | | | | Coefficient of variation (%) | | | | | | | |
| | (5–15 cm) | Exch. Na | Exch. K | Exch. Ca | Exch. Mg | Ext. Al | Ext. Fe | Ext. Mn | Ext. P | Ext. Si | N | C | C/N |
| Valley | Unit A | 17.9 | **60.3** | 63.4 | 30.8 | 24.0 | 62.1 | 40.5 | 17.6 | 42.3 | **24.8** | 31.8 | 11.7 |
| ↓ | Unit B | 18.5 | 38.7 | 53.1 | 29.0 | **33.3** | 75.1 | 33.7 | **27.5** | 55.9 | 18.5 | 27.4 | 14.9 |
| | Unit C | 17.1 | 42.2 | 60.5 | 39.4 | 25.3 | 60.3 | 36.5 | 21.1 | 41.3 | 14.4 | 26.6 | 16.3 |
| | Unit D | 13.6 | 39.3 | 52.5 | 35.3 | 19.2 | 35.1 | 54.1 | 17.6 | 42.8 | 21.0 | 35.0 | 20.5 |
| | Unit E | 20.4 | 41.1 | 74.6 | 39.9 | 17.8 | 53.6 | 68.6 | 15.8 | 54.5 | 21.3 | **35.8** | **21.0** |
| | Unit F | **21.0** | 46.4 | **94.8** | **45.1** | 25.7 | **77.2** | **99.7** | 21.7 | **58.0** | 18.7 | 25.9 | 15.4 |
| ↓ | Unit G | 10.0 | 31.9 | 80.8 | 37.9 | 20.2 | 40.7 | 98.4 | 16.3 | 33.5 | 13.6 | 27.7 | 18.6 |
| Ridge | Unit H | 13.9 | 30.2 | 82.9 | 35.7 | 22.7 | 9.1 | 29.4 | 17.9 | 10.9 | 19.0 | 31.2 | **21.7** |
| | Average | 16.5 | 41.3 | 70.3 | 36.7 | 23.5 | 51.6 | 57.6 | 19.4 | 42.4 | 18.9 | 30.2 | 17.5 |
| | S.D. | 3.7 | 9.3 | 15.3 | 5.2 | 4.9 | 22.7 | 28.5 | 3.9 | 15.4 | 3.7 | 3.8 | 3.5 |
| | Max./Min. | 2.1 | 2.0 | 1.8 | 1.6 | 1.9 | 8.5 | 3.4 | 1.7 | 5.3 | 1.8 | 1.4 | 1.9 |

crown projection, the biggest tree, Swintonia schwenchii, occupied the biggest area, about 30 x 30 m size (Hotta, 1984). Soil characteristics were also shown using the moving mean of 9 consecutive subplots (30 x 30 m). Horizontal variations of exchangeable cations were evaluated using the values of standard deviations of 9 consecutive subplots.

The areas with higher variation (higher standard deviations) of exchangeable Ca seem to have greater species diversity. The areas with lower variation of exchangeable Ca seem to have less species diversity. On the other hand, the areas with higher concentrations of exchangeable Ca and extractable Si showed less species diversity. Distribution patterns of exchangeable Mg seem to have no clear relation to the species diversity. According to the results of *Masunaga* et al. (1997), soil edaphic level especially the nutrient status of K, Ca and Sr in top 0–15 cm depth and distribution of some kinds of tree species, such as accumulators and excluders, have close association. The trees with enormously diverse absorption characteristics of mineral elements might contribute to create such variation of exchangeable Ca in soil. As a reward this soil variation may contribute to support the coexistence nutritionally diverse various tropical trees in the plot.

Table 3 shows the results of calculation of ω index to quantify the degree of similarity of distribution pattern of soil characteristics to the tree species diversity. For the calculation of ω between higher diversity area and higher variation or higher concentration area, we gav 2 for the subplots with higher than mean + S.D., 1 for the subplots with higher than mean but lower than mean + S.D., and 0 for all subplots with less than mean. On the other hand, for the calculation of ω between lower diversity area and lower variation or lower concentration areas, we gave 2 for the subplots with less than the mean – S.D., 1 for the subplots with lower than the mean but higher than the mean – S.D., and 0 for all the subplots with higher than the mean. In order to emphasize the degree of matching of distribution, bold-letters represent the ω values greater than 0.3, of which negative values were underlined (Table 3).

Great species diversity showed great positive ω values, i.e. greater correlation with greater horizontal variations of exchangeable Ca, Mo, S and Zn. Less species diversity showed greater positive ω values with less horizontal variations, such as exchangeable Al, B, Ca, K, Mo, Na, P and Zn. On the other hand, high species diversity showed negative ω value with greater concentration of exchangeable K, extractable Mn, Si and Zn. *Masunaga* et al. (1997) showed some accumulator trees absorbed specific elements actively from the soils even if the soils were poor, creating the diverse soil edaphic characteristics. This may be important to support the coexistence of the nutritionally diverse tree species. On the other hand, nutritional absorption characteristics of trees may have a relatively small influence on

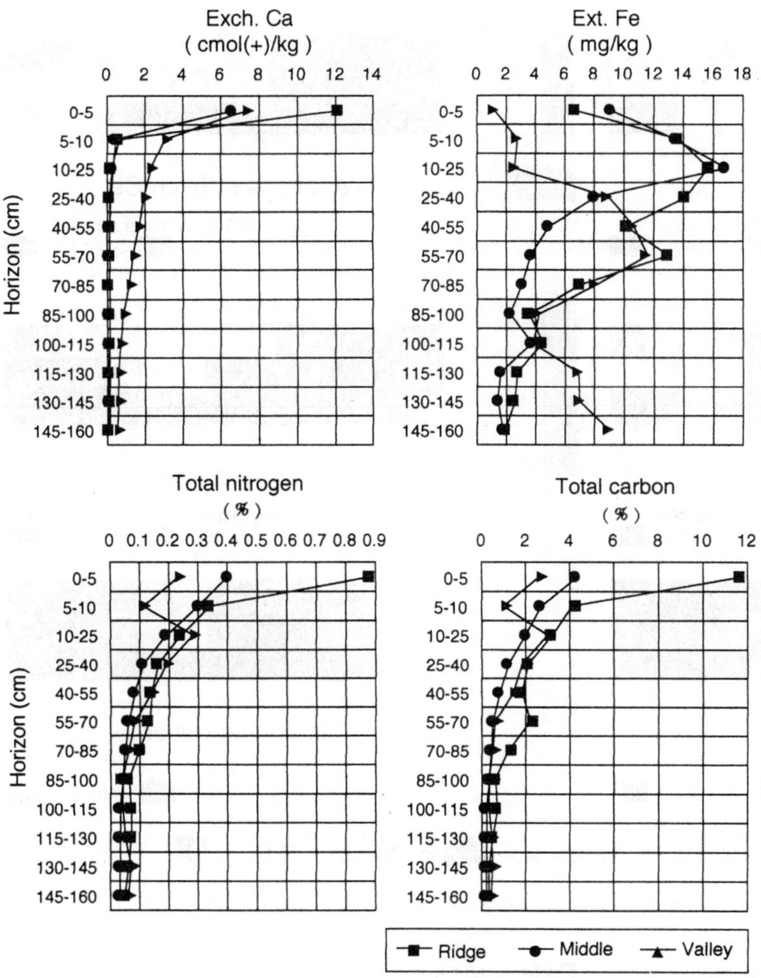


**Figure 6:** Exchangeable Ca, Extractable Fe, Total carbon and Total nitrogen variation with depth in three soil pits on the ridge, middle and valley positions in a toposequence of PAT plot.

**Table 3:** The results of calculation* of ω index between the areas of higher diversity and higher variations or higher concentrations and between the areas of lower diversity and lower variations or lower concentrations in Pinang Pinang Plot.

| Elements | Low species diversity vs. Low variation | High species diversity vs. High variation | Low species diversity vs. Low concentration | High species diversity vs. High concentration |
|---|---|---|---|---|
| Al | **0.39** | 0.24 | −0.24 | −0.12 |
| B | **0.43** | 0.05 | **0.39** | **0.33** |
| Ca | **0.50** | **0.60** | 0.20 | **0.40** |
| Cu | | −0.11 | **0.48** | 0.02 |
| Fe | 0.23 | 0.16 | **0.50** | **0.46** |
| K | **0.54** | −0.05 | 0.11 | **−0.31** |
| Mg | −0.13 | −0.09 | −0.16 | −0.12 |
| Mn | −0.03 | 0.25 | 0.03 | **−0.42** |
| Mo | **0.50** | **0.57** | 0.22 | 0.24 |
| Na | **0.37** | 0.02 | **0.52** | **0.39** |
| P | **0.36** | 0.25 | −0.09 | 0.26 |
| S | 0.25 | **0.55** | −0.27 | 0.12 |
| Si | **0.35** | −0.26 | −0.26 | **−0.50** |
| Sr | | 0.26 | **0.38** | 0.24 |
| Zn | **0.31** | **0.35** | −0.21 | **−0.57** |

* The calculations were made using consecutive 9 subplots, 30 x 30 m.
Positive indices ≥ 0.30 are shown by bold-figure and negative indices ≤ −0.30 are shown by bold-figure with underlines.

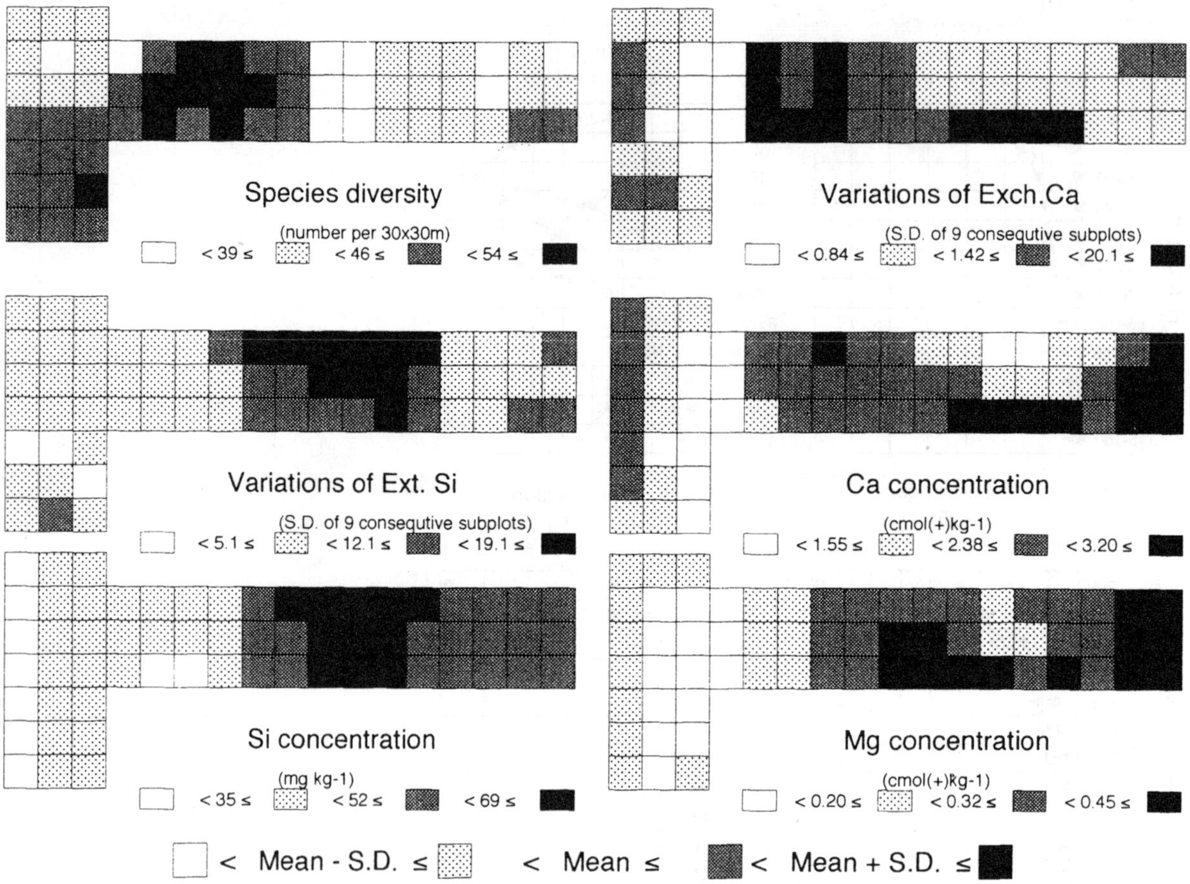


**Figure 7:** The pattern of species diversity, concentration of Exch. Ca, Mg and Ext. Si, and variation of Exch. Ca and Ext. Si

the soils of greater fertility. This makes the soil environment become relatively homogenous. In this case, a relatively few amount of species which have adapted well to the homogenous environment may be dominant.

However, less species diversity shows negative ω values with lower concentrations of B, Cu, Fe, Na and Sr.

Although it is not easy to single out the soil character relating to the tree species diversity, it may conclude from the results (Table 3) that low spatial, in both vertical and horizontal, variability of soil characters and higher soil fertility trend to decrease the diversity of tree species.

## Acknowledgements

We very much appreciated the FBRT project (Field Biology, Research and Training), lead by Prof. S. Kawamura and Fachri Ahmad, which is financially supported by JICA, Japan International Cooperation Agency. The assistance of field sampling and scientific suggestions from Ir. Junaidi, Ir. Armon, Ir. Mssa Suarbaini and Dr. Djohan Aliusiusi, Department of Soil Science, Andalas University, Padang Indonesia, were deeply appreciated. Prof. K. Ogino, Ehime University, (who established the plots) to him we are deeply grateful.

## References

*Allison, L. E.* 1965: Organic carbon. In C. A. Black et al.: Methods of soil analysis, Part 2. Am. Soc. of Agron., Agronomy, 9, p. 1367–1378 Inc., Madison, Wis.

*Ashton, P. S.* 1982: Dipterocarpaceae. In: van Steenis, C. G. G. J. (gen. ed.), Flora Malesiana ser. I 9, 237–552.

*Baillie, I. C., Ashton, P. S., Court, M. N., Anderson, J. A. R., Fitzpatrick, E. A.* and *Tinsley, J.* 1987: Site characteristics and the distribution of tree species in Mixed Dipterocarp Forest on Tertiary sediments in central Sarawak, Malaysia. Journal of Tropical Ecology, 3, 201–220.

*Gartlan, J. S., Newbery, D. M., Thomas, D. W.* and *Waterman, P. G.* 1986: The influence of topography and soil phosphorus on the vegetation of Korup Forest Reserve, Cameroun. Vegetatio, 65, 131–148.

*Gentry, A. H.* 1988: Tree species richness of upper Amazonian forests. Proceedings National Academy of Sciences USA, 85, 156–159.

*Holdrige, L. R., Grenke, W. C., Hatheway, W. H., Liang, T.* and *Tosi, J. A.* 1971: Forest Environment in Tropical Life Zones: A Pilot Study, New York: Pergamon.

*Hotta, M.* 1984: Forest ecology and flora of G. Gadut, Sumatra Nature Study (Botany), p. 220 Kyoto University, Kyoto

*Hotta, M.* 1986: Diversity and dynamics of plant life in Sumatra, Sumatra Nature Study (Botany) Part 2, p. 128. Kyoto University, Kyoto

*Hotta, M.* 1989: Diversity and plant-animal interaction in equatorial rain forests, Sumatra Nature Study (Botany), p. 302 Kagoshima University, Kagoshima

*Hotta, M.* 1994: Recent move on international tropical forest research 3. Species richness and dynamics of the tropical rain forest, an example of West Sumatra. JIRCAS Workshop Paper 2, 64–75.

*Huston, M.* 1980: Soil nutrients and tree species richness in Costa Rican forests. J. Biogeogr., 7, 147–157.

*Huston, M.* 1994: Biological Diversity, The coexistence of species on changing landscapes. p. 483–557, Cambridge University Press, Cambridge

*Iwao, S.* 1977: Analysis of spatial association between two species based on the interspecies mean crowding. Researches on Population Ecology, 18, 243–260.

*Korning, J., Thomsen, K., Dalisgaard, K.* and *Nørnberg, P.* 1994: Characters of three Udults and their relevance to the composition and structure of virgin forest of Amazonian Ecuador. Geoderma, 63, 145–164.

*Khoyama, T., Suzuki, E.* and *Hotta, M.* 1994: Spatial distribution pattern of representative tree species in a foothill rain forest in West Sumatra. Tropics, 4(1), 1–15.

*Masunaga, T., Kubota, D., Hotta, M.* and *Wakatsuki, T.* 1997: The nutritional characteristics of mineral elements of tree species of tropical rain forest, West Sumatra, Indonesia. *Soil Sci. Plant Nutr.*, (in print)

*Nagano, M.* and *Kira, T.* 1978: Aboveground biomas. In: *Kira, T., Ono, Y.* and *Hosokawa, T.* (eds.) Biological production in a warm-temperate evergreen oak forest of Japan, JIBP-PT-F Progress report for 1967, p. 45–52, Osaka City University, Osaka

*Newbery, D. M.* and *Procter, J.* 1984: Ecological studies in four contrasting lowland rain forests in Gunung Mulu national park, SARAWAK IV. Associations between tree distribution and soil factors. Journal of Ecology, 72, 475–493.

*Newbery, D. M., Gartlan, J. S., McKey, D. B.* and *Waterman, P. G.* 1986: The influence of drainage and soil phosphorus on the vegetation of Douala-Edea Forest Reserve, Cameroon. Vegetatio, 65, 149–162.

*Richards, P. W.* 1996: The tropical rain forest (2nd edn.), p. 289–294, Cambridge University Press, Cambridge

*SYSTAT Inc., 1992: SYSTAT for Windows Statistics, Version 5 Edition, Evanston, IL, 750 pp. SYSTAT, Inc.*

*Wakatsuki, T., Saidi, A.* and *Rasiydin, A.* 1986: Soils in toposequence of the Gunung Gadut tropical rain forest, West Sumatra. Southeast Asian Studies, 24, 243–262.

*Whitmore, T. C.* 1984: Tropical rain forests of the Far East, (2nd edn.), Clarendon, Oxford

*Whitmore, T. C.* and *Sidiyasa, K.* 1986: Composition and structure of a lowland rain forest at Toraut, northern Sulawesi., 41, p. 747–756, kew Bulletin

*Whitmore, T. C., Sidiyasa, K.* and *Withmore, T. J.* 1987: Tree species enumeration of 0.5 ha on Halmahera. 40, p. 31–34, Graden's Bulletin, Singapore

*Whittaker, R. H.* 1956: Vegetation of the Great Smoky Mountains. Ecological Monographs, 26, p. 1–80.

*Yoneda, T., Ogino, K., Khoyama, T., Tamin, R., Syahbuddin* and *Rahman, M.* 1994: Horizontal Variance of Stand Structure and Productivity in a Tropical Foothill Rain Forest, West Sumatra, Indonesia, Tropics, 4 (1), 17–33.

# Distribution characteristics of mineral elements in trees of tropical rain forest, West Sumatra, Indonesia

**Tsugiyuki Masunaga, Daisuke Kubota, Mitsuru Hotta*, Yoshiaki Shinmura** and **Toshiyuki Wakatsuki**

Faculty of Life and Environmental Science, Shimane University, Matsue 690, Japan
*Faculty of Science, Kagoshima University, Kohrimoto, Kagoshima 890, Japan

## Abstract

The distribution of mineral elements in trees and soils in a 1 ha permanent observation plot on mixed dipterocarp forest in West Sumatra, Indonesia was studied. Seven elements, including S, K, Al, Cl, Si, G and Sr, revealed significant correlation between leaves and barks, suggesting that barks can also be used to study the nutritional characteristics of trees instead of leaves as the first approximation. The concentrations of elements in 457 different samples of bark showed quite a wide range, indicating the enormous diversity of nutritional characteristics of mineral elements among these tree species. The concentration ranges in mg kg$^{-1}$ were Mn (0.1–3800), Sr (0.1–1070), Si (2–1700), Fe (0.6–4100), Al (10–46000), Mg (10–7200), Zn (0.4–290), K (60–22000), Cu (0.8–150), P (19–2540), Cl (110–9200), S (170–13600), Ca (2300–100000), B (4–70) and Na (130–700).

The accumulators and excluders were defined for each element as the trees which were the top 5% and the bottom 5% respectively (i.e. top and bottom 23 in elemental concentrations in their barks among the 457 tree samples). The soil edaphic factor appears to regulate the distribution patterns of some accumulators and excluders. The accumulators of K, Ca, B and Sr showed a tendency to distribute in the areas rich in those extractable elements in soils. On the other hand, the excluders of K, Al, Na, Cu, Mn and Sr showed a tendency to distribute in the areas poor in those extractable elements in soils. The accumulators and excluders of the other elements, however, did not show such a relationship to the soil edaphic characteristics. Genetic character of trees and/or other environmental factors including complex biological interaction may superimpose in edaphic factors in the distribution of these trees.

## Introduction

In primary lowland tropical rain forests, the number of tree species bigger than 10 cm in trunk diameter at breast height (DBH) is generally 100 to 150 per ha; in very rich areas, it can be more than 200 or even up to 300 per ha (*Huston*, 1994; *Richards*, 1996). This vast amount of tree species are coexisting and forming very complex tree communities (*Huston*, 1994), where the tree species may interact with each other in various ways and degrees. These interactions may be affected by several environmental factors, such as light, carbon dioxide, water and nutrients, through the resource competition among trees (*Tilman*, 1982). The distribution characteristics of mineral elements in both trees and soils are regarded as one of the major factors controlling the tree communities. There have been several reports on the relation between the elemental conditions in soil and the vegetation (*Gratin* et al., 1986; *Newbery* et al., 1986; *Baillie* et al., 1987, *Pastor* et al., 1984). Although it may be very difficult to show unambiguously some general principles to explain the observed variations in tree community composition or species distribution (*Richards*, 1996), we made an attempt to correlate the edaphic factors to explain the composition and distribution of the trees in a small scale plot. On small scale, within a few hectares or less, it can be assumed to be climatically uniform, therefore environmental factors become more important with regards to the composition of forest (*Richards*, 1996).

In this study, we analyzed the various mineral elements, such as N, S, K, Al, Ca, Mg, Cl, Si, Na, B, Cu, Fe, Mn, P, Sr, Zn in samples of 42 leaves, 457 barks and soils from 1 ha plot, West Sumatra (*Hotta*, 1984), focusing on the characteristics of mineral nutrients in both in the trees and the soils. In order to understand the association between edaphic factors and plant distribution pattern, this paper study on the relationships between the distribution pattern of the specific accumulator and excluder trees and the status of mineral nutrients in the soils. The analyses on bark were used to define accumulator trees which showed the concentration in any mineral elements higher than those of the top 5% among 457 samples, i.e. 23 trees, irrespective of species, whereas for excluder trees the concentrations lower than the bottom 5% were adapted. Although *Chenery* and *Sporne* (1976) defined an accumulator in family level with a criterion of concentration, it is difficult to evaluate the edaphic and genetic factors separately which contribute to satisfy such a criterion. In order to close up the relation between the soil edaphic characteristics and the vegetation, only top soils, 0–15 cm depth, were sampled and analyzed (*Jordan*, 1985).

## Materials and methods

### Site description

This study was conducted in a 1 ha long-term observation plot, Pinang Pinang, established by *Hotta* and *Ogino* et al. in 1980 (*Hotta*, 1984, 1986, 1989). The plot was located at a foothill dipterocarp forest in Mt. Gadut area 18 km east of Padang, West Sumatra. The plot has divided into 115 subplots with the measurements of 10 m x 10 m. The trees having over 8 cm in trunk diameter at breast height, DBH, were numbered and recorded. There were 231 species (not included undefined species) and 892 trees in this plot. Various ecological observations, including DBH and tree height measurement were started in 1981. *During* 1981 to 1987, tree mortality (% per year) and recruitment (% per year) were 0.95 and 1.92. Thus, turnover rate was 1.44 (*Phillips* and *Gentry*, 1994, *Hotta*, 1984, 1989). The Pinang Pinang plot is located on a hill top, which is partly narrow with partly broad ridges at 550 m altitude. Annual rainfall exceeds 5000 mm. Soils are relatively young Typic Dystropepts, developed from andesite and limestone. Detail of soil characerization were conducted by *Wakatsuki* et al. (1986) and *Kubota* et al. (1995).

### Samples

Barks for all trees having over 10 cm in DBH in the plot, 457 trees were sampled during June–July 1990 and September–October 1991. Barks of 293 trees were identified from 113 species while 164 trees were not yet identified. The number of tree species bigger than 10 cm DBH in this plot exceeds 180. Bark samples included both the inner and outer bark. Leaf samples, however, could only be collected from 42 trees which have some of their branches lower than 10 m above ground. Soil samples were collected at 0–15 cm depth from five points of each subplot and 115 composite samples were made for analyses.

### Analytical methods

Samples of plant materials were dried for more than 24 h in an oven at 80°C, and then ground into a powdery form by tungsten carbide vibrating mixer mill. Total elemental concentrations of S, K, Al, Ca, Mg, Si, Na, B, Cu, Fe, Mn, P, Sr and Zn were measured by inductively coupled plasma atomic emission spectrometry (ICP) using Shimazu ICPS-2000 after digestion with nitric acid in a high pressure Teflon vessel (*Quaker* et al., 1970, *Koyama* and *Sutoh*, 1987). The chlorine ion concentration was measured by thiocyanate method (*Adriano*, 1982). A soil extraction was made by 0.1 N HCl (soil to solvent ratio of 1:10), and analyzed by ICP.

## Results and discussions

### Leaves and barks

Table 1 shows the mean concentrations, standard deviations (S.D.), concentration ranges and coefficient of variations (C.V.) of samples for 42 leaves and 457 barks in the plot. Although the concentrations of Al, Ca and Sr in the bark samples were slightly higher than those in the leaves, most of the other elements had a higher representation in the leaves. This trend was same, when the 42 sets of sample leaves and barks were compared. In general, leaves are considered more appropriate than the bark in studying the characteristics of mineral characteristics of the tree. However, trees located in tropical rain forest are too tall to collect sufficient samples of leaves. It will require enormous labor and skill to collect enough leaves. Therefore, in order to validate the use of bark samples instead of

**Table 1:** The elemental composition and correlation coefficients of the leaves and the barks

| | g $kg^{-1}$ S | K | Al | Ca | Mg | Cl | mg $kg^{-1}$ Si | Na | B | Cu | Fe | Mn | P | Sr | Zn |
|---|---|---|---|---|---|---|---|---|---|---|---|---|---|---|---|
| leaves (n = 42) | | | | | | | | | | | | | | | |
| Mean | 3.41 | 8.95 | 1.31 | 21.9 | 3.08 | 2.61 | 831 | 225 | 28.7 | 4.3 | 350 | 786.5 | 731 | 14.8 | 23.0 |
| S.D. | 1.93 | 4.96 | 2.33 | 8.5 | 1.29 | 2.07 | 2338 | 166 | 30.7 | 6.8 | 203 | 710.2 | 319 | 16.6 | 29.3 |
| Min. | 1.11 | 2.98 | 0.01 | 7.0 | 1.05 | 1.01 | 2 | 131 | 3.9 | 0.8 | 77 | 44.0 | 207 | 0.1 | 2.2 |
| Max. | 11.50 | 26.27 | 9.54 | 50.4 | 6.03 | 11.90 | 13810 | 674 | 136.1 | 39.8 | 925 | 3035.0 | 1619 | 80.5 | 167.5 |
| C.V. (%) | 57 | 55 | 177 | 39 | 42 | 79 | 281 | 74 | 107 | 159 | 58 | 90 | 44 | 113 | 127 |
| barks (n = 457) | | | | | | | | | | | | | | | |
| Mean | 1.44 | 4.10 | 2.01 | 26.9 | 1.16 | 1.24 | 804 | 168 | 9.3 | 9.7 | 120 | 301.3 | 139 | 65.9 | 17.1 |
| S.D. | 1.26 | 2.89 | 4.78 | 15.0 | 1.09 | 1.00 | 2200 | 68 | 9.1 | 10.8 | 227 | 429.4 | 181 | 86.2 | 24.7 |
| Min. | 0.17 | 0.06 | 0.01 | 2.3 | 0.01 | 0.11 | 2 | 131 | 3.9 | 0.8 | 1 | 0.1 | 19 | 0.1 | 0.4 |
| Max. | 13.58 | 22.12 | 45.89 | 100.3 | 7.24 | 9.22 | 16960 | 700 | 69.6 | 149.2 | 4126 | 3797.8 | 2540 | 1071.5 | 285.0 |
| C.V. (%) | 88 | 70 | 238 | 56 | 94 | 81 | 274 | 40 | 98 | 111 | 189 | 143 | 131 | 131 | 144 |
| Corelation coefficient# | | | | | | | | | | | | | | | |
| | 0.69*** | 0.43** | 0.8*** | 0.02 | 0.04 | 0.57*** | 0.91*** | −0.31* | 0.59*** | 0.11 | 0.03 | −0.07 | −0.12 | 0.6*** | 0.21 |

*** > 0.1% > ** > 1% > * > 5% significance.
#42 sets of leaves and barks were used for calculation.
Number of samples is in parenthesis.

leaves, we calculated the correlation coefficients of elements in the barks with that of the leaves (Table 1). The concentrations of S, K, Al, Cl, Si, B and Sr showed significant correlation, and Na had a negative correlation. Calcium, Mg, Cu, Fe, Mn, P and Zn, however, did not show any significant correlation. These results suggest, at least as our first approximation, the composition in the bark can be used for the characterization of mineral elements in tropical rain forest trees.

The order of the arithmetic mean concentrations of the elements in the bark samples showed that Ca was the highest and B was the lowest, i.e., Ca > K > Al > S > Cl > Mg > Si > Mn > Na > P > Fe > Sr > Zn > Cu > B. The order of the coefficients of variation were Si > Al > Fe > Zn > Mn > Sr > P > Cu > B > Mg > S > Cl > K > Ca > Na. Those of Al and Si were beyond 200%. The concentration range in mg $kg^{-1}$ were Mn (0.1–3800), Sr (0.1–1070), Si (2–1700), Fe (0.6–4100), Al (10–46000), Mg (10–7200), Zn (0.4–290), K (60–22000), Cu (0.8–150), P (19–2540), Cl (110–9200), S (170–13600), Ca (2300–100000), B (4–70) and Na (130–700) respectively. These results showed the extremely high variation in the concentrations, indicating the enormous diversity of the nutritional characteristics of mineral element of the trees even in merely a 1 ha plot.

Table 2 shows the correlation coefficient of elemental concentrations in the bark samples and DBH. Correlations of Ca-Sr and B-P were extremely high. The high correlation between Ca and Sr in soils is well known (*Wakatsuki*, 1978). An effect of B on P uptake at the root tips of faba bean and maize has been reported (*Pollard* et al., 1977). Similar physiological mechanisms may apply in tropical trees as well. Phosphorus showed strong correlation with Zn, K and Cu. Potassium showed significant correlations with S, B and Mg. Correlations among Zn-Ca, Zn-Mg, Zn-B and Zn-Mn were strong. The DBH showed negative correlations with the concentrations of most elements, especially S, Ca, Mg and Zn at 1% level of significance. This negative correlation was probably due to the dilution in concentration of elements in the bark as the tree was growing.

Fig. 1 shows the frequency distributions of several elements in the bark samples. The distribution pattern of Ca showed 3 peaks at around 30, 50 and 90 g $kg^{-1}$, and the frequency distribution of logarithmic concentration of Si also showed 3 peaks around 3.5, 2 and less than 1. Those peaks may suggest that there are 3 groups of tree species with respect to the requirement Ca or Si. Not only for Ca and Si but also for most of the other elements showed extremely high concentration ranges. This indicated that some tree species had trends to accumulate specific elements.

The tree species which showed a concentration higher than mean + 2 S.D. in any elements were listed as tree species showing accumulative characteristics in Table 3. The concentration ranges of accumulation of each element were up to 80 g $kg^{-1}$ for Ca, up to 40 g $kg^{-1}$ for Al, up to 20 g $kg^{-1}$ (20000 mg $kg^{-1}$) for K and Si, up to 10 g $kg^{-1}$ (1000 mg $kg^{-1}$) for S, Mg, Cl, Fe, Mn and P, below 1000 mg $kg^{-1}$ for Na, Sr and Zn. Those concentration ranges of accumulation of most elements may be higher in leaves considering that the elemental concentrations in leaves were higher than those in bark samples as shown in Table 1. Aluminium accumulation of *Aporusa sp.* (Euphorbiaceae) and *Memecylon sp.* (Melastomaceae) correspond to Chenery's findings (*Chenery*, 1950, *Chenery* and *Sporne*, 1976). *Maschalocorymbus corymbosus* (Rubiaceae) and *Eurya acuminata* (Theaceae) also correspond to Chenery's finding, but only in terms of family level. The accumulative characteristics of *Aporusa sp.* trees for Al and *Ficus sp.* trees for Ca and Cl correspond in genera levels, whereas most species possessed their own accumulative characteristics. Although

**Table 2:** Correlation coefficients matrix of 457 barks and DBH.

| | S | K | Al | Ca | Mg | Cl | Si | Na | B | Cu | Fe | Mn | P | Sr | Zn |
|---|---|---|---|---|---|---|---|---|---|---|---|---|---|---|---|
| K | 0.38** | | | | | | | | | | | | | | |
| Al | 0.28** | −0.03 | | | | | | | | | | | | | |
| Ca | 0.13** | 0.09 | 0.15** | | | | | | | | | | | | |
| Mg | 0.19** | 0.30** | −0.05 | 0.15** | | | | | | | | | | | |
| Cl | 0.11 | 0.17** | −0.04 | 0.01 | 0.11 | | | | | | | | | | |
| Si | 0.13** | 0.00 | 0.00 | −0.10 | 0.02 | 0.03 | | | | | | | | | |
| Na | 0.00 | 0.02 | 0.06 | 0.00 | 0.01 | −0.01 | −0.04 | | | | | | | | |
| B | 0.11 | 0.37** | −0.02 | 0.13** | 0.22** | 0.07 | 0.05 | 0.15** | | | | | | | |
| Cu | 0.00 | 0.04 | −0.05 | −0.03 | 0.09 | 0.02 | −0.01 | 0.04 | 0.11 | | | | | | |
| Fe | 0.11 | 0.09 | 0.11 | 0.14** | 0.12 | 0.16** | 0.10 | −0.03 | 0.11 | 0.04 | | | | | |
| Mn | 0.10 | −0.01 | −0.03 | 0.13** | 0.15** | 0.08 | 0.06 | 0.04 | 0.07 | −0.05 | 0.02 | | | | |
| P | 0.15** | 0.33** | 0.18** | 0.12** | 0.22** | 0.06 | −0.06 | 0.12** | 0.53** | 0.27** | 0.12 | 0.04 | | | |
| Sr | −0.03 | 0.05 | 0.02 | 0.56** | 0.07 | 0.02 | −0.03 | −0.01 | 0.13** | 0.00 | 0.02 | 0.00 | 0.07 | | |
| Zn | 0.09 | 0.15 | 0.09 | 0.27** | 0.29** | 0.07 | 0.05 | 0.10 | 0.36** | 0.10 | 0.17** | 0.32** | 0.39** | 0.12 | |
| DBH | −0.16** | −0.11 | −0.09 | −0.15** | −0.16** | −0.06 | −0.06 | 0.02 | −0.10 | −0.09 | −0.07 | −0.08 | −0.08 | −0.11 | −0.14** |

** Significant at 1% level

DBH, trunk diametre at breast height

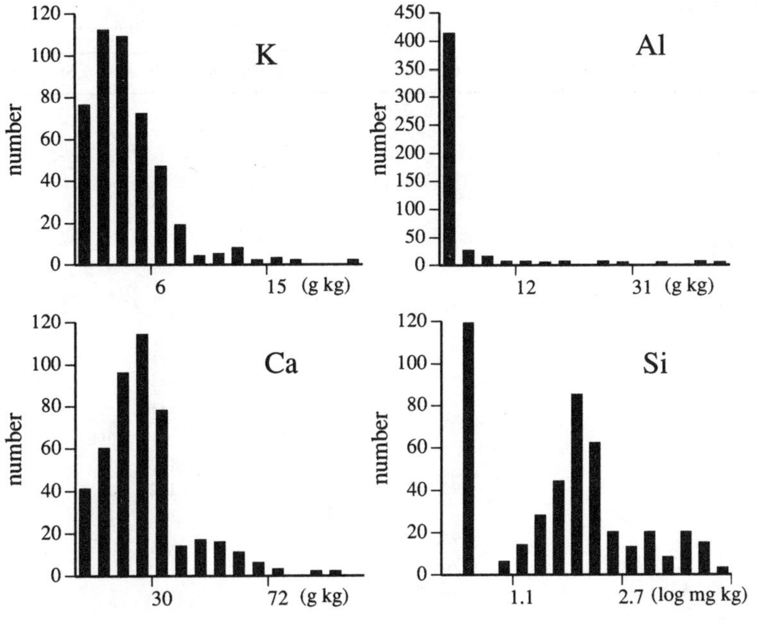


**Figure 1:** Frequency distribution of mineral elements in the barks.

data are not shown, the mineral compositions of many species varied even in the same species. These observation indicate that tree species and genotypes within species differ widely in nutritional characteristics as described by *Foy* (1983) on the tolerance to various mineral stress. Plants adapt their nutritional characteristics to mineral nutrient availability in soils (*Chapin*, 1980).

### Spatial distribution of accumulators and excluders, and soil edaphic level

In order to elucidate the effect of availability of mineral elements in soil on the nutritional characteristics and the distribution patterns of the tree species, accumulators and excluders were defined tentatively for each element as trees which were included in the top 5% and the bottom 5% in elemental contents of bark samples respectively, i.e. 23 out of 457 trees each, irrespective of tree species. However, there were more than 23 trees which had a concentration lower than detection limits for Si, Na, B, P and Zn. All trees lower than detection limit for Si (119 trees), Na (249), B (241), P (79 trees) and Zn (93 trees) were regarded as excluder trees. Those accumulators and excluders are considered to be more influenced by soil than the other trees. We divided all the subplots into two categories: high and low with regards to the mean value of extractable elements. The subplots higher than the means are represented as the shaded areas in Fig. 2–1 to 2–15, however, Cl in soil was not analyzed and therefore not shown in Fig. 2–6. The distribution patterns of accumulators (black squares) and excluders, (white circles) are also shown in Fig. 2–1 to 2–15. With regards to the distributing area of accumulators and excluders on the plot, the accumulators of K, Ca, Mg, Cl, Na, B, Sr seem to concentrate in the middle part of the plot, while the excluders of Cu, Sr and Zn seem to distribute mainly on both right and left ends of the plot. Other accumulators and excluders seem to distribute throughout the plot. The relationships between the distribution of accumulators and excluders, and soil edaphic levels are shown in Table 4 which shows the relative distribution percentages of accumulators and excluders on the subplot higher than the mean of each extractable element. If the percentage is higher than 50%, this suggests edaphic interactions for accumulators and negative interaction for excluders may exist. However, if lower than 50%, negative interaction for accumulators and edaphic interaction for excluders is possible, although explaining the negative interaction is difficult. But, those values around 50% seem to show a weak interaction between the distribution of accumulators and excluders, and soil edaphic levels. Therefore, we regarded that values over 60% indicate edaphic interactions for accumulators and negative interaction for excluders and that those lower than 40% indicate negative interaction for accumulators and edaphic interaction for excluders. The distribution patterns of some accumulators and excluders were related to the distribution patterns of the high elemental extractable areas. The elements which revealed over 60% (Table 4) in relative distribution percentage of accumulators on the subplots higher than the mean of each extractable element in soils include K, Ca, B and Sr. The elements which showed less than 40% for excluders include K, Al, Na, Cu and Sr. These results indicate that there are interactions between the distribution of accumulators and excluders, and soil edaphic levels. The soil edaphic levels can affect the distribution of accumulators. The accumulators, in turn, can contribute to the increase of the available

**Table 3:** List of tree species showing an accumulative characteristics.

| | | | | g $kg^{-1}$ | | | | | | mg $kg^{-1}$ | | | | | | | | |
|---|---|---|---|---|---|---|---|---|---|---|---|---|---|---|---|---|---|---|
| family | genus | species | n | S | K | Al | Ca | Mg | Cl | Si | Na | B | Cu | Fe | Mn | P | Sr | Zn |
| Bruseraceae | *Canarium* | sp | 1 | | | | | | | 7660 | | | | | | | | |
| Celastraceae | *Euonymus* | *javanicus* | 1 | | | | | | | | | | | 485 | | | | |
| Dipterocarpaceae | *Parashorea* | *lucida* | 7 | | | | | | | | | | | | | | 256 | |
| Dipterocarpaceae | *Vatica* | *umbonata* | 1 | | | | | | | 5550 | | | | | | | | |
| Ebenaceae | *Diospyros* | sp. | 1 | | | 41 | | | | | | | | | | | | |
| Euphorbiaceae | *Aporusa* | *frutescens* | 1 | | | 43 | | | | | | | | | | | | |
| Euphorbiaceae | *Aporusa* | *maingayi* | 1 | 5.5 | | 25 | | | | | | | | | | | | |
| Euphorbiaceae | *Macaranga* | *triloba* | 5 | | | | | 3.9 | | | | | | | | | | |
| Euphorbiaceae | *Mallotus* | *affinis* | 3 | | | | | | | 10400 | | | | | | | | |
| Euphorbiaceae | *Mallotus* | *glaberrimus* | 9 | 4.9 | | | | | | | | | | | | | | |
| Fagaceae | *Castanopsis* | sp. | 2 | | | | | | | 6895 | | | | | | | | |
| Fagaceae | *Lithocarpus* | *elegans* | 1 | | | | | | | | 700 | | | | | | | |
| Guttiferae | *Garcinia* | *parvifolia* | 1 | | | | | | | | | | | | 3219 | | | |
| Icacinaceae | *Gonocaryum* | *macrophyllum* | 2 | | | | | | | | 369 | 34 | | | | 1011 | | |
| Icacinaceae | *Platea* | *excelsa* | 2 | | | | | | | | | | | | 1222 | | | |
| Lauraceae | *Barringtonia* | *pendula* | 1 | | | | | | 9.2 | | | | | | 1923 | | | |
| Lauraceae | *Crptocarya* | *scortechinii* | 1 | | | | | | | | 382 | | | | | | | |
| Lecythidaceae | *Barringtonia* | *pendula* | 1 | | | | | | 9.2 | | | | | | | | | |
| Lecythidaceae | *Barringtonia* | sp. | 1 | 5.3 | 13 | | | | 7.3 | | | | | | | | | |
| Leguminosae | *Pithecellobium* | *microcarpium* | 1 | 4.1 | | | | | 3.2 | | | | | | | | | |
| Melastomaceae | *Memecylon* | *laurinum* | 2 | 7.2 | | 23 | | | | | | | | | | | | |
| Melastomaceae | *Memecylon* | sp. | 1 | | | 25 | | | | | | 38 | | | | | | |
| Melastomaceae | *Pternandra* | *caerulescens* | 2 | | | | | | | | | | | 495 | | | | |
| Meliaceae | *Aglaia* | *argentea* | 1 | | | | | 4.7 | | | | | | | | | | |
| Meliaceae | *Aglaia* | sp. | 3 | | 13 | | | 3.4 | | | | | | | | | | |
| Meliaceae | *Didymocheton* | sp. | 1 | | | | | 4.1 | | | | | | | | | | |
| Meliaceae | *Dysoxylum* | sp. | 2 | | | | | 5.9 | | | | | | | | | | 143 |
| Moraceae | *Artocarpus* | *heterophyllus* | 2 | | | | | | | 7275 | | 29 | | | | | | |
| Moraceae | *Ficus* | *lepicarpa* | 1 | | | | 78 | | 3.3 | | | 28 | | 4126 | | | | |
| Moraceae | *Ficus* | *ribes* | 2 | | | | 82 | | 3.3 | | | | | | | | 747 | |
| Moraceae | *Streblus* | *elongata* | 1 | | | | | | | 8580 | | | | | | | | |
| Myrtaceae | *Eugenia* | *decipiens* | 1 | | | | | | | | 325 | | | | | | | |
| Myrtaceae | *Eugenia* | *opaca* | 1 | | | | | | | | | 64 | 62 | | | 2540 | | |
| Opilaceae | *Champereia* | *manillana* | 1 | | | | 65 | | | | | | | | | | | |
| Rhizophoraceae | *Pellacalyx* | *axillaris* | 2 | 4.6 | 12 | | | | | | | | | | | | | |
| Rubiaceae | *Anthocephalus* | *chinensis* | 1 | | | | | 3.9 | | | | | | | | | 350 | |
| Rubiaceae | *Maschalocorymbus* | *corymbosus* | 1 | | | 25 | | 5.2 | | | 380 | | | | 2020 | | | |
| Rubiaceae | *Spathostemon* | *javensis* | 1 | 4.8 | 11 | | | | | | | | | | | | | |
| Rubiaceae | *Uncaria* | sp. | 1 | | | | | | | | | | | | | | 396 | |
| Rubiaceae | *Urophyllum* | *macrophyllum* | 2 | | | | | | 5.2 | | | | | 411 | | | | |
| Rutaceae | *Evodia* | *latifolia* | 2 | | 12 | | | | | | | 47 | | | | | | |
| Sabiaceae | *Meliosma* | *lepidota* | 3 | | | | | 3.9 | | | | | | | | | | |
| Sapindaceae | *Nephelium* | *dasyneurum* | 1 | | | | | | | | | | | | | | 319 | |
| Sapotaceae | *Madhuca* | *sericea* | 1 | 5.2 | | | | | | 16960 | | | | | | | | |
| Sapotaceae | *Madhuca* | sp. | 1 | | | | | | 4.7 | | | | | | | | | |
| Styracaceae | *Styrchnos* | sp. | 1 | | | | | | | | | | | 496 | 1192 | 672 | | |
| Theaceae | *Eurya* | *acuminata* | 4 | | | 17 | | | | | | | | | | | | |
| Thymelaeaceae | *Aquilaria* | *malaccensis* | 2 | | | | | | | | | | | | 3186 | | | 200 |
| Urticaceae | *Dendrocnide* | *stimulans* | 1 | | 15 | | | 4.8 | | | | | | | | | | |
| Urticaceae | *Villebrunea* | *rubescens* | 1 | | 12 | | 63 | | 3.5 | | | | | | | | | |
| Verbnaceae | *Callicarpa* | *arborea* | 1 | | 11 | | | | | | | 51 | | | | | | |

nutrient levels in soils, especially topsoils. With regards to the influence of the soil edaphic characteristics on the vegetation in rain forests in thousands ha scales, *Gartan* et al. (1986) described the influence of the soil phosphorus and *Baillie* et al. (1987) described that of magnesium. Those reports also showed that the soil elements can affect the vegetation in forests. Copper, Mn and P are the elements which showed less than 40% (Table 4) with regards to the relative distribution percentage of accumulators on the subplots higher than the mean of each extractable element in soils. The excluders of Mg, Si and P revealed over 60% in those relative distribution. Other

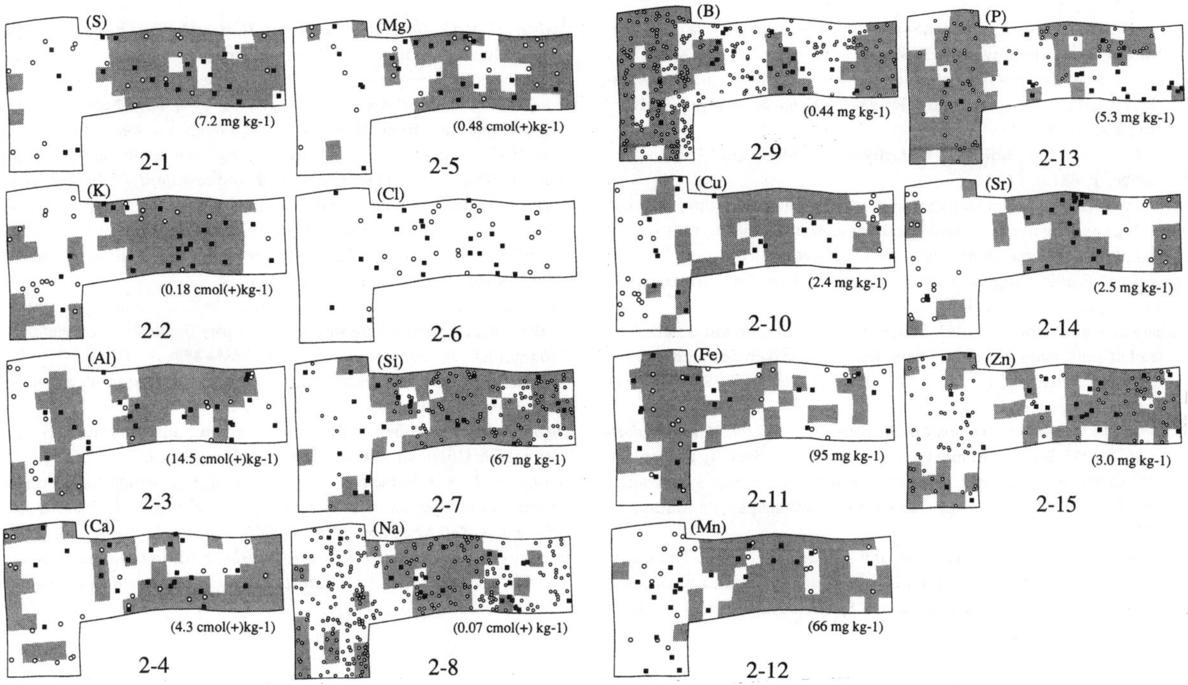


**Figure 2:** Spatial distribution of accumulators, excluders and extractable elements in soil (0–15 cm) in the plot. Black squares: accumulators, White circles: excluders, Shaded area: higher than mean concentrations (in parenthesis).

**Table 4:** Relative distribution percentages of accumulators on the subplots higher than mean elemental concentration in soil

| | (%) S | K | Al | Ca | Mg | Cl | Si | Na | B | Cu | Fe | Mn | P | Sr | Zn |
|---|---|---|---|---|---|---|---|---|---|---|---|---|---|---|---|
| accumulators | 57 | 78 | 48 | 65 | 52 | – | 57 | 48 | 70 | 17 | 52 | 22 | 35 | 87 | 48 |
| excluders | 57 | 39 | 35 | 48 | 65 | – | 64 | 40 | 52 | 9 | 57 | 17 | 80 | 26 | 44 |

accumulators and excluders which showed between 40% to 60% in relative distribution percentage in Table 4 were less related to the soil edaphic levels in terms of the tree distribution. These results suggest that some tree species have the ability to accumulate or exclude a particular element rather independent of its edaphic status in the soil. The nutritional characteristics of these trees may be affected by more a genetic nature than the soil edaphic nature.

## Acknowledgements

We greatly appreciate the efforts of Dr. S. Kawamura, Dr. R. Ogushi, Prof. T. Yoneda, Prof. A. Bakar and Fachri Ahmad for the successful operation of FBRT project (Field Biology, Research and Training) which is financially supported by JICA (Japan International Cooperation Agency). Ir Armon, MS. Junaidi, Ir. Maas Syarbaini and Dr. Djohan Aliusius, Department of Soil Science, Andalas University, Padang, Indonesia, are appreciated for their assistance in field sampling and scientific suggestions. The contribution of Prof. K. Ogino, Ehime University, is also deeply appreciated for establishing the plot investigated in this study.

## References

*Adriano, D. C.* and *Doner, H. E.* 1982. Bromine, chlorine and fluorine. In *Method of Soil Analysis*. Part 2, Ed. Page, A. L., Miller, R. H. and Keeney, D. R., 449–483. pp., ASA, SSSA, Madison, Wisconsin

*Baillie, I. C., Ashton, P. S., Court, M. N., Anderson, J. A. R., Fitzpatrick, E. A.* and *Tinsley, J.* 1987. Site characteristics and the distribution of tree species in mixed dipterocarp forest on tertiary sediments in central Sarawak, Malaysia. *Journal of Tropical Ecology* 3, 201–220.

*Chapin, F. S. III* 1980. The mineral nutrition of wild plants. *Annu. Rev. Ecol. Syst.* 11, 233–260.

*Chenery, E. M.* 1950. Some aspect of the aluminium cycle. *Jurnal of Soil Science* 2, 97–109.

*Chenery, E. M.* and *Sporne, K. R.* 1976. A note on the evolutionary status of aluminium-accumulators among dicotyledons. *New Phytol.* 76, 551–554.

*Foy, C. D.* 1983, The physiology of plant adaptation to mineral stress. Iowa State Journal of Research 57: 355–391.

*Gartlan, J. S., Newbery, D. McC., Thomas, D. W.* and *Waterman, P. G.* 1986. The influence of topography and soil phosphorus on the vegetation of Korup forest reserve, Cameroun. *Vegetation* 65, 131–148.

*Hotta, M.* 1984. Forest ecology and flora of G. Gadut, Sumatra Nature Study (Botany). 220 pp. Kyoto University, Kyoto.

*Hotta, M.* 1986. Diversity and dynamics of plant life in Sumatra, Sumatra Nature Study (Botany) Part 2. 128 pp. Kyoto University, Kyoto.

*Hotta, M.* 1989. Diversity and plant- animal interaction in equatorial rain forests, Sumatra Nature Study (Botany). 302 pp. Kagoshima University, Kagoshima.

*Huston, A. M.* 1994. Biological diversity. Cambridge Univ. Press, Cambridge, P. 681.

*Iwao, S.* 1977. The m*-m statistics as a comprehensive method for analyzing spatial patterns of biological population and its application to sampling problems. *JIBP Synthesis* 17, 21–46.

*Jordan, C. F.* 1985. *Nutrient Cycling in Tropical Rain Forest Ecosystems.* 189 pp. John Wiley & Sons, Chichester.

*Koyama, T.* and *Sutoh, M.* 1987. Simultaneous multielement determination of soils, plant and animal samples by inductively coupled plasma emission spectrometry. *Jpn. J. Soil Sci. Plant Nutr* 58, 578–585. (in Japanese)

*Kubota, D., Masunaga, Hermansyah, Rasydin, A., T., Hotta, M.* and *Wakatsuki, T.* 1995. Soil envirnmental and tree species diversity in tropical rain forest, West Sumatra, Indonesia. In 3rd international symposium on "Soils of Tropical Forest Ecosystems", Balikpapan, Indonesia. 6, 122–139.

*Masunaga, T., Kubota, D., Hotta, M.* and *Wakatsuki, T.* 1996. The nutritional characteristics of mineral elements of tree species of tropical rain forest, West Sumatra, Indonesia. *Soil Sci. Plant Nutr.* submitted

*Newbery, D. McC., Gartlan, J. S., Mckey, D. B.* and *Waterman, P. G.* 1986. The influence of drinage and soil phosphorus on the vegetation of Douala-Edea forest reserve, Cameroun. *Vegetation* 65, 149–162.

*Ogino, K., Hotta, M., Rusjdi, T.* and *Yoneda, T.* 1984. Forest ecology of G. Gadut area. In: Hotta, M. (ed.), *Forest ecology and flora of G. Gadut*, pp. 15–37. Sumatra Nature Study (Botany). Kyoto University, Kyoto.

*Pastor, J., Aber, J. D., McClaugherty, C. A.* and *Melillo, J. M.* 1984. Aboveground production and N and O cycling along a nitrogen mineralization gradient on Blackhawk island, Wisconsin, *Ecology* 65, 256–268.

*Phillips, O. L.* and *Gentry, A. H.* 1994. Increasing turnover time in tropical rain forests. *Science* 263, 954–958.

*Quaker, N. R., Klucker, P. D.* and *Chang, G. N.* 1970. Calibration of inductively coupled plasma emission spectrometry for analysis of the environmental materials. *Anal. Chem.* 51, 885–895.

*Richards, P. W.* 1996. *The Tropical Rain Forest.* 575 pp. Cambridge University Press, Cambridge.

*Tilman, D.* 1982. Resource competition and community structure. 296 pp. Princeton University Press, Princeton New Jersey.

*Wakatsuki, T.* 1978. Behaviors of elements during weathering of transported soil materials (part 7), Leacheate elements – Mg, Ca, Na and Sr. *Jpn. J. Soil Sci. Plant Nutr.* 49, 270–274. (in Japanese)

*Wakatsuki, T., Saidi, A.* and *Rasydin, A.* 1986. Soils in the toposequence of the G. Gadut topical rain forest, West Sumatra. *South Asian Studies* 24, 23–262.

# Forest soils under alternatives to slash-and-burn agriculture in Sumatra, Indonesia

**Meine van Noordwijk**[1], **Daniel Murdiyarso**[2], **Kurniatun Hairiah**[3], **Upik Rosalina Wasrin**[4], **Achmad Rachman**[5] and **Thomas P. Tomich**[1]

[1]ICRAF-S.E. Asia, P.O. Box 16001, Bogor, Indonesia
[2]Bogor Agricultural University, Bogor, Indonesia
[3]Brawijaya University, Malang, Indonesia
[4]SEAMEO-BIOTROP, Bogor, Indonesia
[5]Center for Soil and Climate Research, Bogor, Indonesia

## Abstract

A global project on 'Alternatives to Slash and Burn' agriculture was initiated by a consortium of international and national research institutes to facilitate intensification of the use of converted forest land, in order to help alleviate poverty and protect the remaining forest areas for their biodiversity values and their role in mitigating greenhouse gas emissions.

Data for the Indonesian benchmark areas in the lowland peneplain, piedmont and mountain zone of Sumatra are presented. A significant amount of forest land, especially in the lowland peneplain, has been converted in the last ten years into agricultural use, usually following logging concessions. Soils on the peneplain are poor (oxi- and ultisols) and current intensive crop based production systems are not sustainable. In the piedmont zone on better soils (inceptisols), rubber agroforests (still) characterize the area. Agroforests have emerged during the 20'th century as the major alternative to slash-and-burn agriculture, based on a shift of emphasis from food crops to cash-earning tree crops. Emphasis on food crops, however, continues in government resettlement schemes.

Differences in organic C content of the topsoil between forests and crop land are about 0.5 % C, with agroforests and tree crop plantations in an intermediate position. A new size-density fractionation scheme for soil organic matter demonstrated larger changes in light and intermediate fractions. Forest soils can be significant sinks for methane and thus partly compensate for the methane emissions in lowland rice production.

Overall, the Sumatra benchmark areas demonstrate the need to combine intensification of land use at the field/household level with effective protection of remaining forest areas at the community level and reducing other driving forces of deforestation at the national level.

## Introduction

### The need for alternatives to slash-and-burn agriculture

One-and-a-half century before *Nye* and *Greenland* (1960) wrote their seminal 'the soil under shifting cultivation', *Marsden* (1811) described swidden practices in Sumatra. His contemporary sources in Sumatra were well aware of the importance of soil organic matter as the source of soil fertility in the few years after opening a piece of forest by slash-and-burn methods and the fact that depletion of this fertility necessitated abandoning the plot to a period of fallow regrowth. Marsden was early in expressing reverence for the rain forest so destroyed 'I could never behold this devastation without a strong sentiment of regret', 'it is not difficult to account for such feelings on the sight of a venerable wood, old, to appearance, as the soil it stood on, and beautiful beyond what pencil can describe, annihilated for the temporary use of the space it occupied'. Nowadays concern about forest destruction expressed by outside interest groups is usually articulated in terms of biodiversity conservation, watershed protection and greenhouse gas emissions, but the underlying emotions may be close to the ones formulated by Marsden.

In the early part of the 20'th century the debate on the role of shifting cultivation in the stil forested landscape of Sumatra, focussed on the need for technical developments allowing a more intensive land use, partly necessitated by the increasing allocation of land to large scale plantation crops (*Koens*, 1925; *Hagreis*, 1930). Intensification of land use for crop production was expected to depend on better management of soil organic matter and nutrient content. At the same time the development of *Imperata* grasslands was attributed to overexploitation of the soil in failed attempts at intensification (*Danhof*, 1941). The global project on 'Alternatives to Slash-and-Burn' at the end of the 20'th Century thus has well-developed roots in Sumatra. Global concerns about the fate of tropical rain forests should be linked to the opportunities and constraints perceived by local land users.

Human exploitation of forests can be based on above as well as belowground resources (Fig. 1). Aboveground the usual division is between non-timber forest products

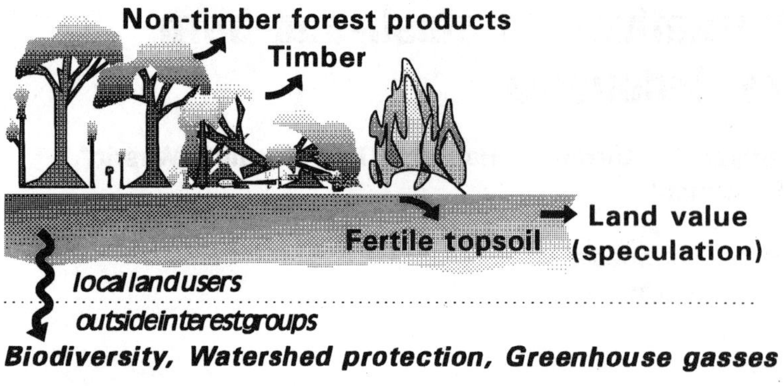


**Figure 1:** Functions of forest soils to local land users as well as external interest groups.

(NTFP), which may include medicinally used roots, and timber, where the first may be collected in less destructive ways then timber harvesting (logging), although this is not true for all products. Belowground we can distinguish between the short term benefits of soil fertility after slash-and-burn conversion of forests in a long-fallow rotation and the value of the deforested land for permanent agricultural or plantation use. This latter may include a speculative element. Apart from these values of the forest for local users and newcomers, there is increasing attention for its value to the outside world for biodiversity conservation, watershed protection and as mitigator of greenhouse gas emissions. These 'environmental service functions' refer to the soil as well as aboveground part of the forest ecosystem.

The damage done to the forest ecosystem and its 'environmental service functions' increases from collecting NTFP, via logging and long-fallow rotations to conversion into non-forest land uses, either consisting of short-fallow rotations, semipermanent agriculture or non-agricultural land use. Interactions between these activities are common. For example, logging roads provide easy access to farmers opening the remaining forest by slash-and-burn methods. Slash-and-burn methods may be used both for the short term exploitation of soil fertility (in the 'slash-and-burn agricultural system') and for (semi)permanent use of the land for farming or other types of activity.

Compared to other methods of forest clearing, slash-and-burn methods have less impact on soil physical conditions of forest soils and they should be preferred over bulldozer land clearing (*Alegre* and *Cassel*, 1996). Losses of nutrients due to leaching may be high in the first few years after burning (*Juo* and *Manu*, 1996), however, and this may lead to a gradual decline of soil fertility with each slash-and-burn cycle. The soil organic matter content of ex-forest soils is of crucial importance to the productivity of subsequent crops (*Nye* and *Greenland*, 1960; *Palm* et al., 1996).

Slash-and-burn farming systems exploiting the forest for its soil fertility effects are generally viewed as having a low productivity relative to the amount of damage they do to forest resources (*Sanchez* et al., 1990; *Brady*, 1996). The global project on 'Alternatives to Slash-and-Burn' (ASB) is built on the hypothesis that: 'Intensifying land use as alternative to slash-and-burn farming can help to reduce deforestation, conserve biodiversity, reduce net emission of greenhouse gasses and alleviate poverty'. The hypothesis thus implies (semi)permanent activities on a small area as alternative to extensive slash-and-burn activities on a large area.

If and where this central ASB hypothesis is true, research and development efforts should be aimed at supporting farmers in developing land use technology, in which agroforestry options may play a central role. For conditions where the hypothesis does not seem to apply, we may need different types of activities to achieve the aims of reducing deforestation.

In the first phase of the ASB project a broad ranging 'characterization and diagnosis' activity was initiated in Brazil, Cameroon and Indonesia to collect baseline data on the nature of current slash-and-burn conversion of tropical forests in the three continents and judge the relevance of this intensification hypothesis in the local context of farmers, other users of forest resources and government institutions. Guidelines and procedures were developed for 'Characterization and Diagnosis' for the global project (*Izac* and *Palm*, 1994). We will summarize results for the Indonesian sites here, with particular attention to the soil-related aspects. For a broader account, including socio-economic and policy aspects, the reader is referred to *Tomich* and *Van Noordwijk* (1996) and *Van Noordwijk* et al. (1995).

After describing the various land use options which exist, we will focus on soil organic matter (SOM) and its changes during conversion of forests to other land uses, as it is important for at least three of the functions of forest soils:

– SOM is important for the **sustainability** of food crop production in slash and burn (fallow rotation) systems via its role in: (a) N and P mineralization, (b) Al detoxification, (c) maintenance of soil structure,

– SOM is relevant to **greenhouse emissions** and sinks, based on total C stocks in the soil and possibly via a link with $CH_4$ and $N_2O$ sinks and sources,

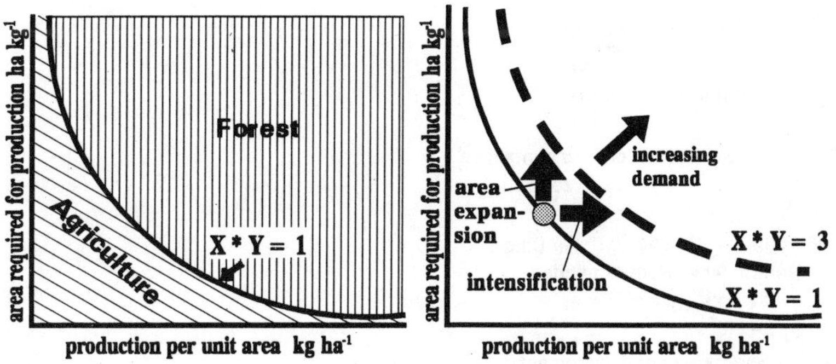


**Figure 2A:** The amount of land needed for agricultural production is the inverse of the productivity per ha; thus 'land use intensification' can contribute to forest protection. **B.** Increase in production can be achieved by both area expansion and intensification.

– SOM is related to soil **biodiversity**, via functional groups of soil organisms such as mycorrhizae, rhizobia, methanotrophe bacteria etc.

Measurement of total soil-C is adequate for evaluating C-stocks in the soil, but not for studying soil-C dynamics on a short term, as only a small part of the total C is responding rapidly. Thus methods are needed for quantifying specific fractions of soil organic matter (*Hairiah* et al., 1995).

We will first discuss the theoretical basis of the intensification hypothesis, then summarize the characterization and diagnosis data for Sumatra, focussing on land use patterns and soil organic matter and then return to the intensification hypothesis and its relevance for Sumatra.

### Intensification hypothesis

The underlying model for the intensification hypothesis is: $Y = 1/X$, where Y is the amount of land needed for a given amount of agricultural products (ha per kg of product) and X is the productivity per unit land of the agricultural system (kg of product per ha) (Fig. 2A). If we compare the rice yields in a true shifting cultivation system (say 1.5 and 0.5 Mg $ha^{-1}$ in two years of cultivation, alternating with 28 years of fallow: this leads to $2/30 = 0.067$ Mg $ha^{-1}$) with the 10 Mg $ha^{-1}$ that is possible in intensive irrigated rice fields (at least two crops of rice per year a 5 Mg $ha^{-1}$ each, no fallow periods), we can easily see that intensifying rice from shifting cultivation to intense paddy fields allows a 150-fold increase in production per ha and a reduction of the amount of land needed to feed one person (say 250 kg per person per year) from 3.7 to 0.025 ha. This type of intensification, provided that it is technically possible, can thus reduce the land claims for agriculture and allow more forest to be conserved. But will it happen that way?

The first thing to note is that Fig. 2A assumed a constant 'demand' for agricultural products. Fig. 2B indicates that increased demand for production, e.g. due to population growth, can be met by area expansion (the vertical arrow), intensification (the horizontal arrow) or by mixed strategies. Intensification may thus help to keep up with growing demands (population size and demands per person) as alternative to area expansion, rather than actually allowing currently used land to go back to more natural systems. There is an important school of theory stating that in fact intensification of land use will only start when all opportunities for area expansion have been utilized (*Boserup*, 1965).

In the project documents ASB has emphasized two situations: 'forest margins' where active forest conversion occurs and 'degraded lands'. The links between these two situations are manifold:

- forest conversion for unsustainable land use systems can lead to land degradation and continued 'land hunger' for the remaining forest,
- more sustainable land use systems directly following forest conversion may thus reduce the rate at which degraded lands are formed and slow down forest conversion,
- intensified land use on degraded lands may be an alternative to forest conversion, but only if the remaining forests are effectively protected.

Landscape level processes and driving forces of human migration may be at least as important as field level technical aspects of the land use system.

## Sites and methods

### Benchmark areas

The global guidelines and procedures for 'Characterization and Diagnosis' for the ASB project (*Izac* and *Palm*, 1994) specify a stepwise approach to the choice of study sites based on stepwise stratified sampling, in order to extrapolate results in a later stage to the strata identified beforehand (Fig. 3).

Benchmark areas were chosen in the lowland peneplain, piedmont and mountain zone of central Sumatra (in the provinces of Jambi and West Sumatra) and in Lampung (Table 1; Fig. 4). The latter represents higher population densities. Existing data were summarized for the purpose of the project and new data were collected on: vegetation, land use, soil type at family level, soil organic carbon fractions, net flux of methane and a number of socio-economic indicators.

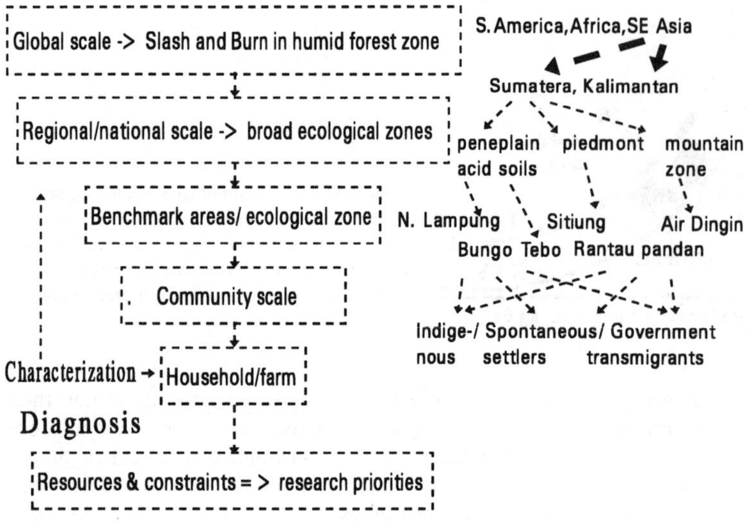


**Figure 3:** Stepwise choice of research sites and extrapolation domains.

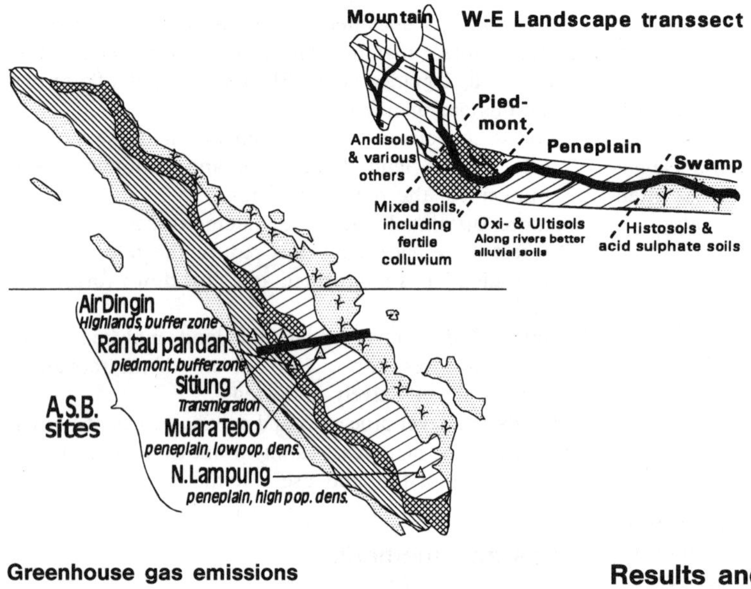


**Figure 4:** Ecological zones of Sumatra according to Scholz (1983) and benchmark areas characterized for the ASB project.

### Greenhouse gas emissions

Preliminary measurements were made on methane ($CH_4$) and nitrous oxide ($N_2O$) sources and sinks in various land use types derived from forest soils. Gas samples from a small sampling chamber (*Murdiyarso* et al., 1994) were transported to Bogor in airtight bottles for analysis with a Gas Chromatograph. From a linear regression of concentration against time, the source (positive slope) or sink (negative slope) strength were derived.

### Soil organic matter fractions

A size-density fractionation method for soil organic matter was used (*Meijboom* et al., 1995; *Hairiah* et al., 1995) to separate macro-organic matter of the 150–2000 µm size class into light, intermediate and heavy fractions, using Ludox (silica) suspensions of 1.13 and 1.3 g $cm^{-3}$ for the density distinctions.

## Results and discussion

### Indonesia as benchmark area

Indonesia was chosen to represent the humid tropical forest zone in Asia for the global ASB project. Indonesia still has large forest areas, but forest conversion to other land uses is rapid. The transformation from primary to secondary forest types is largely due to timber extraction, with a smaller role for traditional shifting cultivation systems. Subsequent transformation of secondary and logged over forest types generally is based on 'slash-and-burn' practices, by a variety of actors for a variety of reasons. Part of the forest is converted to (temporary) crop land, either in government sponsored schemes or by spontaneous migrants. These lands can evolve into *Imperata* grasslands (alang-alang) or into permanent treebased production systems (agroforests or tree (crop) plantations). Both the

**Table 1:** Site selection for characterization and diagnosis activities by ASB-Indonesia

| Benchmark Area | Ecological Zone | Main Focus in ASB | Population density relative to resources |
|---|---|---|---|
| Air Dingin, W. Sumatra | Mountain | Buffer zone of National Park (KSNP) in highlands | High, emigration |
| Rantau Pandan, Jambi | Piedmont | Buffer zone of National Park (KSNP) in piedmont, rubber agroforests, traditional shift. cult. | Intermediate |
| Sitiung, W. Sumatera | Piedmont/ peneplain | Transmigration villages interacting with local farmers | Intermediate, recent immigration |
| Bungo Tebo, Jambi | Peneplain | Forest margin: spont. settlers, transmigrants | Low, immigration |
| N. Lampung | Peneplain | Degraded land rehabilitation as alternative to migration | High, immigration + emigration |

KSNP = Kerinci Seblat National Park

'forest margin' and the 'degraded land' focus of the global ASB project are relevant in Indonesia.

Characterization at the regional/national scale should identify broad agroecological-economic areas. The historical transformation of 'shifting cultivation' to 'permanent agriculture' has occurred at different rates in various provinces of Indonesia. Broadly speaking four groups can be distinguished (*Richards* and *Flint*, 1993):

I. Java + Bali, where the transformation to permanent agriculture occurred before 1880

II. North and West Sumatra and South Kalimantan, where the transformation was nearly complete by the middle of the 20'th century,

III. Most of Sumatra, where most of the transformation took place during the middle of the 20'th century,

IV. The rest of Kalimantan and Irian Jaya which are still in the early stages of the transformation.

It was decided to start the ASB project in Sumatra (group III), but Kalimantan and Irian Jaya may offer other perspectives in a later stage.

The next step was to identify 'benchmark areas', defined as 'homogenous areas in terms of the biophysical and general socioeconomic factors that influence slash and burn activities'. Sumatra is 350 km at its widest, almost 1700 km long, and is cut in two roughly equal parts by the equator; its total land area is 480 000 km$^2$. Five major agro-ecological zones (*Scholz*, 1983) can be identified with boundaries running from N.W. to S.E. approximately parallel to the coast (Fig. 4):

1. a narrow Western coastal zone, the lower slopes of the mountain zone on the S.W. side, with various soil types,
2. a mountain zone, dominated by andisols and latosols of reasonable to high soil fertility,
3. a narrow piedmont (foothill) zone, the lower slopes of the mountain range on the N.E. side, dominated by latosols and red-yellow podzolics,
4. a broad peneplain zone, almost flat land with Tertiary sediments, deposited in the sea; at present its altitude is less than 100 m above sea level and it consists for about 10% of river levees and flood plains with more fertile alluvial soils and for 90% of uplands with a gently undulating landscape and mostly red-yellow podzolic soils,
5. a coastal swamp zone with peat and acid sulphate soils.

The zones 1, 2 and 3 contain the most fertile soils and have been inhabited for long periods of time. The coastal swamps and the peneplain were inhabited sparsely as human population was traditionally concentrated along the river banks on relatively favourable sites.

Since the beginning of 20th Century, population density in Sumatra increased also in the peneplain by (trans)migration from Java both spontaneously and sponsored by the government. A clear gradient in population density exists from the South (Lampung) to the central part (Jambi, Riau) of the island. Although the major part of the land in Sumatra is considered to be government forest land, a substantial part of this land is no longer under forest cover.

## Forest conversion and slash-and-burn as a land clearing method

A forest map by *Van Steenis* (1935) shows that although the major part of the island was still under forest cover by that time, it started to look like an Emmenthaler cheese with big holes. Forest conversion by that time had taken place mainly in a) coastal zones, esp. Aceh, W. Sumatra, Bengkulu and Lampung, b) close to the major rivers in the eastern peneplain, esp. the Musi river in S. Sumatra and the Batanghari river in Jambi and c) N. Sumatra, the area of the plantation boom in the late 19'th, early 20'th century (tobacco, rubber, oil palm).

In 1982 forest conversion had affected most of the remaining forest in Lampung and South Sumatra, but in Jambi had not changed much in comparison with 1932 (FAO/*MacKinnon*, 1982). The completion of the Transsumatra Highway and associated Transmigration projects in the early 1980's would soon make their presence felt, however. The ASB-benchmark areas in Jambi are thus located

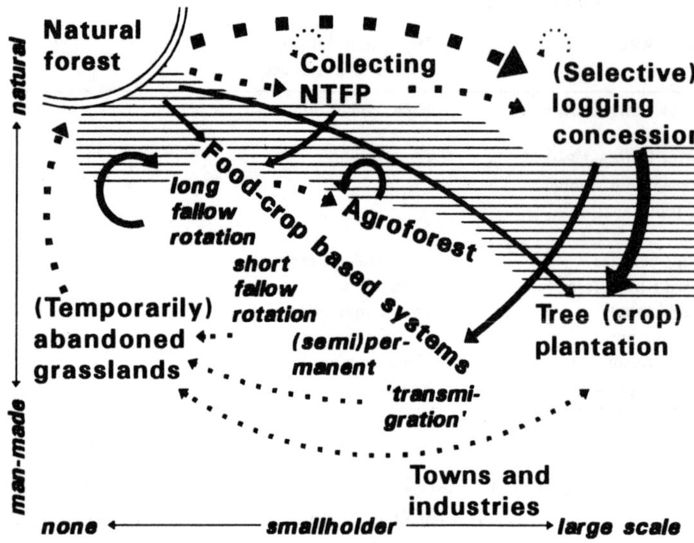


**Figure 5:** Transformations of forest soils to other land use types as found in Sumatra; all transformations in the shaded areas use 'slash-and-burn' as a method for land clearing; the 'agroforests' are the major farmer-developed alternative to slash-and-burn based food crop production systems.

in an area where forest conversion along the major rivers took place before the 1930's but which remained under forest cover at least until the early 1980's. The N. Lampung benchmark area neighbours on one of the few forest patches left in the Lampung-S. Sumatra part of the E. peneplain.

Slash-and-burn is the common method for land clearing in Sumatra. It is applied to a large range of forest types, however, by a range of land users and leading to a range of new land use systems (indicated by solid arrows in Fig. 5). The connotation of 'slash-and-burn' as a conversion of natural forest into temporary food-crop based land use systems in not the most common type. Slash-and-burn methods are used for converting logged-over forests into tree crop (oil palm) plantations, industrial timber estates or (typically crop-based) government resettlement (transmigration) schemes. Logged-over forests are supposed to get the chance to recover, but in practice large areas have been converted to such other land uses based on (local) government plans. A substantial part of the remaining logged forests were converted by spontaneous migrants, usually into rubber gardens. The negative sloping diagonal in fig. 5 shows the various types of 'shifting cultivation', 'long rotation fallow' and 'short rotation fallow', where forest or shrub land is opened to grow food crops. Intensive food crop based systems have been attempted by the transmigration program, but have been much less successful than the rubber agroforests of the local population. Where tree crops have been used for transmigration sites, poverty alleviation can be successful (Levang, *pers. comm.*). Intensification of the original long-fallow agricultural system by local farmers since the beginning of the 20'th century has led to rubber agroforests, rather than more intensive food-crop based systems (*Gouyon* et al., 1993). Elsewhere in Sumatra, on the better soils of the western coastal strip, the even more forest-like damar agroforests developed (*Torquebiau*, 1984). Agroforests, man-made forests, with a large share of directly useful trees, are thus the major 'alternative to slash and burn' in Sumatra. These agroforests can be seen as the ultimate form of 'enriched fallow systems', in the sense that the trees planted in the fallow are the major source of income for the farmers and the food crops grown in the initial years are no longer the major 'raison d'être' of the land use system. When the rubber agroforests are to be renewed, slash-and-burn methods are still used, however.

The (*Imperata*) grass fallows which can be formed after prolonged cropping, tend to be perpetuated by fire and can lead to an 'arrested succession' in the form of large ('sheet') alang-alang (*Imperata cylindrica* grasslands). Sometimes these grasslands are abandoned and farmers try their luck elsewhere, as happened in the early stages of the transmigration settlements in the benchmark areas (especially Lampung), but otherwise they can be reclaimed at a later stage. *Imperata* grasslands can also originate from failed attempts to develop tree (crop) plantations.

## Current land use

The 'agroforest' land use type has not been recognized in many of the previous descriptions. For example, the land use classification system proposed for Indonesia by *Malingreau* and *Christiani* (1981) and used in the LREP (Land Resources Evaluation and Planning) project, recognizes 'taungya' type tree plantations with food intercrops as 'agroforestry'. In analyzing remote sensing images a distinction between 'rubber agroforests' and 'secondary forest' is difficult.

A vegetation map (scale 1: 250 000) for Sumatra was published in three map sheets by SEAMEO-BIOTROP, based on Landsat MSS satellite data for the period

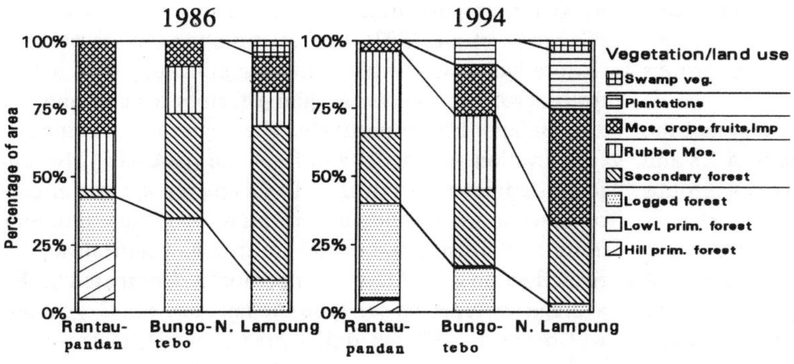


**Figure 6A:** Land use in three benchmark areas according to the 1986 vegetation map; **B.** *idem* for 1994, interpreted from Landsat TM images.

1983–1985. A description of the various vegetation types is given by *Laumonier* (1997). The vegetation description is based on natural vegetation (100 legend units, 82 for uplands and 18 for swamp vegetation) and cultivated types (21 legend units, often 'mosaics' of one or more crops and secondary vegetation). Agroforests are not identified as a separate class on this map either.

Fig. 6A shows the vegetation/land use types of the three benchmark areas according to the 1986 vegetation map. Figure 6B shows the situation in 1994 and attempts were made to separate the 'agroforest' category. A substantial increase occurred in 'plantations' (oil palm, rubber, sugar cane and industrial timber). The 'mosaic of crops, fruit trees and *Imperata*' category mainly refers to transmigration settlements in the peneplain.

### Soils in the benchmark areas

For the ASB project about 5000 ha in each benchmark area was mapped at the 1: 50 000 scale. Field soil characterization was conducted along transects, based on the differences on land form, soil catena, and land use. Undisturbed soil and bulk samples were collected to determine the soil physical and chemical properties respectively.

The Bungo Tebo and North Lampung sites are dissected peneplains, consisting of acid tuffaceous sediments. Sitiung is also a dissected peneplain and consists of acid clayey sediment alternating with acid volcanic tuff cover. The Rantau Pandan site represents a piedmont area which was built mainly by granite and andesitic lava.

Soils in Bungo Tebo and Sitiung predominantly are very deep, well drained, very acid, and low soil fertility status. Soils in Rantau Pandan are more varied and complex than the ones in Bungo Tebo. The soils range from shallow to very deep, moderate to fine in texture, well to moderately excessive drained, very acid, and low soil fertility status.

Soils in North Lampung are very deep, well drained and very acid, with a low soil fertility status; iron concretions are often found within the soil profiles. Soil erosion has occurred throughout the area with various intensities depending on land management.

Dominant soil types are: Typic Hapludox in Sitiung, Typic and Oxic. Dystropept (Rantau-Pandan), Typic Kandiudox (Bungo-Tebo) and Plinthic and Typic Hapludox (N. Lampung).

### Estimates of net C release due to land use change

The estimated land use change obtained from the overlays of land use maps at two time periods was used to estimate the changes in the above-ground biomass per unit area (Murdiyarso and Wasrin, 1995). The largest change occurred in the conversion of secondary forests and logged-over forests into cultivated areas, involving areas of 4,386,110 ha and 2,363,000 ha, respectively, or 43.5 and 21.2 percent of the initial areas. During the six year period 1986–1992 the amount of carbon released from land use changes in Jambi was estimated as 176,403,000 ton or 29,400,000 ton/yr and 2.9 t Ca $ha^{-1}$. In comparison with the countrylevel estimate of 0.19 Gt C $yr^{-1}$ for Indonesia (IPCC, 1990), Jambi may have contributed some 16 percent. This is much more than proportional to its surface area.

In the period 1986–1994 the Rantau Pandan benchmark may have been a net C sink (3.1 t C $ha^{-1}$ $yr^{-1}$), as the jungle rubber agroforests matured. The North Lampung and Bungo-Tebo benchmark areas emitted considerable amounts of C as forest was converted to vegetation types with lower C stocks (6.8 and 9 t C $ha^{-1}$ $yr^{-1}$). These estimates of changes in average C-stock of land use systems do not yet include changes in belowground C stocks.

### Effects of Land Use on Soil Carbon Content on Sumatera

*Van Noordwijk* et al. (1997) analyzed soil data for Sumatra obtained in the 1980's in the context of the LREP project. About 2800 profile data were found with complete records of soil type (Soil Taxonomy), land use, texture and $C_{org}$. Five broad soil groups emerged from the analysis with significant between group differences in $C_{org}$:
Histosols (peat, covering about 10% of Sumatra, but which may contain more than 90% of all C stored in Sumatran soils,
Andisols and wetland soils (Aquic groups) both contain about 10% of $C_{org}$. On the Andisols C is intimately bound to clay complexes (allophane), while in wetland soils, the

C is partially protected from decomposition by anaerobic conditions,

Among the remaining mineral soil types, two groups could be identified:

relatively fertile upland soils (mainly Inceptisols) and the Oxisols + Ultisols, with an average $C_{org}$ content of 3.8 and 3.2%, respectively. The differences between all groups were statistically significant in a t-test.

In general, the $C_{org}$ content decreases in the order: primary forest > secondary forest > tree crop plantations > areas used for food crops or covered by *Imperata*. On the major upland soils, the difference in $C_{org}$ content between land use types is about 0.5% C. At an average bulk density of 1.25 g $cm^{-3}$, this represents 10 Mg $ha^{-1}$ for a 15 cm top soil layer. Changes in deeper layers may be expected to be less, and the total change between natural forest and 'degraded' soils is probably in the range of 15–20 Mg $ha^{-1}$; belowground effects may thus be only 10% of aboveground changes in C stocks. On the Andisols and the wetland soils, larger differences in $C_{org}$ content are observed between land use types, but the smaller number of observations makes comparisons less certain. Potentially, land use effects on $C_{org}$ may be more pronounced on these soils as management reduces the protection of $C_{org}$ when Andisols are tilled and wetland soils drained.

A comparison can be made with an analysis made in the 1930's of a large data set obtained by *Hardon* (1936) from Lampung on the southermost corner of the island. Lampung was then under transformation from forest to agricultural land, a change which today has been virtually completed. For nearly all land use categories, Hardon's data fell within range for LREP data for major upland soils. Hardon's average topsoil content over all land uses (3.53%) is close to the present average of 3.46% for these soil groups. We conclude that the average $C_{org}$ content of the topsoil in Lampung/S. Sumatera in the early 1930's was similar to the average for the whole of lowland Sumatera, excluding volcanic, wetland and peat soils in the late 1980's. There is no indication of any change in soil C storage under forests in the 50 year time span during which atmospheric $CO_2$ concentration increased by 20% in this period, from 0.29 to 0.35%.

The data set for the 1980's confirms a relation between soil pH and $C_{org}$ established in the 1930's by *Hardon* (1936). The combined data show that the lowest $C_{org}$ content can be expected in the pH range 5.0–6.0. Below a pH of 5.0 reduced biological activity may slow down the break down of organic matter. Interestingly, most agricultural research recommends lime applications to the range 5.0–6.0; this may stimulate breakdown of organic matter and thus contribute to crop nutrition, but possibly at the costs of maintaining the soil organic matter content. By selecting acid soil tolerant germplasm, adequate crop production can be obtained in the pH range 4.5–5.0, with higher $C_{org}$ levels.

In a multiple regression analysis the relative importance for $C_{org}$ was estimated of soil pH, texture, elevation (a proxy for temperature), slope, land use and soil type (*Van Noordwijk* et al., 1997). The quantitative factors pH and texture had a slope which differs significantly ($p < 0.001$) from zero. The relative weighing factors for clay and silt are 1.4 and 1.0, respectively. The regression coefficient for elevation ($p < 0.01$) and for slope ($p < 0.05$) also differed significantly from zero. In this regression, the effects of elevation are studied separately from the different altitudinal distribution of soil groups. They indicate a positive effect on $C_{org}$ of lower temperatures. Compared to the average contents per soil type and land use, the $C_{org}$ conten will decrease 15% per unit increase in pH, increase 1% and 0.7 per percent increase in clay and silt content, respectively, increase by 4% per 100 m increase in elevation and decrease by 0.3% per percent increase in slope. The following equation emerged for predicting $C_{org}$ in the upper 15 cm of mineral soil from Sumatra under forest cover:

$C_{org}$ = exp (1.256 + 0.00994 * Clay% + 0.00699 * Silt% – 0.156 * pH + 0.000427 * Elevation + 0.834 (if soil is Andisol) + 0.363 (for swamp forest on wetland soils)

As relative C-levels for non-forest land uses, assuming no change in soil pH, the data set suggests 91% for upland crops, 83% for land under tree crops of various types, 85% for *Imperata* grasslands and 81% for young secondary vegetation (shrub). The relative values for crop land become 77 and 65% if pH is increased by 1 or 2 units, respectively. No indication was obtained that tree-based production systems in plantations differ in $C_{org}$ content from land used for annual crops. The agroforests, however, were not recognized as a separate group.

These changes in soil C stocks may be smaller than often assumed. Contributing factors may be the fact that on Sumatra prolonged cropping of a soil, depleting its C stocks is not common as *Imperata* takes over in such cases and the use of livestock for ploughing is limited. For research concerned with the global C budgets and the effects of land use change on C emissions, priority should be given to the peat and wetland soils; drainage of the Histosols may release more $CO_2$ into the atmosphere from current soil sources than the belowground effects of transforming all remaining forests into *Imperata* grasslands.

## Effect of burning on nutrients and soil organic matter

In the North Lampung benchmark area data were collected on the direct effects of burning on ash and soil nutrient content (Table 2). The ash layer was sampled separately for the top 3 cm and the 3–5 cm layer. It consisted of burned plant material, fine charcoal and as well as true ash. The table shows that soil pH was increased by at least two points, due to accumulation of base cations came from burnt above biomass. Exchangeable cation content as well as available P content increased dramatically; these data, once again, demonstrate that slash-and-burn methods are

**Table 2:** Chemical properties of forest soil before and after burn in N. Lampung.

| Layer cm | pH $H_2O$ | pH KCl | $C_{org}$ % | P-Olsen, mg $kg^{-1}$ | $K^+$ | $Ca^{2+}$ | $Mg^{2+}$ |
|---|---|---|---|---|---|---|---|
| | | | | | cmol. $kg^{-1}$ | | |
| Before burn: | | | | | | | |
| 0– 5 | 6.2 | 4.7 | 2.44 | 5.0 | 0.20 | 1.44 | 0.62 |
| 5–10 | 5.6 | 4.6 | 2.12 | 2.0 | 0.20 | 1.85 | 0.52 |
| After burn: | | | | | | | |
| 0– 3 | 8.1 | 7.5 | 7.15[1] | 51.4 | 5.37 | 25.5 | 4.47 |
| 3– 5 | 8.3 | 7.2 | 4.28 | 25.6 | 2.02 | 14.8 | 3.46 |
| 5–10 | 7.2 | 6.0 | 1.94 | 6.7 | 0.29 | 3.12 | 0.63 |
| Soil surface ash | | | | 384 | 176 | 23.6 | 17.6 |

1. This probably includes partly burnt charcoal

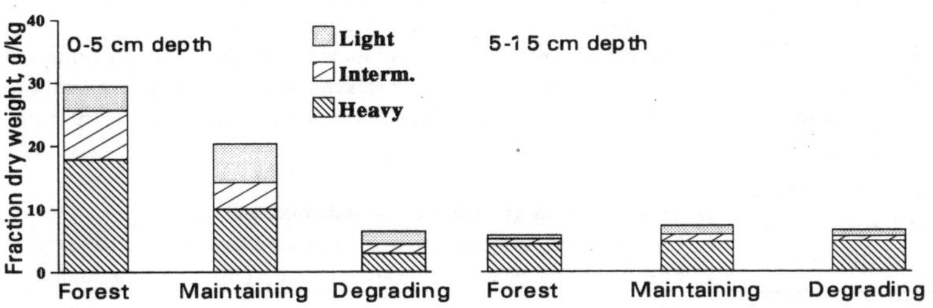


**Figure 7:** Fractionation of soil organic matter for the 0–5 and 5–15 cm layer of soils under forest, under the best local practices and under soil degrading conditions (including frequent fire and/or tillage) in the North Lampung benchmark area (*Hairiah* et al., 1995).

an effective way of supplying available nutrients to the next vegetation.

## Soil organic matter fractionation

In a survey of the organic matter content of topsoil in the N. Lampung benchmark area (*Hairiah* et al., 1995), three groups of land use practices could be differentiated on the basis of size-density fractions of the top 5 cm of mineral soil (Fig. 7):

a. Forest (remnants of logged-over primary and various types of secondary forest)

b. SOM-maintaining practices: woodlots, forest plantations established with slash-and-burn land clearing, home gardens and unburnt *Imperata* grasslands,

c. Degrading lands: burnt *Imperata*, sugar cane plantations with annual burning of residues and forest plantations established with bulldozer land clearing.

For the second category of land use systems the sum of the Ludox fraction (g $kg^{-1}$) in the top 5 cm of soil may still decrease by about 20–30% from the forest level. Under degrading situations, the data suggested that 8–10 years after opening the forest, the sum of the Ludox fraction decreased by 70–80%. In the 5–15 cm depth layer, however, the converted forest sites exceeded the forest. Total content of the Ludox fractions (in g $kg^{-1}$ of soil) for this second layer is only 20–50% of that in the top 5 cm. In the 5–15 cm soil layer the heavy fraction is dominant over the light and intermediate fraction in dry weight.

The size-density fractionation scheme also showed large differences between forest soil (both before and after a burn) and soil under 9 year of continuous cropping. Intermediate values were obtained for hedgerow intercropping plots, where frequent application of tree prunings was slowed down the soil degradation. The data show, however, that the degradation of soil organic matter after forest conversion can not be completely avoided by this practice. The trees with the highest polyphenolic content of the prunings (*Calliandra* and *Peltophorum*) have the largest heavy pool, which indicates soil carbon closely linked with mineral particles.

## Methane ($CH_4$) oxidation and nitrous oxide ($N_2O$) emission in forest soils

Secondary and logged-over forest as well as rubber agroforests can act as sink for methane; values were in the range of 0.1 mg $CH_4$ $m^{-2}$ $hr^{-1}$. A much lower sink strength was shown by newly burnt forest soil, which might be caused by less active methanotrophe bacteria due to higher soil temperature, or as a direct effect of the burning. If the methane consumption rates found here for forest soils are extrapolated to 24 hours per day on a yearly basis an estimated oxidation occurs of 9 kg $CH_4$ $ha^{-1}$ $yr^{-1}$. As the net effect on global warming is supposed to be 25 times larger for $CH_4$ than for $CO_2$, this methane sink is equivalent to an annual sink of 165 kg C $ha^{-1}$ $yr^{-1}$. This is equivalent to the changes in C stock from forest soils to *Imperata* grasslands if this change is allocated to a period of 60–100 years. Net emissions from rice-fields (*Nugroho* et al., 1996) are about 200 times higher per ha than the oxidation in forest soils, so a forest-paddy rice ratio of 200 would have to be maintained to potentially offset net emissions. Estimates of the total $CH_4$ release of smouldering logs after burning effects are difficult to obtain, but air samples next to such a site showed a $CH_4$ concentration of about 15 times as high as in normal air.

Initial data on $N_2O$ emission show an opposite trend to those for $CH_4$, with the biologically more active forest soils emitting more $N_2O$ than the more depleted rice and alang-alang soils.

## Concluding remarks

The characterization results show that rapid conversion of forests takes place in the peneplain of Sumatra, but the initial expectation was not confirmed that this conversion is due to smallholders opening natural forests for food crop production systems. This type of conversion appears to be largely restricted to transmigration sites and normally the most valuable timber is removed by logging before such a conversion occurs. Widespread conversion of forests takes place for large scale plantation crops, often in a 'nucleus estate smallholder' project linked with transmigration villages. Spontaneous migrants, who may originate from transmigration villages, moved from elsewhere in Sumatra or came straight from Java, tend to convert logged-over forests primarily for growing rubber. Rubber agroforests have been developed as a major land use system on the peneplain. As prolonged cropping is rare, the average content of soil organic matter in agricultural soils is not as low as found elsewhere in the tropics. Yet a decline in 'active' soil organic matter is indicated by the results of the size-density fractionation scheme used. The main escape from negative soil degradation effects is apparently a shift to tree-based production systems in the form of rubber (and fruit tree) agroforests. Although in a biophysical sense the data for Sumatra may not be very different from the globally expected trends, the land use patterns differ substantially and the agroforests developed by farmers on Sumatra may be a major model as alternative-to-slash-and-burn elsewhere in the tropics.

We will now come back to the central hypothesis that 'Intensifying land use as alternative to slash-and-burn farming can help to reduce deforestation, conserve biodiversity, reduce net emission of greenhouse gasses and alleviate poverty'.

Traditional 'shifting cultivation' systems hardly exist any more in Sumatra, but slash-and-burn is still widely used as a land clearing technique. The intensification model of Fig. 1 probably holds if 'all other things are equal'. But not necessarily if total demand is increasing. An important reason that local demand can increase is migration, either spontaneous or government sponsored. This leads to the main lesson from Sumatra for the global ASB project. The experience of the rubber agroforests in Sumatra, which go back to the start of the 20'th Century, shows that:

**Table 3:** Hypotheses for phase-2 research of ASB-Indonesia

**I. Hypotheses on dynamics of land use change and poverty**

1. Intensified land use by local farmers in already converted forests can, in the absence of immigration, alleviate pressure to convert remaining local forest,
2. Intensified land use in already converted forests or degraded lands can reduce the 'poverty push' to migrate to new forest margins,
3. Local community involvement in forest management, including logging, can reduce the 'pull' attracting migrants ('forest squatters'),
4. The combination of cheap labour, via recent (trans)migrants, abundant land access to local people and profitable (tree) crops accelerates forest conversion
5. Where communal forest land has to be cleared before it can be claimed by individual families, this tenure arrangement accelerates forest conversion,
6. Technical options for intensification exist for all but the most marginal soils,
7. Improved market access, physical as well as economical, is a key to intensified land use, esp. on 'marginal' soils,
8. 'Protection' and 'conservation' forests have to be actively protected and conserved by (external) stakeholders from local exploitation; 'bufferzone agroforestry' by itself is not enough to achieve such protection.

**II. Hypotheses on greenhouse gas emissions and sinks**

1. Upland forest soils can be a considerable sink for methane,
2. Burning forest vegetation can produce considerable amounts of methane,
3. The source/sink relationships of soils for methane and nitrous oxide are as important for the global climate as that for carbondioxide.

**III. Hypotheses on biodiversity conservation**

1. The biodiversity values of low-management intensity systems, such as jungle rubber can be largely maintained while productivity is increased, if more productive tree *germplasm* is introduced,
2. Intensifying *management* in agroforests such as jungle rubber systems leads to a large loss of biodiversity and only moderate gains in productivity,
3. Soil biodiversity and active soil organic matter fractions can fall below critical levels during intensification of land use, especially for annual food crops.

a) economically attractive tree-based production systems exist as alternatives to extensive food-crop based systems,
b) they do help to alleviate poverty, but
c) they speed up rather than slow down the rate of natural forest conversion, especially because they attract an inflow of migrants who want to share the benefits of such systems.
d) the rate of forest conversion can be further increased by easy access via logging roads and temporary off-farm employment opportunities in the logging and lumber industry which helps to bridge the unproductive period of the tree crops.

Other reasons under which the central ASB hypothesis would not hold true are that other factors than the need for agricultural products may be the main driving force of forest conversion, such as logging, mining or land speculation. Privatization of formerly communal land is based on slash-and-burn practices followed by tree planting in the Minangkabau area of West Sumatera and Jambi (and possibly elsewhere). This may be a major reason for the existence of extensively managed 'jungle rubber' (*Gouyon* et al., 1983), as it stakes claims recognized by the local community, though not necessarily by the government.

In summary, we can conclude that three requirements can be formulated for the intensification hypothesis:

1. The intensification techniques must be ecologically and agronomically sound, socially and economically feasible and lead to marketable products; agroforests are a promising model,
2. The forest which is supposed to be saved from conversion by this intensification must be effectively protected and boundaries enforced,
3. The inflow of people from elsewhere must be brought under control.

Increased involvement of local communities in forest management may help to address both issue 2 and 3. Hypotheses for further research in the next phase of the ASB project are formulated (Table 3) and will be tested in the coming years.

## Acknowledgements

The ASB project is financially supported by the Global Environment Facility (GEF) with UNDP sponsorship; thanks are due to Dr. Mike Swift for comments on an earlier draft and to many colleagues in the project for sharing ideas.

## References

*Alegre, J. C.* and *Cassel, D. K.*, 1996. Dynamics of soil physical properties under alternatives to slash-and-burn. Agric. Ecosyst. and Environm. 58: 39–48

*Boserup, E.*, 1965. Conditions of Agricultural Growth. Aldine, Chicago, USA.

*Brady, N. C.*, 1996. Alternatives to slash-and-burn: a global imperative. Agric. Ecosyst. and Environm. 58: 3–11

*Danhof, G. N.*, 1941. Tweede bijdrage tot oplossing van het alang-alang vraagstuk in de Lampongse Districten. [Second contribution to solving the alang-alang problem in the Lampung Districts]. Tectona 34: 67–85.

*FAO/MacKinnon, J.*, 1982. National Conservation Plan for Indonesia, Vol. II. Sumatra. FAO, Bogor. (10. 2. 16).

*Gouyon, A., De Foresta, H.* and *Levang, P.*, 1993. Does 'jungle rubber' deserve its name? An analysis of rubber agroforestry systems in southeast Sumatra. Agroforestry Systems 22, 181–206.

*Hagreis, B. J.*, 1930. Ladangbouw [Shifting cultivation]. Landbouw 6: 43–78.

*Hairiah, K.., G. Cadisch, M. van Noordwijk, A. R. Latief* and *G. Mahabharata, Syekhfani*, 1995. Size-density and isotopic fractionation of soil organic matter after forest conversion. In: A. Schulte and D. Ruhiyat (eds.) *Proc. Balikpapan Conf. on Forest Soils* Vol. 2: 70–87

*Hardon, H. J.*, 1936. Factoren, die het organische stof- en het stikstofgehalte van tropische gronden beheerschen [Factors, controlling the organic matter and the nitrogen content of tropical soils]. Landbouw XI (12): 517–540.

*IPCC*, 1990. Climate change: the IPCC scientific assessment. Cambridge University Press, Cambridge (UK)

*Izac, A. M.* and *C. A. Palm*, 1994. Guidelines for characterization and diagnosis for the global project on alternatives to slash-and-burn. ICRAF, Nairobi.

*Juo, A. S. R.* and *Manu, A.*, 1996. Chemical dynamics in slash-and-burn agriculture. Agric. Ecosyst. and Environm. 58: 49–69

*Koens, A. J.*, 1925. Ladangbouw [Shifting cultivation]. Landbouw 1: 334–340.

*Laumonier, Y.*, 1997. The vegetation and Physiography of Sumatra. Kluwer, Dordrecht, the Netherlands, 215 pp.

*Malingreau, J. P.* and *R. Christiani*, 1981. A land cover/ land use classification for Indonesia. First revision. Indonesian J. of Geog. 11 (41): 13–47.

*Marsden, W. H.* 1811. The History of Sumatra. Reprinted from 3rd edition, by Oxford University Press, Oxford.

*Meijboom, F. W., Hassink J.* and *Van Noordwijk, M.*, 1995. Density fractionation of soil macroorganic matter using silica suspensions. Soil Biol. Biochem. 27: 1109–1111.

*Murdiyarso, D. M., Hairiah, K.* and *Van Noordwijk, M.* (Eds.) Modelling and Measuring Soil Organic Matter Dynamics and Greenhouse Gas Emissions after Forest Conversion. Proceedings of Workshop/ Training Course 8–15 August 1994, Bogor/Muara Tebo. ASB-Indonesia publication No. 1.

*Murdiyarso, D.* and *U. R. Wasrin*, 1995. Estimating land use change and carbon release from tropical forest conversion using remote sensing techniques. Journal of Biogeography 22: 715–722.

*Nugroho, S. G., Lumbanraja, J., Suprapto, H., Sunyoto, Ardjasa, W. S., Haraguchi, H. and Kimura, M.*, 1996. Three-year measurement of methane emission from an Indonesian paddy field. Plant and Soil 181: 287–293.

*Nye* and *Greenland*, 1960 The Soil under Shifting Cultivation. Commonwealth Bureau of Soils Tech. Comm. 51, Harpenden, UK.

*Palm, C. A., Swift, M. J.* and *Woomer, P. L.*, 1996. Soil biological dynamics in slash-and-burn. Agric. Ecosyst. and Environm. 58: 61–74.

*Richards, J. F.* and *Flint, E. P.*, 1993. Historic land use and carbon estimates for South and Southeast Asia, 1880–1980. Carbondioxide information analysis center, Oak Ridge National Laboratory, Environmental Sciences Division Publication No. 4174. 326 pp.

*Sanchez, P. A., Palm, C. A.* and *Smyth, T. J.*, 1990. Approaches to mitigate tropical deforestation by sustainable soil management practices. In: Scharpenseel, H. W., Schomaker, M. and Ayoub A. (eds.) Soils on a Warmer Earth. Elsevier, Amsterdam, p. 211–220.

*Scholz U.*, 1983. The natural regions of Sumatra and their agricultural production pattern. A regional analysis. Central Research Institute for Food Crops (CRIFC). Bogor.

*Tinker, P. B., Ingram, J. S. I.*, and *Struwe, S.*, 1996, Effects of slash-and-burn agriculture and deforestation on climate change. Agric. Ecosyst. and Environm. 58: 13–22.

*Tomich, T. P.* and *M. van Noordwijk*, 1996. What drives deforestation in Sumatra? in: B. Rerkasem (ed.) Montane Mainland Southeast Asia in Transition, Chiang Mai University, Thailand, pp 120–149.

*Torquebiau, E.*, 1984. Man-made Dipterocarp forest in Sumatra. Agroforestry Systems 2(2): 103–128.

*Van Noordwijk, M., T. P. Tomich, R. Winahyu, D. Murdiyarso, S. Partoharjono* and *A. M. Fagi* (editors) 1995. Alternatives to Slash-and Burn in Indonesia, Summary Report of Phase 1. ASB-Indonesia Report Number 4, Bogor, Indonesia

*Van Noordwijk, M., Cerri, C., Woomer, P. L., Nugroho, K.* and *Bernoux, M.*, 1997 Soil carbon dynamics in the humid tropical forest zone. Geoderma (in press)

*Van Steenis, C. G. G. J.*, 1935. Maleische vegetatieschetsen [Sketches of Malaysian vergetation]. Tijd. Kon. Ned. Aard. Gen. 52: 25–67, 171–203, 363–390.

# Fine root inventories in Dipterocarp forests and plantations in East Kalimantan, Indonesia

**Dieter Murach**[1], **Daddy Ruhiyat**[2], **E. Iskandar**[2] and **Andreas Schulte**[3]

[1]Institute of Silviculture, Buesgenweg 1, D-37077 Goettingen, Germany
(E-mail: dmurach@gwdg.de)
[2]Faculty of Forestry, Mulawarman University, Samarinda 75123, Kaltim, Indonesia
[3]University of Paderborn, Forest Ecology, An der Wilhelmshöhe 44; D-37671 Höxter, Germany

## Abstract

Results of fine root inventories in Dipterocarp forests and plantations of Eucalyptus deglupta Bl. and Paraserianthes falcataria (L.) Niel. in East Kalimantan, Indonesia are presented. The influence of different forest land use on the amount and vertical distribution of fine roots in the soil is examined. Possible consequences for physical soil parameters related to root growth and morphology of roots are discussed. Differences in nutrient acquisition between tree species are indicated by the chemical compositon of fine roots. Fine root development and turnover in the Dipterocarp forests and the plantations is assessed by the comparison of samplings taken during dry and wet season.

## 1 Introduction

Human impact is changing the environment of forest ecosystems on a global scale. Chemical properties of soils are affected by anthropogenic changes of climate, by forest management and by air pollution. The maintenance of the chemical state of the soil is one of the most important preconditions for a sustainable use of forests (Ulrich, 1987; Schulte, 1996). While there have been a number of investigations on nutrient budgets in temperate forests during the last decade, comparable research is scarce for tropical forest ecosystems (Ruhiyat, 1989) despite its particular importance for forest management in the humid tropics, where tree growth is mainly limited by unfavourable soil conditions (Jordan, 1985). The understanding of site related plant growth is depending on informations about the response of root growth to changing soil conditions as roots are the interface between soil and plants. Total biomass and distribution of fine roots and mycorrhizae are regulating water and nutrient uptake. Root turnover is an important sink in carbon allocation. Root growth is influenced by soil chemistry and there is a feed-back between root growth and physical soil properties (Glinski and Lipiec, 1990).

The most sensitive approach assessing the chemical stress acting in an ecosystem is the analysis of the soil solution and of those parts which are in interaction with it, like plant roots (Ulrich, 1987). Ecosystem-level research on nutrient cycling is incomplete without data on belowground processes.

We here present the first results of fine root inventories from research and demonstration plots in East Kalimantan, Indonesia.

## 2 Site descsription and methods

The effect of forest land use on nutrient cycling and root growth has been studied in four plots in the lowland evergreen rain forest area of East Kalimantan, Borneo: a primary growth Dipterocarp forest, a logged Dipterocarp forest and two plantations of Eucalyptus deglupta Bl. and Paraserianthes falcataria (L.) Niel. The logged Dipterocarp forest has been treated according to the Indonesian selective cutting and replanting system (Tebang Pilih dan Tanam Indonesia, TPTI). With this system only trees of more than 50 cm in diameter are cut, with the condition that 25 trees with a diameter of more than 20 cm of those species which are cut have to be left on the plot. These can be used during the next cutting rotation after 35 years.

A short description of the plots is listed in table 1. Detailed soil data are given by Wenzel et al. (this volume). The classification of the soil types and the data on soil chemistry indicate comparable soil conditions for all plots. The higher base saturation of the uppermost soil layers of the plantations may be explained by the burning of the areas before replanting.

In each of the four plots, 20 soil cores were taken with an soil auger to a depth of 100 cm at the end of the dry season in September 1994 and at the end of the wet season in March 1995. The roots were washed out using sieves and tap water with a final rinse in demineralized water. To store the cleaned roots for later sorting they were dried at a temperature of about 40°C for a period of 3 days.

For the differentiation of living and dead fine roots the dried samples were soaked in demineralized water. Under a microscope the rootlets were cut into pieces of about 2 cm in length and small cross-sections of these pieces

**Table 1:** Stand description of the experimental plots at PT. Kiani Lestari, in Batu Ampar, region of Muara Benskal, Kutai, East Kalimantan, Indonesia

| | | Dipterocarp forest | | logged Dipterocarp | | Eucalyptus plantation | | Paraserianthes plantation | |
|---|---|---|---|---|---|---|---|---|---|
| altitude [m] | | 250 | | 250 | | 250 | | 250 | |
| main tree species | | Dipterocarps | | Dipterocarps | | Eucalyptus deglupta | | Paraserianthes falcataria | |
| age in 1995 [y] | | old growth | | 4 years after logging | | 9 | | 9 | |
| mean height of trees [m] | | | | | | 12 | | 15 | |
| relief | | crest | | upper slope | | middle slope | | middle slope | |
| exposition | | SE | | WNW | | ENE | | E | |
| type of soil | | acid Haplic-Acrisol | | acid Haplic Acrisol | | acid Haplic Acrisol | | acid Haplic Acrisol | |
| pH ($H_2O$) | 0– 5 cm | 4,1 | | 4,2 | | 4,6 | | 4,5 | |
| | 60–80 cm | 5,0 | | 4,8 | | 4,7 | | 5,1 | |
| base saturation “BS” [%]* and cation exchange capacity “CEC” [μmol IE/g soil] | | CEC | BS | CEC | BS | CEC | BS | CEC | BS |
| | 0– 5 cm | 80 | 8 | 80 | 9 | 90 | 39 | 95 | 38 |
| | 20–30 cm | 110 | 3 | 75 | 4 | 95 | 6 | 90 | 17 |
| | 60–80 cm | 130 | 2 | 95 | 2 | 135 | 3 | 120 | 7 |

*: share of Na, K, Ca and Mg in exchangeable cations of $NH_4Cl$-extract

were used for the classification of vitality and plant species. The main criterion for vitality was the coherence of the tissues within the central cylinder of the rootlets. The appearance of the primary cortex was not used for roots which already showed a secondary thickening. Because of missing references for all the different species on the plots the colour of the central cylinder was excluded as a differentiation criterion.

Rootlets were related to different plant categories by comparison with reference samples, which were taken from excavated roots of selected plant species on the plots. Small roots of ferns, grasses (Zingiberaceae), Eucalyptus deglupta and Paraserianthes falcataria were cut off from bigger roots traced back to plant individuals, cleaned and fixed in AFE (acetic-acid: formaldehyde: ethanol 70% = 5:5:90). Characteristic features of the cross sections of such roots, like the size and distribution of vessels in the central cylinder or the ratio of the area of the central cylinder to the cortex were used for classification of plant category.

Parts of the living fine roots were photographed for digital image analysis by which root tips, length and volume of the samples were measured automatically. After drying the roots at a temperature of 40°C to 50°C they were weighed. For statistical analysis of biomass data the Kruskal-Wallis-test was used.

For the chemical analysis, fine roots were crushed in an agate mill. The samples were dissolved in nitric acid using small pressure tanks (Heinrich et al., 1986). The solution was analysed for P, K, Ca, Mg, AL, Fe, Mn und Na by ICP-atomic emission spectrometry. N was analysed by gas chromatography (Koenig et al., 1996).

Exchangeable cations were estimated according to Ulrich by $NH_4Cl$-extract (Meiwes et al., 1984).

## 3 Rooting patterns in Dipterocarp forests and plantations

The plantations and the logged Dipterocarp forest show significantly fewer fine roots compared to the primary growth forest (table 2). Fine root biomass of about 4000 kg $ha^{-1}$ in the primary growth forest is in the range of values reported for other tropical forests (Cannell et al., 1995). Data for the species investigated are not available.

Differences between plots are proved to be significant for both sampling dates. The effect of selective cutting is visible in the gaps even 4 years after treatment. Fine root biomass is reduced in the logged stand to about 40% of that in the primary growth plot. There was no significant seasonal difference of fine root biomass between the end of the dry and the wet season in either Dipterocarp forest, but an increase in the dead fine roots in March ’95 is indicating a higher regeneration rate of fine roots during the wet season in both stands.

In the two plantations, seasonal variations seem to be greater. Living fine roots of Eucalyptus increase during wet season while those of Paraserianthes decrease. Thus, there is evidence that the two tree species respond quite differently to changing water availability.

Seasonal variations in the amount of grass and fern roots are not significant. The reduced root biomass of other plant species in the plantations, mainly shrubs, at the end of the wet season may be related to the cutting of the understory vegetation during the first sampling. Nevertheless, data of the first sampling indicate that the amount of roots of other species in the plantations is high compared to that of planted trees.

**Table 2:** Means (standard deviation) of different fine root parameters at two sampling dates in September 1994 and March 1995

| | Dipterocarp | | Logged Dipterocarp | | Eucalyptus | | Paraserianthes | |
|---|---|---|---|---|---|---|---|---|
| | IX '94 | III'95 | IX '94 | III '95 | IX '94 | III '95 | IX '94 | III '95 |
| Living fine roots of main tree species [kg/ha] | 4100 (1300) | 4000 (1700) | 2300 (800) | 2400 (1200) | 600 (800) | 1150 (400) | 1300 (600) | 750 (400) |
| Living fine roots of grass and fern [kg/ha] | 50 (70) | 0 | 50 (130) | 10 (20) | 250 (300) | 100 (200) | 550 (200) | 250 (200) |
| Living fine roots of other species [kg/ha] | 0 | 0 | 0 | 0 | 900 (450) | 100 (100) | 750 (250) | 450 (700) |
| Living fine roots of all species [kg/ha] | 4150 | 4000 | 2350 | 2400 | 1750 | 1350 | 2600 | 1450 |
| Share of living fine roots | | | | | | | | |
| • to a depth of 5 cm | 0.3 | 0.4 | 0.4 | 0.4 | 0.4 | 0.4 | 0.4 | 0.6 |
| • to a depth of 10 cm | 0.4 | 0.5 | 0.5 | 0.5 | 0.5 | 0.6 | 0.5 | 0.7 |
| • to a depth of 40 cm | 0.8 | 0.8 | 0.8 | 0.8 | 0.7 | 0.8 | 0.8 | 0.8 |
| in fine root biomass of total profile | | | | | | | | |
| Dead fine roots of all species [kg/ha] | 700 (300) | 1350 (900) | 800 (450) | 1200 (750) | 400 (200) | 650 (450) | 700 (200) | 2000 (600) |

The relative amount of dead fine roots is highest in the Paraserianthes plot, where it exceeds the values of the living fine roots during the wet season. Assuming that decomposition rate in the Paraserianthes stands is not lower than in the other plots the greater amount of dead fine roots is indicating a higher fine root turnover in this plantation.

Roots of ferns and grasses are mainly growing within the uppermost soil layer (fig. 1). Although absolute amounts of biomass are different, the other species in the plantations show nearly the same vertical gradients as the planted trees. There are sharp gradients in fine root concentration between the uppermost 10 cm of soil profil and the deeper soil horizons.

## 4 Influence of root growth on soil conditions

Most publications on root ecology are concerned with the influence of soil properties on root growth but little is known about the changes in soil properties due to the growth of roots. Roots may influence soil chemical and physical properties in different ways. The absorption of nutrients and water is creating rhizosphere effects. Well known are the effects of nitrate and ammonium uptake on rhizosphere pH-values (Marschner et al., 1991) or the accumulation of discriminated ions like aluminum near the root cortex.

Of particular importance for tropical soils is the influence of roots on soil physical properties. Clay soils in the humid tropics, exposed to high precipitation may become water-logged, which is unfavourable for plant growth and increases the risk of erosion.

Roots need space to grow, space which was previously occupied by soil particles and soil pores. Thus, they are compacting the soil in the rhizosphere, especially when secondary thickening starts. There are reports on increased bulk density of the soil near the roots (Greacen et al., 1968).

On the other hand, root penetration can perforate compact soil layers by creating macropores after decomposition of dead roots. Those macropores improve water movement and gaseous diffusion. They may also be pathways for the roots of other plant species (Heinonen, 1986). Thus a high turnover of deep growing roots and not high stocks of roots will increase soil permeability.

To assess the possible influence of root growth on soil permeability in the different plots, the volume of the rootlets in the different soil layers has been calculated by image analysis for a random sample. A good allometric relationship exists between the biomass of fine roots and their volume (tab. 3).

Differences in equation parameters between the plots are reflecting the differences in wood density and root morphology. Roots of ferns and Paraserianthes occupy nearly double the volume per unit biomass compared to roots of the Dipterocarp forests. Eucalyptus roots also seem to have a lower density compared to roots in Dipterocarp forests. The average concentration of root volume for the different plots and plants has been calculated for the subsoil from 40 to 100 cm (tab. 4) from data in table 3.

These values can be related to the volume of macropores ($> 10$ μm diamter). If macropores are estimated to amount to 10% of the total soil volume for the clayey horizons in the profil (see Wenzel et al., this volume, tab. 4), the amount of fine root volume may be in the range of about 1% of macropore volume for the subsoil layer.

Data of the Eucalyptus plantation in Sept. '94 (tab. 4) indicate that the root volume in the subsoil of plantation may be reduced to about 50% or more compared to pri-

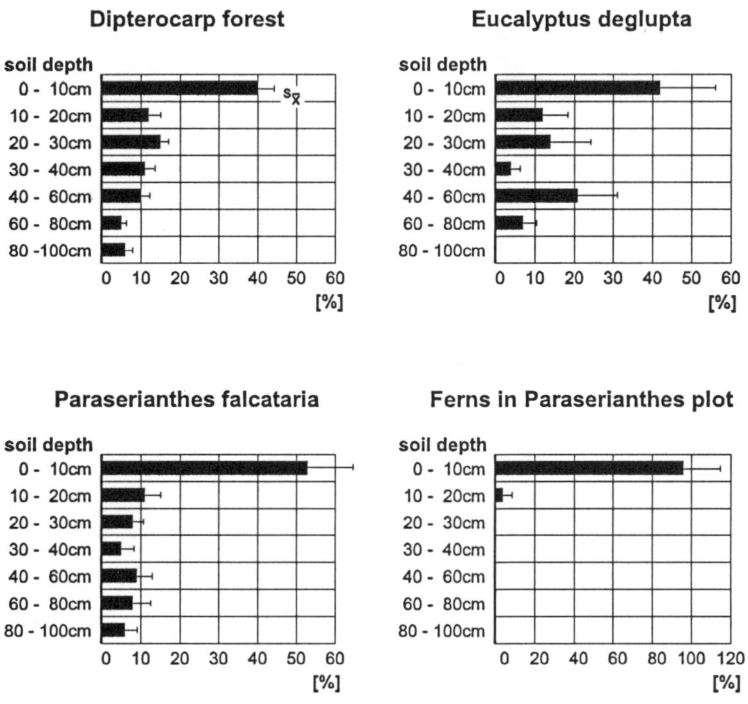


**Fig. 1:** Vertical gradients of fine root biomass for different plant species (biomass in different soil layers as percentage of biomass of total profile)

**Table 3:** Regression equations and correlation coefficients for the estimation of fine root volume y [μ] by fine root biomass x [mg]

| | Equation | Corr.-coefficient |
|---|---|---|
| Dipterocarp | y = 4.53 x | $r^2 = 0.86$ |
| Logged Dipterocarp | y = 5.09 x | $r^2 = 0.94$ |
| Eucalyptus | y = 5.56 x | $r^2 = 0.94$ |
| Paraserianthes. | y = 8.50 x | $r^2 = 0.83$ |
| other plants in plantations | y = 7.42 x | $r^2 = 0.93$ |
| fern | y = 9.26 x | $r^2 = 0.95$ |

**Table 4:** Average concentration of fine root volume (living and dead fine roots) [μl/100 ml soil] in the subsoil below 40 cm depth

| | Sept. '94 | March '95 |
|---|---|---|
| Dipterocarp | 80 | 80 |
| Logged Dipterocarp | 50 | 60 |
| Eucalyptus | 40 | 50 |
| (other plants) | (50) | (0) |
| Paraserianthes | 60 | 120 |
| (other plants) | (40) | (15) |

mary growth forests. But differences may be compensated if understory vegetation is growing within the plantations. For the other plantation, the root volume in the subsoil may be enhanced compared to primary growth forests as the data for the Paraserianthes plantation in March '95 show. Although fine root mass for the Paraserianthes plantation and the Dipterocarp primary forest are similar at that date, the root volume of the Paraserianthes plantation is greater due to the differences in root-wood density (tab. 3).

Two samplings do not give a reliable estimate of fine root turnover. Nevertheless, according to the amount of dead fine roots of table 2 and the observed changes in bio- and necromass an indication is obtained that fine root turnover in plantations may be of the same magnitude or even higher as in primary growth Dipterocarp forests. Since root turnover creates porosity, a reduced porosity of the subsoil is not a necessary consequence of tree plantations. The influence of plantations on soil permeability depends on tree species and on the management of the understory vegetation.

## 5 Nutrient storage in fine roots

The highest concentrations of N are found in roots of the legume Paraserianthes falcataria (tab. 5). There is clear difference to the concentrations of N in the roots of the other plants. Extremely low N concentrations are found in roots of Eucalyptus.

There is little difference between the P and Mg concentrations of the different plant species. P concentrations are very low compared to temperate forest ecosystems in Europe where values of less than 0.5 mg/g are found only

**Table 5:** Element concentrations in living fine roots

| | N [%] | P | K | Ca [mg/g DS] | Mg | Al |
|---|---|---|---|---|---|---|
| Dipterocarp | | | | | | |
| 0– 5 cm | 1.1 | 0.4 | 1.5 | 1.8 | 1.1 | 4.8 |
| 5–10 cm | 0.8 | 0.3 | 1.8 | 1.0 | 1.2 | 6.4 |
| 60–80 cm | 0.7 | 0.3 | 2.1 | 0.8 | 1.3 | 12.7 |
| logged Dipterocarp | | | | | | |
| 0– 5 cm | 1.3 | 0.4 | 1.6 | 2.9 | 1.3 | 3.7 |
| 5–10 cm | 1.2 | 0.3 | 1.9 | 1.6 | 1.2 | 4.6 |
| 60–80 cm | 0.9 | 0.3 | 1.9 | 1.3 | 1.2 | 8.4 |
| Eucalyptus | | | | | | |
| 0– 5 cm | 0.6 | 0.3 | 2.5 | 4.5 | 1.4 | 3.3 |
| 5–10 cm | 0.5 | 0.3 | 2.1 | 4.0 | 1.0 | 4.6 |
| 30–80 cm | 0.5 | 0.3 | 2.0 | 1.4 | 0.9 | 13.8 |
| Paraserianthes | | | | | | |
| Of/Oh | 1.4 | 0.3 | 1.9 | 10.3 | 1.2 | 1.6 |
| 0– 5 cm | 1.4 | 0.3 | 2.5 | 7.8 | 1.1 | 2.2 |
| 5–10 cm | 1.5 | 0.3 | 2.0 | 8.5 | 1.2 | 3.9 |
| 60–80 cm | 1.4 | 0.3 | 3.2 | 2.4 | 1.4 | 9.5 |
| fern in Paraserianthes plantation | | | | | | |
| Of/Oh | 0.7 | 0.4 | 3.4 | 3.1 | 1.8 | 1.4 |
| 0– 5 cm | 0.8 | 0.4 | 2.4 | 2.6 | 1.9 | 2.9 |
| 5–100 cm | 0.9 | 0.5 | 2.3 | 2.6 | 2.3 | 6.6 |
| other plants in Paraserianthes plantation | | | | | | |
| Of/Oh | 1.0 | 0.3 | 0.9 | 9.9 | 2.1 | 1.9 |
| 0– 5 cm | 0.9 | 0.3 | 2.0 | 6.6 | 2.3 | 3.5 |
| 5–10 cm | 0.9 | 0.5 | 1.0 | 7.7 | 1.4 | 4.2 |
| 60–80 cm | 1.3 | 0.3 | 2.4 | 3.4 | 2.2 | 9.6 |
| Spruce (Solling, Germany) | | | | | | |
| 0–05 cm | 1.2 | 1.3 | 2.3 | 3.4 | 0.8 | 4.5 |
| 30–40 cm | 0.7 | 0.9 | 2.0 | 1.4 | 0.8 | 14.4 |
| Beech (Solling, Germany) | | | | | | |
| 0– 5 cm | 1.3 | 0.7 | 1.6 | 0.6 | 0.4 | 4.5 |
| 30–40 cm | 0.7 | 0.6 | 2.4 | 0.7 | 0.6 | 6.4 |

in reforestations of degraded soils of former heath lands. Magnesium concentrations are high compared to broad-leaved stands in Europe. Roots of beech stands in temperate forests may have only 10–50% of the Mg concentrations listed in tab. 5 (Murach 1988).

Calcium concentrations of the roots in the top soil of the plantations are high, especially with Paraserianthes. This corresponds to the high base saturation of the soil in those soil layers (tab. 1).

There are only small differences between the element concentrations of living and dead fine roots except with aluminum, which is clearly accumulated in dead fine roots.

Total nutrient storage in the root compartment is highest in the primary growth dipterocarp stand. But in relation to nutrient storages in the leaves, the amount of elements in fine roots is significantly lower in Dipterocarp forest compared to the plantations (tab. 6). Data for element storage in leaves are taken from Ruhiyat (1989), who has investigated the element budget of Dipterocarp forests and plantations in East Kalimantan.

In Dipterocarp forest, only N and Mg stocks in roots account for up to 10% of stocks in leaves. With the other elements the amounts in fine roots are of minor importance. In contrast to the Dipterocarps, the element stocks in fine roots of the plantations are in the same range as those in the leaves except for K.

## 6 Management of the "hidden half" in plantations

An important aspect of sustainable forestry in the humid tropics is the management of the nutrient budges of the forest ecosystems (Jordan 1985). Especially the management of plantations has to focus on the nutrient stocks in vegetation and soil (Ruhiyat 1989) in order to avoid nutrient deficiencies for the next rotation.

The presented results indicate that fine roots in plantations are as important as the leaves for element storage. Roots of understory plants also may have an important part in storage capacity.

The maintenance of soil physical properties also is essential for sustainable wood production. In areas where clayey soils are exposed to high precipitation, a compac-

**Table 6:** Element storage in fine roots and leaves

| | N | P | K | Ca | Mg |
|---|---|---|---|---|---|
| | | | kg/ha | | |
| Dipterocarp | | | | | |
| living fine roots | 36 | 1.2 | 6.5 | 5 | 4.5 |
| dead fine roots | 15 | 0.5 | 3 | 1.5 | 1.5 |
| leaves* | 350 | | 180 | 200 | 40 |
| logged Dipterocarp | | | | | |
| living fine roots | 26 | 0.8 | 4.5 | 5 | 3 |
| dead fine roots | 20 | 0.6 | 2.5 | 2 | 1.5 |
| Eucalyptus | | | | | |
| living fine roots | 8 | 0.5 | 3 | 4.5 | 1.5 |
| dead fine roots | 16 | 0.7 | 4.5 | 10.5 | 3 |
| leaves* | 40 | | 35 | 17 | 5 |
| Paraserianthes | | | | | |
| living fine roots | 17 | 0.5 | 3 | 10 | 2.5 |
| dead fine roots | 29 | 0.7 | 4.5 | 12.5 | 3.5 |
| leaves* | 60 | | 40 | 25 | 7 |

* data of comparable stands in East Kalimantan from Ruhiyat (1989)

tion of the subsoil should be avoided to minimize the risk of water logging. The porosity of the subsoil is improved by high root turnover in these layers. Our results show that the creation of macropores in clayey subsoil horizons by root penetration need not be decreased with tree plantations compared to natural growth Dipterocarp forests. However, the effect of root growth on compaction of the subsoil depends on the planted tree species and the management of the understory vegetation.

The importance of root growth for both aspects of plantation management in the humid tropics, nutrient conservation and maintenance of soil physical properties, means that sustainable management of plantations should include the management of the "hidden half" of the plants.

Further root research in plantations should improve knowledge of root ecology of important plantation tree species, and screen suitable auxiliary plants for their ability to store nutrients above and below ground and to increase soil porosity by high root turnover in dense subsoil horizons.

## References


*Cannell, M. G. R.; N. M. J. Crout; R. C. Dewar* (1994): Agroforestry modelling and research co-ordination. Annual report June 1993 – June 1994, ODA forestry research programme RS 651. Institute of Terrestrial Ecology, Penicuik, Midlothian, Scotland.

*Glinski, J.* and *J. Lipiec* (1990): Soil physical conditions and plant roots. CRC press. Boca Raton, Florida.

*Greacen, E. L.; K. P. Barley* and *D. A. Farrell* (1968): The mechanics of root growth in soils with particular reference to the implications for root distribution. In W. J. Whittington (ed.): Root growth, Butterworths, London.

*Heinonen, R.* (1986): Alleviation of soil compaction by natural forces and cultural practices. In R. Lal et al. (eds.): Land clearing and development in the tropics, Balkema, A. A., Rotterdam, 285.

*Heinrichs, H.; H.-J. Brumsack; N. Loftfield* and *N. König* (1986): Verbessertes Druckaufschlußsystem für biologische und anorganische Materialien. Z. Pflanzenern. Bodenkd. 149, 350.

*Jordan, C. F.* (1985): Nutrient cycling in tropical forest ecosystems. John Wiley & Sons, Chichester.

*Koenig, N.* und *H. Fortmann* (1996): Probenvorbereitungs-, Untersuchungs- und Elementbestimmungs-Methoden des Umweltanalytik-Labors der Nieders. Forstl. Versuchsanstalt und des Zentrallabors II des Forschungszentrums Waldökosysteme. Berichte des Forschungszentrums Waldökosysteme, Reihe B, Band 46/47.

*Marschner, H.; M. Häussling* and *E. George* (1991): Ammonium and nitrate uptake rates and rhizosphere pH in non mycorrhizal roots of Norway spruce (Picea abies (L.) Karst.), Trees 5, 14.

*Meiwes, K.-J.; N. König; P. K. Khanna; J. Prenzel* und *B. Ulrich* (1984): Chemische Untersuchungsverfahren fuer Mineralboden, Auflagehumus und Wurzeln zur Charakterisierung und Bewertung der Versauerung in Waldboeden. Berichte des Forschungszentrums Waldökosysteme Bd. 7.

*Murach, D.* and *B. Ulrich* (1988): Destabilization of forest ecosystems by acid deposition. GeoJournal 17, 253.

*Ruhiyat, D.* (1989): Die Entwicklung der standörtlichen Nährstoffvorräte bei naturnaher Waldbewirtschaftung und im Plantagenbetrieb, Ostkalimantan (Indonesien). Goettinger Beitraege zur Land- und Forstwirtschaft in den Tropen und Subtropen, 35.

*Schulte, A.* (1996): Dipterocarp forest ecosystem theory based on matter balance and biodiversity. In: Schulte, A. and D. Schöne (eds.). Dipterocarp Forest Ecosystems: Towards Sustainable Management. World Scientific Publishing, Singapore, London, New Jersey 3–28.

*Ulrich, B.* (1987): Stability, elasticity, and resilience of terrestrial ecosystems with respect to matter balance. Ecological Studies 61, 11.

# Vesicular-arbuscular mycorrhizas of durian and other plants of forest gardens in West Kalimantan, Indonesia

**Hilary F. Smith**[1], **Patrick, J. O'Connor**[2], **Sally E. Smith**[2] and **F. Andrew Smith**[3*]

Departments of [1]Geography, [2]Soil Science and [3*]Botany, The University of Adelaide, Adelaide SA 5005, Australia

## Abstract

Mycorrhizal symbioses are very widespread and often result in improved growth and nutrition of plants. In the lowland tropics, the vesicular-arbuscular (VA) mycorrhizal symbiosis is the most common form and ectomycorrhizas occur in some forest trees. The mycorrhizal status of many tropical plants is not known and in some cases there have been conflicting reports. *Durio* species, which are ecologically and economically important in S.E. Asia, have been claimed by different authors to have VA mycorrhizas, ectomycorrhizas, or to be non-mycorrhizal.

We have found that young roots of *D. zibethinus* (durian) from a village forest garden (complex agroforest) in West Kalimantan, Indonesia, form extensive VA mycorrhizas. Cells of the root cortex contained many coils of hyphae, arbuscules and infrequent vesicles. Intercellular hyphae were not seen. VA mycorrhizal fungi were less frequently found in older roots but other fungi were present, some of which colonized the root surface. *Nephelium lappaceum* (rambutan) and *Artocarpus integer* (cempedak), cultivated in association with durian, formed similar VA mycorrhizas. Roots of *Lansium domesticum* (langsat) only showed initial stages of colonization – i.e. entry-points and coils of hyphae, but this may be a function of the age of the roots that were sampled. Although we did not survey the many other species in the forest gardens, work elsewhere by others on the same species or their relatives suggests that most plants will also be VA mycorrhizal.

The possible nutritional significance of the VA mycorrhizal symbiosis in this forest garden agroecosystem is briefly discussed and suggestions are made for further study.

## 1 Introduction

The roots of most terrestrial plants are colonized by mycorrhizal fungi that often improve the growth and nutrition of the plants, compared to similar but non-mycorrhizal plants. In addition, root diseases can be suppressed and external hyphae of mycorrhizal fungi have an important role in stabilizing soil structure. Mycorrhizas thus have a major ecological significance in the overall protection of the soil resource base and this protection is an important contributing factor for the sustainability of any ecosystem, natural or agricultural. In natural ecosystems, mycorrhizas ensure efficient recycling of nutrients. Agricultural plant production removes nutrients at each harvest and sooner or later can lead to soil degradation. Mycorrhizas have a range of beneficial effects in agricultural ecosystems, including reducing the need for input of fertilizer and amelioration of undesirable side effects on soil conditions: see *Sieverding* (1991) and *Smith* and *Read* (1997) for further discussion.

Mycorrhizas are receiving increasing attention in tropical ecosystems, particularly because of their potential economic significance in agriculture and forestry. Alternatives are being sought to intensive agricultural practices in the tropics. Agroforestry in its various forms is one possible option. It has potential for low input, continuous crop production with a reduced risk of soil degradation. Agroforestry technology is still being developed world-wide and there is also interest in traditional practices, of which there are many kinds (e.g. see *Nair*, 1989, for references). Traditional agroforestry in Indonesia includes forest garden systems, often called complex or multi-storied agroforests (*Michon* et al. 1986). These systems have high species diversity which is managed both temporally and spatially. They often focus on the growth of cereals (rice and maize) and vegetables in conjunction with trees for fruit, nuts and timber. However, there is variation in practice even locally, as shown by studies in Sumatra (e.g. *Michon* et al., 1986; *Gouyon*, 1996), East Kalimanan (e.g. *Sardjono*, 1996; *Sorensen*, 1996) and West Kalimantan (e.g. *Padoch* and *Peters*, 1992; *Salafsky*, 1994). Most recently, *Smith* (1996) has conducted an intensive study of forest gardens near Sukadana, W. Kalimantan, in which cultivation does not include a cereal crop but is centred around commercial production of bananas and durian.

Studies on traditional agroforests have concentrated mainly on above-ground tree-plant interactions, with some attention paid to nutrient cycling but very little to mycorrhizas. For example, *Schroth* (1995) reviewed belowground interactions in agroforestry and the importance of tree root characteristics as criteria for species selection and system design, but even in this extensive review mycorrhizas received only passing mention. Studies of mycorrhizal dependency of some individual agroforest tree species in Africa have been made by *Nouaim* et al. (1994), *Nouaim* and *Chaussod* (1994) and *Tomlinson* et al. (1995).

Little is known about the occurrence of mycorrhizas in many important plants in the tropics, particularly in trees. Vesicular-arbuscular (VA) mycorrhizas – sometimes now simply called arbuscular mycorrhizas – are the most widespread type in tropical forest ecosystems (*Janos*, 1987), although ectomycorrhizas can also be common, as in mixed dipterocarp forests (*Alexander*, 1988; *Smits*, 1992). VA mycorrhizas show significant diversity in structure that may relate to differences in function, as discussed in detail by *Smith* and *Smith* (1996).

The information about occurrence of mycorrhizas in some individual plant species or genera appears contradictory. This is the case with species of *Durio*, which have been variously recorded as non-mycorrhizal (*Shamsuddin*, 1979), VA mycorrhizal (*Chong*, 1989) and 'possibly ectomycorrhizal' (*Smits*, 1992). *D. zibethinus* (durian) is economically very important in many Southeast Asian countries and is the major source of income in the forest garden system studied by *Smith* (1996). We had the opportunity to examine roots of *D. zibethinus* and associated woody perennials collected from these forest gardens for the presence of mycorrhizal fungi. This has led us to evaluate the possible significance of mycorrhizal associations in this system and suggest further possible avenues for study that might be beneficial.

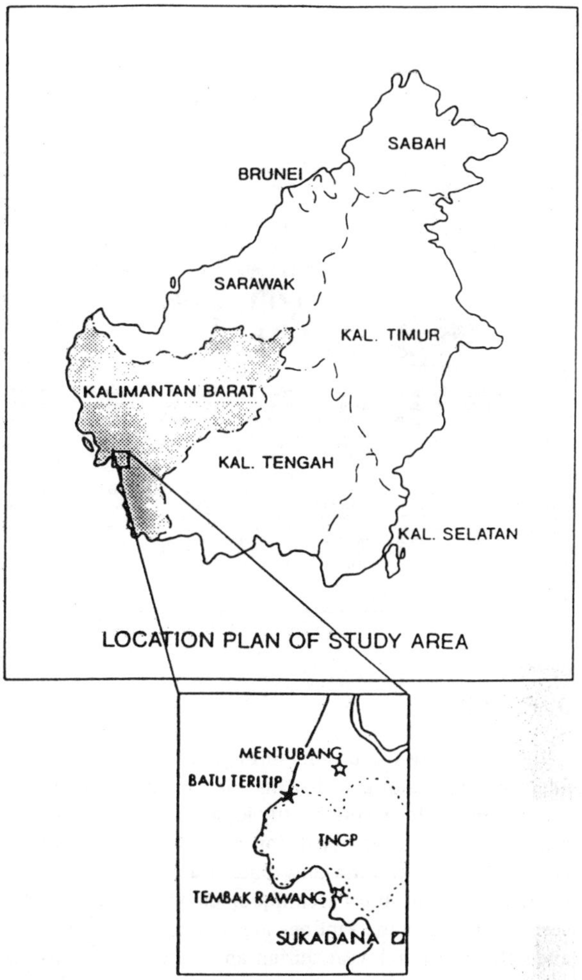


**Figure 1:** Location of the study site at Batu Teritip, W. Kalimantan, Indonesia. From Smith, 1996.

## 2 Methods

The study was conducted in forest gardens in the coastal village of Batu Teritip, about 10 km north-west of Sukadana, W. Kalimantan, and near Gunung Palung National Park. The location is shown in Figure 1.

Extensive inventories were made of the age of the gardens, their species composition, structure and function (*Smith*, 1996). Soils in the forest gardens were classified using the current United States Department of Agriculture system as oxisols and ultisols. They are highly weathered, deep and well-drained with loamy-clay texture and acid/low base status (pH probably about 5.5). Further details were not recorded: see *Salafsky* (1993) for more detailed description of similar soils. These soils were distinct from those of the home gardens, which are histosols (red peats) with more than 20% organic matter. They originate from cleared coastal mangrove swamps on which the village is located.

Roots of *D. zibethinus* (durian), *Nephelium lappaceum* (rambutan), *Lansium domesticum* (langsat) and *Artocarpus integer* (cempedak) were collected from forest gardens and were examined later for the presence of mycorrhizas. The youngest durian trees sampled were 1–2 years old and the oldest 30 years, as estimated by the owners of the trees. The age of other trees could not be determined. Samples included young roots and older roots not more than about 3 mm in diameter. These were stored in 45% ethanol and transported to Adelaide for examination. Several methods were tested for staining. The most successful method was to heat segments of roots in 10% KOH at 80°C for 2–3 hours, followed by a rinse in distilled water and bleaching in 1N HCl for 2 minutes. They were then stained in 0.05% trypan blue in lactoglycerol and were stored in 50% glycerol before microscopic examination. Transverse sections of roots were stained and examined without clearing. Background staining made photography of samples difficult.

| ZONE | HOME GARDEN | FOREST GARDEN 20+ years | FOREST GARDEN 15-20 years | FOREST GARDEN 10-15 years | FOREST GARDEN 5-10 years | FOREST GARDEN 5 years | FOREST GARDEN 0.5-1.5 years | FOREST GARDEN 0-0.5 years | PRIMARY FOREST |
|---|---|---|---|---|---|---|---|---|---|
| LAND USE | Gardens, Handicrafts | Durians, Other fruit, Timber | Durians, Other fruit | Fruit & some Durians | Bananas & some fruit | Bananas & vegetables | Bananas & vegetables | Cleared, uncultivated | Minor use of medicines, fruits & timber |
| TREES & (*) OTHER MAJOR SPECIES | *Musa** <br> *Syzygium* <br> *Citrus* <br> *Coffea** <br> *Colocasia** | *Durio* <br> *Lansium* <br> *Artocarpus* <br> *Nephelium* <br> *Mangifera* <br> *Dysoxylum* <br> *Litsea* | *Durio* <br> *Lansium dom.* <br> *Artocarpus* <br> *Nephelium* <br> *Mangifera* <br> *Carica pap.* <br> *Syzygium* | *Carica pap.* <br> *Musa** <br> *Durio* <br> *Syzygium* <br> *Lansium dom.* <br> *Nephelium* <br> *Artocarpus* | *Musa** <br> *Carica pap.* <br> *Syzygium* <br> *Citrus* <br> *Artocarpus* <br> *Nephelium* <br> *Durio* <br> *Coffea** | *Musa** <br> *Lansium dom.* <br> *Durio* <br> *Coffea* <br> *Colocasia** <br> *Allium** <br> *Capsicum** <br> *Piper reto.** <br> *Ocimum bas**. | *Musa** <br> *Capsicum** <br> *Piper reto.** | *Octomeles* <br> *Dysoxylum* | *Octomeles* <br> *Shorea* <br> *Litsea* <br> *Ficus* <br> *Artocarpus* <br> *Nephelium* <br> *Dysoxylum* |

**Figure 2:** Vertical profile of forest gardens and near-by areas at Batu Teritip, W. Kalimantan, showing their age, use and the main associations of plants (summarised mainly as genera) and excluding minor species. Adapted from Smith, 1996.

## 3 Results

Figure 2 shows a profile through forest gardens and adjacent areas along with a list of the major trees and other plants. These were mostly identified to species level but are here summarised for convenience mainly as genera. Initially, the lowland dipterocarp forest is cleared using a slash-and-burn technique, leaving some trees for soil stability. Banana saplings are planted after 6 months, and after another 1 year durian and other fruit trees are planted under the young banana trees. Over 10–15 years the fruit trees mature and stratify, mimicking the structure and function of the nearby forest.

Young durian roots were heavily (> 50%) colonized by VA mycorrhizal (Glomalean) fungi and contained arbuscules in the inner cortex and, very rarely, vesicles. Lengthy intercellular hyphae were not seen and cortical cells contained prominent coils of mainly aseptate hyphae which often made arbuscules difficult to see. Older (thicker) roots were less heavily colonized by VA mycorrhizal fungi. Sometimes their surfaces were colonized by a heavily-branched septate fungus of unknown type. Extensive VA mycorrhizas with intracellular coils were also seen in rambutan and cempedak. Langsat roots, which were more fleshy, had appressoria and hyphal coils in the outer cortex, but no arbuscules or vesicles were seen.

Table 1 summarizes the mycorrhizal status of durian and other fruit trees that we sampled. Previous relevant work with the same species or related plants in the same families is also shown, along with some other examples, and the table shows the widespread occurrence of VA mycorrhizas among these plants.

The degree of colonization of most of the plant species in the forest gardens and adjacent areas (approximately 100 species altogether) is not known. However, their probable mycorrhizal status was checked using a list of publications that describe VA mycorrhizal plants in Indonesia and other tropical and sub-tropical countries in southern Asia (F. A. *Smith*, unpublished). Figure 3 summarises the results of this survey. The results show that most of the species (or at least plants in the same genera) have been found in other studies to form VA mycorrhizas. Relatively few plant species found in the understorey have been studied, so the results do not imply that many of these plants would be expected to be non-mycorrhizal.

## 4 Discussion

*Chong* (1989) found that roots of *Durio* sp. in Sarawak were endomycorrhizal and contained vesicles, but she did not see arbuscules. We have found that young durian roots are endomycorrhizal, with many arbuscules and very infrequent vesicles, and similarly with roots of rambutan and cempedak. The situation with langsat is not yet clear. The extensive intracellular hyphal coils in durian and other plants are definitive '*Paris*-type' structures first described by *Gallaud* (1905), and shown in Figure 4, taken from *Smith* and *Smith* (1996). We did not see lengthy intercellular hyphae that are found in '*Arum*-type' VA mycorrhizas (Fig. 4). Intracellular coils in hypodermes can occur in both types (not shown in Fig. 4).

Most of the herbaceous plants that have been widely studied are *Arum*-types and the frequency of *Paris*-types

**Table 1:** The mycorrhizal status of some major plants from the forest gardens in Batu Teritip, W. Kalimantan, and related tropical species from S.E. Asia.

| Family | Genus/species | Myc. status | Notes and references |
|---|---|---|---|
| Bombacaceae | *Ceiba pentandra* | V | Sarawak; Chong (1989) |
| | *Durio* sp. | V; coils; | Sarawak; Chong (1989) |
| | *Durio griffithii* | ECM? | Kalimantan; Smits (1992) |
| | ***D. zibethinus*** | **–M** | Malaysia; Shamsuddin (1979) |
| | ***D. zibethinus*** | **(V)A; coils** | Kalimantan; Smith et al. this study |
| Meliaceae | *Aglaia* sp | VA | Kalimantan; (Smits, 1992) |
| | ***(Disoxylum sp.)*** | VA; coils | Java; (Janse, 1897) |
| | ***Lansium domesticum*** | **–M/VA?** | Kalimantan; Smith et al. this study |
| | *Swietinia macrophylla* | V | Sarawak; Chong (1989) |
| Moraceae | ***Artocarpus integer*** | **(V)A; coils** | Kalimantan; Smith et al. this study |
| | *A. nitida ssp. griffithii* | VA | Kalimantan; Smits (1992) |
| | *A. dadah* | –M | Kalimantan; Smits (1992) |
| | ***(Ficus sp.)*** | VA | Kalimantan; Smits (1992) |
| Sapindaceae | ***Nephelium lappaceum*** | **(V)A; coils** | Kalimantan; Smith et al. this study |
| Caricaceae | ***(Carica papaya)*** | **VA** | (records from the Americas) |
| Anacardiaceae | ***(Dracontomelum dao)*** | **?** | (no records found) |
| | ***(Mangifera spp.)*** | VA | (records from the Americas) |
| Musaceae | ***(Musa sp.)***[1] | **–M** | Java; Janse (1897) |
| | ***Musa sp.***[1] | **VA** | (records from the Americas) |
| | ***Musa sp.***[1] | **VA** | Costa Rica; Janos, pers. comm. |
| | ***Musa sp.***[1] | **–M** | Costa Rica; Janos, pers. comm. |
| Myrtaceae | ***(Syzygium sp.)***[2] | **?** | See below: note 2 |
| | *Eugenia*, 2 spp.[2] | VA | Kalimantan; Smits (1992) |
| | *Eugenia* sp.[2] | (VA) | Java; Janse (1897) |
| Rubiaceae | ***(Coffea arabica)*** | **VA; coils** | Java; Janse (1897); and others |
| | *Gardenia* sp. | VA | Kalimantan; Smits (1992) |
| | *Ixora* | VA | Kalimantan; Smits (1992) |
| | *Nauclea* sp. | (VA) | Java; Janse (1897) |
| | *Porterandia anisophyllea* | VA | Kalimantan; Smits (1992) |
| | *Urophyllum*, 2 spp | –M | Kalimantan; Smits (1992) |

**Key: Bold type = plants from the garden in W. Kalimantan. (Plants not examined in this study are in brackets).** Normal type = other related plants from S.E. Asia. –M = nonmycorrhizal; V = vesicles but no arbuscules recorded; (V)A = rare vesicles, many arbuscules; VA = vesicles and arbuscules; –M/VA? = entry-points and hypodermal coils only; (VA) = status inferred from description or growth effects; coils = cortical fungal coils (not always recorded in the literature); ECM? = possibly ectomycorrhizal; ? = no records found.

**Notes:**

**1.** Janos (pers. comm.) says that *Musa* spp. have very low dependency on VA mycorrhizas and that there is often low colonization in plantations. Janse (1897) examined *Musa* from the forest at Cibodas, Java.

**2.** Many *Eugenia* spp. have been transferred to *Syzygium* spp.

has probably been greatly underestimated (*Smith* and *Smith*, 1996). *Paris*-type structures can be clearly identified in the seminal survey by *Janse* (1897) of fungal endophytes in roots of plants of Java, and we saw in our material the senescent arbuscules that *Janse* described as 'sporangioles'. It has been suggested that some *Paris*-type VA mycorrhizas lack arbuscules altogether (see *Alexander*, 1988) but this may result from seasonal or other effects (*Smith* and *Smith*, 1996). Absence of arbuscules and, in some cases, vesicles makes recognition of VA mycorrhizas difficult. The problem is worse if the material stains poorly or contains a range of fungi, or both! We did not find that durian can form ectomycorrhizas, as suggested by *Smits* (1992). The fungus that caused external hyphae was only present on older roots, unlike ectomycorrhizas where complete sheaths form over the root tips. As some plant species can form both VA and ectomycorrhizas, the observations by *Smits* may not be incorrect; further work is needed to resolve this issue.

Other plants in the forest gardens also probably have VA mycorrhizas, as suggested by previous work in S.E. Asia and elsewhere (Table 1 and Fig. 3). A very extensive collection would be needed to confirm this. Although we do not know if VA mycorrhizas give significant nutritional benefit to the plants in this agroforestry system, this is strongly suggested by work with other lowland tropical plants (e.g. *Janos*, 1987), including plantation shade trees in Java (*Wibawa* et al., 1995). The VA mycorrhizal symbiosis is very non-specific, so that fungal hyphae growing from one plant can rapidly colonize another, and one root can be colonized by different VA mycorrhizal fungi (see *Smith* and *Smith*, 1996 for discussion). As a large population of fungal propagules develops, the low specificity may be nutritionally beneficial in mixed plantings, as practised in

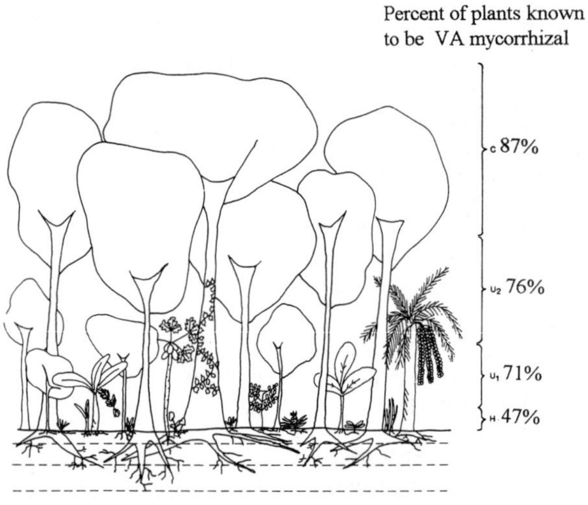


**Figure 3:** Mycorrhizal status of plants in the different storeys of a mature forest garden, showing the percentages (%) of plant genera that are known to be mycorrhizal from previous work by others, and from the present study. From Smith, 1996.

forest gardens in Kalimantan and elsewhere. If fungal hyphae that connect different plants can transfer significant amounts of organic carbon to shaded seedlings or other 'receivers', there may be additional nutritional benefits. There will then also be costs to the 'donor' plants in terms of carbon lost, but these may not be deleterious where photosynthesis is not the limiting factor in plant growth and productivity. The ecological singificance of the linkage of plants by mycorrhizal fungi is as yet unclear. If nutrients such as P and (organic) N can move between plants on a source-sink basis, such linkage would be expected to minimize competition by different species for soil nutrients and promote diversity within ecosystems (*Janos*, 1983, 1987; *Alexander* 1989; *Smith* and *Smith*, 1996). In the absence of fertilizer application, removal of nutrients is an inevitable result of silviculture, agroforest ecosystems and intensive agriculture. In the long term, widespread occurrence of mycorrhizas and resource sharing by mycorrhizal linkages cannot prevent this depletion of nutrients. However, the can promote more efficient nutrient cycling within the soil-plant system and greatly delay the need for application of fertilizers (where the latter is at all economically feasible).

Slash-and-burn (swidden) agriculture is a significant cause of loss of tropical forests. Many argue that leaving forest slash in the field is ecologically sound because organic matter and nutrients are left on site and eventually return to the soil. Burning the forest slash is more common and causes loss of volatile nutrients (e.g. N). High soil temperatures during burning may reduce the number of mycorrhizal propagules (spores and root fragments) but it is difficult to generalize because fire can differ greatly in intensity. Soil temperature near the surface can be greatly increased even by increased radiation input through removal of vegetation cover. The fallow period after burning may result in further loss of propagules if the land is left bare. However, *Sieverding* (1991) believes that slash-and-burn agriculture has little detrimental effect on numbers of VA mycorrhizal propagules and therefore does not threaten mycorrhizally-promoted regeneration of vegetation. In the forest gardens of Batu Teritip, forest clearance is only partial and burning is only light, thus enhancing retention of mycorrhizal propagules and allowing rapid colonization of new plantings. As shown in Table 1, the mycorrhizal dependency of bananas, the first crop planted in these gardens, is variable and colonization in soils of high nutrients status can be light. *Smith* (1996) argues that any temporary decrease in mycorrhizal propagules following slash-and-burn will probably have been reversed by the time that secondary planting of more mycorrhizally dependent species takes place. This situation may be different in systems when the slash process causes complete clearance and the burn is more intensive. This severe clearance may not be deleterious when the first crop is rice, because some cereals show relatively small mycorrhizal dependency (*Smith* and *Read*, 1997). However, prolonged cultivation of sparsely mycorrhizal species may affect secondary regeneration of other species during subsequent fallow periods. In this context, it would also be interesting to know the mycorrhizal status of *Imperata cylindrica* ('alang-alang') grassland, commonly associated with nutrient decline after long-term exploitation of land in the tropics of S.E. Asia. The role of mycorrhizas in plant successions of different kinds is reviewed in Chapter 15 of *Smith* and *Read* (1997).

## 5 Conclusions and recommendations

We believe that, in mimicking the natural processes of succession and forest structure, complex forest gardens are inherently more ecologically sustainable than many more intensive agricultural alternatives. They provide opportunity for low input, small-scale and resource-efficient cultivation under increased pressure from population growth and land shortage. There are many opportunities for future research within the framework covered by *Smith* (1996). In the context of this contribution we recommend especially a detailed inventory of changes in numbers and kinds of mycorrhizal fungal propagules following slash-and-burn and during establishment of agroforest ecosystems of different kinds, along with surveys of the degree of mycorrhizal colonization of the plants that are cultivated. There is increasing interest in this belowground microbial biodiversity in the tropics and we strongly support the extension of experimental studies of effects of mycorrhizas on the growth and productivity of tropical plants.

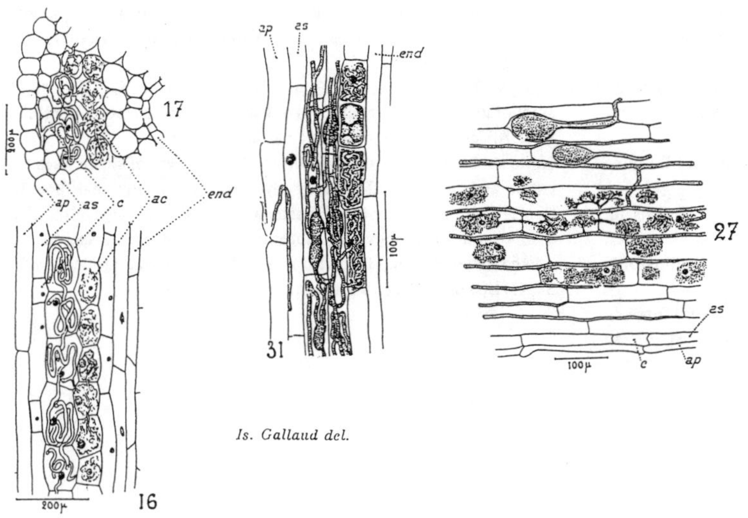


**Figure 4:** Copies of illustrations by *Gallaud* (1905), showing the definitive differences between *Paris*-type and *Arum*-type VA mycorrhizas, as discussed in the text. The original numbers are retained for reference. **Paris-type:** 16 and 17. *Paris quadrifolia*; longitudinal and transverse sections. 31. *Anemone nemorosa*; longitudinal section. ***Arum-type:*** 27: *Allium sphaerocephalum*; longitudinal section. Abbreviations: ap = epidermis; as = hypodermis; c = passage cell; end = endodermis; ac = compound arbuscule. The hyphal coils in the *Paris*-types are like those seen in the present study. Note the intercellular hyphae in *Allium*, and the arbuscules and vesicles. See *Janse* (1897) for comparison and also *Smith* and *Smith* (1996).

## 6 Acknowledgments

The field studies in Kalimantan formed part of a Ph.D project undertaken by Hilary F. Smith, with supervision from Lesley Potter and Ruth Lawrence. Universitas Tanjungpura Pontianak was the Indonesian sponsor for the project. Financial support came from an Australian Postgraduate Research Award and the Australian Vice-Chancellors' Committee Awards for research in Asia. Our mycorrhizal research program is funded by the Australian Research Council. We are very grateful to Dave Janos for advice about tropical mycorrhizas and to the people of Batu Teritip for their help.

## 7 References

*Alexander, I.* (1988). "Mycorrhizas of indigenous tropical forest trees: some research priorities". In F.S.P. NG (ed.): Proceedings of the Asian Seminar, Forest Research Institute Malaysia, Kuala Lumpur, pp. 79–89.

*Alexander, I.* (1989). "Mycorrhizas in tropical forests". In: J. Proctor (ed.): Mineral nutrients in tropical forest and savannah ecosystems. Special Publication no. 9, British Ecological Society. Blackwell Scientific Publications, Oxford.

*Chong, L.* (1989). "Occurrence of mycorrhizae in seedlings of some tree species in Sarawak". In: A. Mahadevan, N. Raman and K. Natarajan (eds.): *Proceedings of the First Asian Conference on Mycorrhizae.* University of Madras, pp. 70–72.

*Gallaud, I.* (1905). "Études sur les mycorrhizes endotrophes". Rév. Gen. Bot. 17: 5–48, 66–83, 123–136, 223–239, 313–325, 425–433, 479–500.

*Gouyon, A.* (1996). "Ecological and socio-economic conditions of rubber agroforestry in the Dipterocarp ecosystems of Indonesia". In: A. Schulte and D. Schöne (eds.): Dipterocarp forest ecosystems: towards sustainable management. World Scientific, Singapore, pp. 369–388.

*Janos, D. P.* (1983). "Tropical mycorrhizas, nutrient cycles and plant growth". In: S. L. Sutton, T. C. Whitmore and *A. C. Chadwick* (eds.): Tropical rain forest: ecology and management. Special Publication no. 2, British Ecological Society. Blackwell Scientific Publications, Oxford.

*Janos, D. P.* (1987). "VA Mycorrhizas in humid tropical ecosystems". In: G. Safir (ed.): VA Mycorrhizae: an ecophysiological approach. CRC Press, Boca Raton, pp. 107–134.

*Janse, J. M.* (1897): "Les endophytes radicaux de quelques plantes Javanaises". Ann. Jard. Bot. Buitenzorg 14: 53–200.

*Michon, G., Mary, F.* and *Bompard, J. M.* (1996). "Multistoried agroforestry garden system in West Sumatra, Indonesia". Agroforestry systems 4: 315–338.

*Nair, P. K. R.*, (ed.). (1989). Agroforestry systems in the tropics. Kluwer Academic Publishers with ICRAF, Dordrecht, The Netherlands.

*Nouaim, R.* and *Chaussod, R.* (1994). "Mycorrhizal dependency of micropropagated agen tree (Argania spinosa): I. Growth and biomass production". Agroforestry systems 27: 53–65.

*Nouiam, R., Lineres, M., Esvan, J. M.* and *Chaussod, R.* (1994). "Mycorrhizal dependency of micro-propagated agen tree (Argania spinosa): II. Mineral nutrition". Agroforestry systems 27: 67–77.

*Padoch, C.* and *Peters, C.* (1992). "Managed forest gardens in West Kalimantan, Indonesia". In: D. Cohen and L. Potter (eds.): Case studies of genetic resource conservation in natural habitats: biological and socioeconomic dimensions. American Association for the Advancement of Science.

*Salafsky, N.* (1993). The forest garden project: an ecological and economic study of a locally developed land-use system in West Kalimantan, Indonesia. Unpublished Ph D Thesis, Duke University, USA.

*Salafsky, N.* (1994). "Forest gardens in the Gunung Palung region of West Kalimantan, Indonesia: defining a locally-developed, market-oriented agroforestry system". Agroforestry systems 28: 237–268.

*Sardjono, M. A.* (1996). "The lembo system: a model for agroforestry in dipterocarp forest ecosystems of East Kalimantan". In: A. Schulte and D. Schöne (eds.): Dipterocarp forest ecosystems: towards sustainable management. World Scientific, Singapore, pp. 354–368.

*Schroth, G.* (1995). "Tree root characteristics as criteria for species selection and system design in agroforestry". Agroforestry systems 30: 125–143.

*Shamsuddin, M. N.* (1979). "Mycorrhizas of tropical forest trees". In J. I. Furtado (ed.): Abstracts, V International Symposium of Tropical Ecology. University of Malaysia, Kuala Lumpur, p. 173.

*Sieverding, E.* (1991). Vesicular-arbuscular mycorrhiza management in tropical agrosystems. Eschborn: Deutsch Gesellschaft für Technische (GTZ) GmbH.

*Smith, F. A.* and *Smith, S. E.* (1996). "Mutualism and parasitism: diversity in function and structure in the 'arbuscular' (VA) mycorrhizal symbiosis". Adv. Bot. Res. 22; 1–43.

*Smith, H. F.* (1996): Semi-commercial agroforestry: a case-study from West Kalimantan, Indonesia. Unpublished Ph. D. Thesis, Department of Geography, The University of Adelaide, Australia.

*Smith, S. E.* and *Read, D. J.* (1997). Mycorrhizal Symbiosis, 2nd Edition. Academic Press, London.

*Smits, W. T. M.* (1992). "Mycorrhizal studies in Dipterocarp forests in Indonesia". In: D. J. Read, D. H. Lewis, A. H. Fitter and I. J. Alexander (eds.): Mycorrhizas in Ecosystems. CAB International, Wallingford, pp. 283–292.

*Sorensen, K. W.* (1996). "Traditional management of dipterocarp forests: examples of community forestry by indigenous communities". In: A. Schulte and D. Schöne (eds.): Dipterocarp forest ecosystems: towards sustainable management. World Scientific, Singapore, pp. 335–353.

*Tomlinson, H., Telehimanot, Z., Traore, A.* and *Olapade, E.* (1995). "Soil amelioration and root symbioses of Parkia biglobosa (Jacq.) Benth. in West Africa". Agroforestry systems 30: 145–159.

*Wibawa, A., Baon, J. B.* and *Nurkholis* (1995). "Growth of shade trees for coffee and cacao as affected by mycorrhizal and rhizobial inoculation". In: Supriyanto and J. T. Kartana (eds.): Proceedings of the 2nd Symposium on Biology and Biotechnology of Mycorrhizae and 3rd Asian Conference on Mycorrhizae. BIOTROP Special Publication 56, SEAMEO/BIOTROP, Bogor, Indonesia, pp. 209–214.

# Natural rehabilitation of disturbed tropical rainforest soils in Sabah, Malaysia

**Anders Malmer**[1], **Eva Johansson**[2] and **Martina Kluge**[1]

[1]Swedish University of Agricultural Science, Faculty of Forestry, Department of Forest Ecology, S-901 83 Umeå, Sweden, E-mail: Anders.Malmer@sek.slu.se.
[2]Swedish University of Agricultural Science, Department of Entomology, P.O. Box 7044, S-750 07 Uppsala, Sweden, E-mail: Eva.Johansson@entom.slu.se.

## Abstract

Physical properties of mineral top soil were investigated on and outside 6 and 13 year old tractor tracks (and outside tractor tracks) on Haplic Acrisol (Ultisol) under selectively logged forest, secondary forest after forest fire and plantation forest. The analysis of possible rehabilitation to pre disturbance soil properties also included earlier published data for new tractor tracks and control forest in the same experimental area. Dry bulk density (DBD), loss on ignition (LOI) and steady state infiltrability (SSI), as well as visual descriptions of vegetation, top soil and erosion, were recorded for 3 positions in slopes as well as for 3 degrees of tractor disturbance to try to resolve the process of top soil rehabilitation.

13 year old tracks under secondary forest grouped significantly together with controls for DBD and LOI, in contrast to nearby 13 year old ones under the canopy of selectively logged forest. Differences in amount and quality of litterfall are discussed to explain this difference. For SSI there was a significant increasing gradient from new to older tracks, but no tracks were as high as control.

From these and other data, it may be argued that DBD decreases and possibly total porosity increases over time with increasing organic content (LOI). Soil permeability is rehabilitated more slowly than DBD and total porosity, as macro porosity is further developed in a later stage with further increased organic content. This later development of macroporosity may also be attributed to changes in the quality of organic material as DBD may not be further lowered.

## Introduction

The use of heavy machinery in mechanical clearing and timber extraction in forests is known to cause disturbance to forest soils (eg. *Nicholson*, 1958; *Van der Weert*, 1974; *Kamaruzaman* and *Nik Muhamad Majid*, 1986; *Malmer* and *Grip*, 1990). Apart from compacting the soil, the machines also remove the looser top soil, thereby exposing the more compact subsoils (*Malmer* and *Grip*, 1990). As a consequence of the disturbance changes of physical properties such as increased bulk density and loss of infiltrability occur (eg. *Van der Weert*, 1974; *Aina*, 1979; *Wierda*, 1989; *Chauvel*, 1991) and are commonly reported to lead to a decrease in site production and an increase in surface runoff and erosion (*Bruijnzeel*, 1990; *Malmer*, 1990 and 1996a). The rehabilitation of the soils back to pre disturbance values seems to be a slow process. Low infiltrability and high dry bulk density have been reported on six year old tracks (*Malmer* and *Grip*, 1990) as well as on twelve year old ones (*Van der Plas* and *Bruijnzeel*, 1993).

Apart from trying to quantify changes of soil properties caused by soil disturbance, few studies have dealt with the process of natural or man made rehabilitation of disturbed rain forest soils. The apparently slow recovery of these soils will be a large problem for future forest production in the humid tropical region, since both selective logging and clear-felling operations disturb up to 30% or more of the soil surface in the above mentioned way (*Malmer*, 1996). *Nussbaum*, et al. (1995) reported on experiments of rehabilitation in Eastern Sabah. However, they gave results after only a six months period of response to different treatments, and they only studied landing sites. For the rehabilitation of tractor tracks in the forest, the time and the (natural or man made) driving force for rehabilitation (loosening of soil, addition of nutrients or organic material, etc.) will be relying on factors like topographical position, degree of disturbance, etc. *Kararuzaman* (1996) reported on a ten year time series of physical parameters for skid trails, bush landings and secondary forest roads in hill forest on Ultisol in Peninsular Malaysia. He showed means of these parameters to undergo a steady change towards values of undisturbed soils. By assuming continued linear rate of change towards control values he estimated time for total rehabilitation of skid trail top soils to 22 years for bulk density and 52 years for saturated hydraulic conductivity.

The objective of this project was to quantify natural rehabilitation of forest soils disturbed by timber extraction, by studying dry bulk density, organic content and infiltrability in crawler tractor tracks of different age. We also tried to learn more about the actual rehabilitation process by studying the influence of degree of disturbance, topography, erosion and vegetation. The intention was also that this new knowledge would be of importance in future experiments with man made rehabilitation of disturbed forest soils (*Johansson* and *Kluge*, 1995).

## Research area

The Mendolong research area is 35 km inland from Sipitang at 115.5°E, 5.0°N on the west coast of Sabah, Malaysia (northern Borneo). The natural vegetation of the research area is hill dipterocarp forest at its elevation of 650–750 m.a.s.l. (*Sim* and *Nykvist*, 1991). The area also includes areas struck by forest fires in 1982/83 (*Woods*, 1989; *Nykvist*, 1996). The research area contains an environmental impact assessment on different ways of conversion od dipterocarp forest to *Acacia mangium* forest plantations in the form of paired catchments with additional process studies (*Nykvist* et al., 1994; *Malmer*, 1996a). Mean yearly precipitation was 3350 mm and mean yearly runoff was 1955 mm for the control catchments in the area (1985–1990, *Malmer*, 1992). Maximum rain intensities of one hour duration in 1988 were generally below 30 mm $h^{-1}$. However, higher rain intensities (> 100 mm $h^{-1}$) of 5–10 min duration were observed at the start of rainstorms (*Malmer*, 1996a). Topography in the area is generally moderately hilly with midslope angles of 25–30°, but single slopes up to 45° also occurs. The dominating soil type in the area was Haplic Acrisol (Ultisol) with clayey top soil (*Malmer* and *Grip*, 1990). As this soil type is common trough out the South East Asian region, it was choosen as the tested soil type in this study. Under forest cover the uppermost 20 cm of this soil was loose with a well developed highly porous structure (*Malmer* and *Grip*, 1990) and typical mineral top soil loss on ignition (LOI) values were in the range of 5–15% (*Malmer*, unpublished).

## Methods

The classes of soils compared in this study are presented in Table 1. Data for control forest top soils and new tractor tracks were extracted from the investigation by *Malmer* and *Grip* (1990). The investigated tracks were chosen within one $km^2$, to minimize effects of areal variation in properties of vegetation, climate and soil. Within this limited area 6 and 13 year old tracks were available. The tracks in the plantation (P6) are the same tracks (P0) that were investigated by *Malmer* and *Grip* (1990). All of the investigated tracks of different soil classes were created at timber extraction from the natural dipterocarp forest using crawler tractors of D4 and D6 classes. The range of maximum slope angles of the tracks were 10–17° (tracks were seldom perpendicular to the slope gradient).

On tractor tracks, the classes of P6, SL6, SL13 and SF13 were divided into three classes of disturbance; 1st order were main tracks connecting to landings, having had heavy traffic, 2nd order were tracks branching from main tracks, having had intermediate traffic and 3rd order tracks were tracks of the last branching, having had only single or few passes by tractors. Furthermore, each disturbance class in each soil class were divided into up slope, mid slope and down slope positions. These divisions resulted in 9 positions of investigation per soil class for P6, SL6, SL13 and SF13. In each position three undisturbed soil core samples for dry bulk density (DBD) and loss on ignition (LOI) were taken at each depth of 0–5, 5–10 and 10–20 cm. Also 2 measurements of infiltration were taken at each of the positions by use of double ring infiltrometer with 156 mm and 256 mm diameters of the inner- and outer rings respectively.

In secondary forest after forest fire, but outside tracks (FF), two sample positions were chosen in each of three plots. One DBD and LOI sample was taken in each of those six positions at each of the three sampling depths. Infiltration was measured once at these six points.

Furthermore, features of vegetation and the soil surface were recorded. This was made every 10 m along stretches of tracks from upsloge to downslope on different soil classes. The features of vegetation recorded were number of trees rooted in the tracks and their basal area as well as degree of cover by ground vegetation. The soil surface was recorded in terms of recent sediment erosion or deposition and presence of leaf wash.

From momentary measurements of infiltration, steady state infiltrability was calculated according to *Collis-George* (1980). Statistical differences ($p < 0.05$) between means of physical parameters were calculated by using non parametric tests according to *Dunn* (1964).

**Table 1:** Soil classes investigated.

| track age | vegetation | abbreviation |
|---|---|---|
| –[1] | undisturbed forest[2] | F |
| 0 | clear-felled forest[2] | P0 |
| 6 | forest plantation | P6 |
| 6 | selectively logged forest | SL6 |
| 13 | selectively logged forest | SL13 |
| 13 | secondary forest 11 years after fire | SF13 |
| –[1] | like SF13 above, but undisturbed soil | FF |

1) Undisturbed control soil.
2) Data from earlier investigation by Malmer and Grip (1990).

## Results

### Dry bulk density

Bulk densities were generally lower on the least disturbed tracks, but these differences were small and non significant. In some cases bulk densities were higher in lower parts of slopes. However, this showed to be related to areas of deposition of sediments and hence enrichment

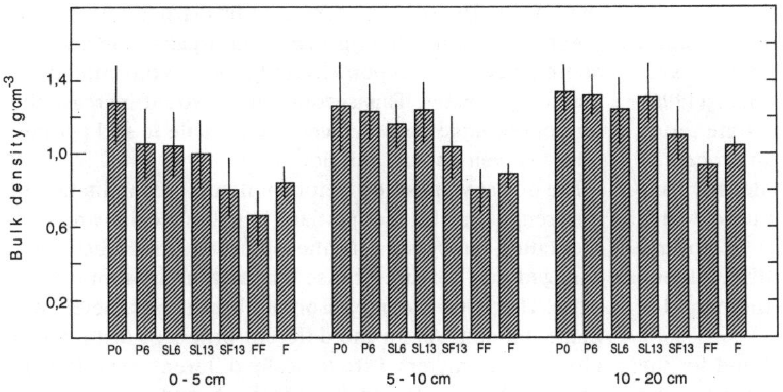


**Figure 1:** Means (and standard deviations) of dry bulk density (g $cm^{-3}$) at different depths for different soil classes. For explanations of abbreviations see Table 1. Data for F and P0 are from Malmer and Grip (1990). (F and P0: n = 7; P6, SL6, SL13 and SF13: n = 27; FF: n = 6).

of sand. Figure 1 shows means of dry bulk densities for different soil types tested. F, FF and SF13 grouped significantly with lower bulk densities than the group of other tested soils for all three soil depths measured. Within these two groups there were no significant differences. The significances apparent were highest in the uppermost 0–5 cm of top soil.

### Loss on ignition

Dry bulk density was negatively related to LOI (Figure 2). Consequently, FF and SF13 again grouped together and had significantly different means of LOI compared to the other soil classes (F was not tested for LOI, *Malmer* and *Grip*, 1990). Also, in accordance with DBD, LOI was generally higher, but not significantly, on tracks with less disturbance.

### Steady state infiltrability

Statistical analysis divided the means of steady state infiltrability of different soil classes (Table 2) in to three groups significantly different from each other. P0 and SL6 showed lowest means, while P6, SL13 and SF13 formed an intermediate level significantly lower than FF and F. In accordance with above, infiltrability were generally higher on less disturbed tractor tracks (Table 3), but the high variability (Table 2) made any significant differences impossible to detect.

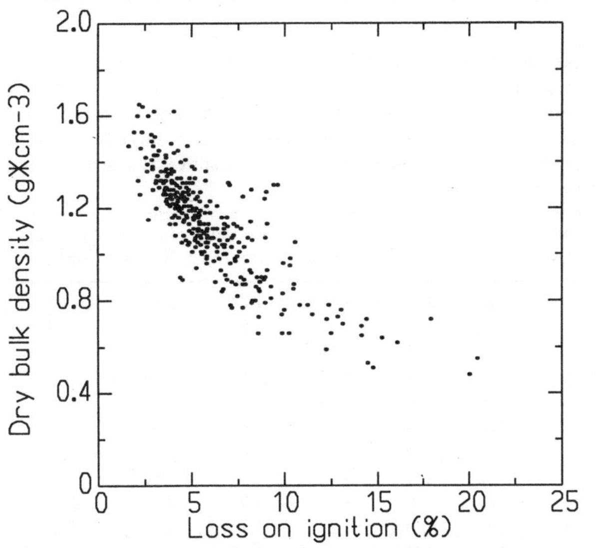


**Figure 2:** Loss on ignition (LOI) versus dry bulk density (DBD) for all soil depths and tested soil classes (P6, SL6, SL13, SF13 and FF).

### Track descriptions

There was a non statistical tendency for litter cover, erosion and leafwash to vary with order of disturbance (Table 4). The exception was SL6 where wild boar had decreased litter cover on the third order tracks described. More trees were found on 13 year old tracks than on 6 year old ones, except for in P6 where the trees in the tracks were planted. However, those planted trees were performing much less well than those on undisturbed soils; basal area of about 5 $m^2$ $ha^{-1}$ on tracks after 6 years compared to about 10 $m^2$ $ha^{-1}$ for the same plantation at less than 3.8 years (*Nykvist* et al., 1996).

**Table 2:** Means (and standard deviations) of steady state infiltrability (mm $h^{-1}$) for different soil classes. For explanations of abbreviation see Table 1. Data from F and P0 are from Malmer and Grip (1990) (P0: n = 8; P6, SL6, SL13 and SF13: n = 18; F: n = 10, FF: n = 6).

| soil class | | | | | | |
|---|---|---|---|---|---|---|
| P0 | P6 | SL6 | SL13 | SF13 | F | FF |
| 0.28 | 56 | 28 | 62 | 54 | 154 | 398 |
| (0.40) | (80) | (44) | (121) | (74) | (90) | (216) |

**Table 3:** Means of steady state infiltrability (m $h^{-1}$) for different soil classes on tracks of different order of disturbance (n = 6). For explanations of abbreviation see Table 1.

| order | soil class | | | |
|---|---|---|---|---|
| | P6 | SL6 | SL13 | SF13 |
| 1st | 26 | 7 | 13 | 65 |
| 2nd | 37 | 38 | 26 | 21 |
| 3rd | 104 | 39 | 147 | 75 |

## Discussion

From these results it is clear that a natural rehabilitation process of tractor track top soils is active in the same manner as to what was shown by *Kararuzaman* (1996). Even though limited to three points in a false time series, the data presented above also describe a linear process of rehabilitation during the first decade for bulk density. This is if the soil classes P0, SL6/P6 and SF13 are used, which means a linear rehabilitation back to DBD of control forest in 13 years. If SL13 is used instead of SF13, rather a slow down in processes in the later part of the period is argumented.

Data on loss on ignition showed a similar trend for the difference between SL13 and SF13, but the change over time can not be evaluated as LOI data for new tracks are not available. However, the lowest level of typical mineral top soil LOI for control dipterocarp forest (5—15%, *Malmer*, unpublished) was reached for all soil classes investigated, even though LOI of SF13 tracks grouped significantly different at the upper limit of that range, in the same group as that of FF (secondary forest after forest fire).

The high variability of steady state infiltrability makes the evaluation of linearity or not meaningless, but clearly the tracks of P6, SL13 and SF13 grouped in an intermediate position between lows of P0 and S6 and highs of FF and F. This also speaks for that the full rehabilitation of soil permeability takes longer time than that of LOI and DBD. This was also estimated by *Kararuzaman* (1996) in terms of top soil saturated hydraulic conductivity. However, it may be noted that tracks in the plantation after 6 years and in selectively logged forest after 13 years are capable of a mean steady state infiltrability that is larger than the one hour maximum rain intensity for the area reported by *Malmer* (1996a).

*Kararuzaman* (1996) showed that rehabilitation of total porosity followed the same pace as DBD and LOI, and that, assumed linearity, rehabilitation of these three properties on tractor tracks would take similar time. In the LOI interval of 2—9%, the DBD decreases rapidly with increased LOI (Figure 2). This is due to the lower specific weight of organic material together with increased porosity from the build up of soil aggregation induced by soil meso- and macro fauna supported by the increase in organic material. Above 10% of LOI the DBD still decreases with increased LOI, but with a less steep slope (Figure 2). This second relation could be connected to only the lower specific weight of organic material and only a minor change in total porosity. In accordance with this, and the data above, on 13 year old tracks there is still a rehabilitation of organic material and soil permeability to take place after the rehabilitation of DBD (and possibly most of total porosity). In other words, the rehabilitation could be divided in two phases: The initial phase contains the build up of soil aggregation and total porosity with addition of organic material, up to about 10% LOI for these soils. The second phase contains structural development of the top soil with increased organic content, not adding much to total porosity, but rather to the build up of macro aggregates. This second phase would increase the number of large pores which is favourable to soil permeability as well as soil aeration.

The objective to study factors of natural rehabilitation in different parts of slopes partially failed as erosion and sedimentation processes on the slopes of the tracks had changed soil textures by sedimentation in some lower parts. Thereby the comparison of physical parameters was made difficult. As for the different orders of disturbance, physical parameters were logically different, even though not significantly. The reason for this have to be sought in the fact that even with only one pass by the tractor, the top soil is often removed and more compact sub soil turned into new track topsoil. The similarity in former subsoil properties would then over ride most of the effect of the differences in the number of passes by the tractors.

With few exceptions, leaf wash was a feature restricted to the first order tracks (Table 4). It was also observed to be most common in midslope positions, with combinations of flat surfaces, suitable for sheet flow, and the highest degrees of slope. Midslope rehabilitation in general was not significantly different from other segments of the slopes, but on the very spots where leaf wash occurred, a slower rehabilitation (in terms of lower organic content in more compact soil) was obvious by visual inspection. Leaf wash and seed wash (*Nussbaum* et al., 1995) have been pointed out as one general factor that prolongs rehabilitation on tractor tracks. In general litter cover were more than 75% on both 6 and 13 year old tracks of all disturbance degrees (Table 4). On the other hand, cover of ground vegetation rooted in the tracks were low (Table 4). This was also the case in SF13 and P6 where

**Table 4:** Summary table of descriptions of full slopes of tractor tracks in terms of cover by litter layer and ground vegetation (classes 1, 2, 3 and 4 representing 0—25, 25—50, 50—75 and 75—100% cover respectively), erosion (0 no occurrence, 1 old signs, 2 recent erosion), leaf wash (0 no occurrence, 1 observed) and basal area of trees rooted in the track ($m^2$ $ha^{-1}$).

| soil class | dist. order | Litter (mean) | Gr. Veg. (mean) | Erosion (mean) | L. wash (mean) | Bas. area ($m^2$ $ha^{-1}$) |
|---|---|---|---|---|---|---|
| P6 | 1st | 3.50 | 1.00 | 1.22 | 0.33 | 4.70 |
| | 2nd | 3.87 | 1.00 | 1.07 | 0.00 | 10.36 |
| | 3rd | 4.00 | 1.00 | 1.03 | 0.16 | 8.34 |
| S6 | 1st | 3.00 | 1.44 | 1.11 | 0.44 | 0.00 |
| | 2nd | 3.50 | 1.50 | 0.33 | 0.17 | 1.12 |
| | 3rd | 2.88 | 1.13 | 0.63 | 0.00 | 1.26 |
| SL13 | 1st | 3.11 | 1.00 | 1.78 | 0.44 | 0.00 |
| | 2nd | 3.67 | 1.00 | 0.33 | 0.00 | 4.44 |
| | 3rd | 3.67 | 1.17 | 1.00 | 0.00 | 16.26 |
| SF13 | 1st | 3.33 | 1.14 | 1.14 | 0.24 | 9.83 |
| | 2nd | 3.36 | 1.79 | 1.00 | 0.29 | 6.16 |
| | 3rd | 3.75 | 1.50 | 0.00 | 0.00 | 6.47 |

more light could be expected to reach the tracks. However, even if canopy closure was incomplete, tracks were often effectively shaded by extensive vegetation like ferns, rooted on the side of the tracks and expanding into the open tracks.

In this study, 13 year old tracks on similar soils and similar slopes in selectively logged forest and in secondary forest differed significantly in natural rehabilitation in terms of most variables observed. The investigated tracks of SL13 and SF13 is less than 500 m apart and are the result of the same logging operation. Possibly the difference in rate and quality of litterfall between the pioneer trees in SF13, and the mix of high canopy and few pioneer trees of SL13, may be the reason. The plantation of *Acacia mangium* in P6 can also be regarded as pioneer vegetation in terms of quick accumulation of biomass and nutrients in trees (*Sim* and *Nykvist*, 1991; *Nykvist* et al., 1996). However, there were little difference in rehabilitation between P6 and S6. The *Acacia mangium* leaves have been observed to be prone to slower decomposition. This may indicate that this organic material is less positive for the soil faunal activity and consecutive soil aggregation. That may be one reason for the difference beteen SL13 and SF13 not to be apparent between SL6 and P6. Changes in composition of soil organic matter is still poorly understood but may play a key role in soil degradation and reclamation (*Martins* et al., 1991).

## Conclusions

The amount of soil organic matter, measured as LOI, is concluded to play a key role for the rehabilitation of soil physical properties. It may be possible to distinguish between two stages where the major part of DBD (and possibly most of total porosity) is first restored with increased LOI. In a later stage macro porosity and soil permeability will further develop with further increase in LOI.

Furthermore, also the importance of the quality of organic material is underlined by the results of this study. If addition of organic material should be used for man made rehabilitation of disturbed soils, care should be put on choosing such materials that have indications to be more attractive for the soil fauna.

## Acknowledgements

Resources for this study are gratefully acknowledged from Sabah Forest Industries and from Sida trough the International Rural Development Centre at Swedish University of Agricultural Science.

## References

*Aina, P. O.* (1979): Soil changes resulting from long-term management practices in Western Nigeria. Soil Sci. Soc. Am. J., 43: 173–177.

*Bruijnzeel, L. A.*, (1990): Hydrology of moist tropical forests and effects of conversion: A state of knowledge review, UNESCO – IHP, Humid Tropics Programme, Paris, 224 pp.

*Chauvel, A., Grimbaldi, M.* and *Tessier, D.* (1991): Changes in pore-space distribution following deforestation and revegetation: An example from the Central Amazon Basin, Brazil. Forest Ecology and Management, 38: 259–271.

*Collis-George, N.* (1980): A pragmatic method to determine the parameters that characterize ponded infiltration. Aust. J. Soil Res., 18: 111–117.

*Dunn, O. J.* (1964): Multiple comparisons using rank sums. Tecnometrics, 6(3): 241–252.

*Grip, H., Malmer, A.* and *Wong, F. K.* (1994): Converting tropical rainforest to forest plantation in Sabah, Malaysia. 1. Dynamics and net losses of nutrients in control catchment streams. Hydrological Processes, 8: 179–194.

*Johansson, E.* and *Kluge, M.* (1995): Natural rehabilitation of tractor disturbed soils in tropical rainforest in Sabah, Malaysia. Swedish University of Agricultural Sciences, International Rural Development Centre, Working paper no. 285, 27 pp.

*Kararuzaman, J.* and *Nik Muhamad Majid* (1986): The impact of skid trails on the physical properties of tropical hill forest soils. Pertanika, 9(3): 311–321.

*Kararuzaman, J.* (1996): Estimation of rate of recovery of disturbed soils from ground-based logging in Peninsular Malaysia. Journal of Tropical Forest Science, 9(1): 88–100.

*Malmer, A.* (1990): Stream suspended sediment load after clear-felling and different forestry treatments in tropical rainforest, Sabah, Malaysia. IAHS Publ., 192: 62–71.

*Malmer, A.* (1992): Water yield changes after clear-felling tropical rainforest and establishment of forest plantation in Sabah, Malaysia. Journal of Hydrology 134: 77–94.

*Malmer, A.* (1996a): Hydrological effects and nutrient losses of forest plantation establishment on tropical rainforest land in Sabah, Malaysia. Journal of Hydrology, 174: 129–148.

*Malmer, A.* (1996b): Observations on slope processes in a tropical rain forest environment before and after forest plantation establishment. In: Anderson, M. G. and Brooks, S. M. (eds); Advances in hillslope processes. John Wiley & Sons, Chichester, UK, 961–973.

*Malmer, A.* (1996c): Phosphorus loading to tropical rain forest streams after clearfelling and burning in Sabah, Malaysia. Water Resources Research, 32(7): 2213–2220.

*Malmer, A.* and *Grip, H.* (1990): Soil disturbance and loss of infiltrability caused by mechanized and manual extraction of tropical rainforest in Sabah, Malaysia. Forest Ecology and Management, 38: 1–12.

*Malmer, A.* and *Grip, H.* (1994): Converting tropical rainforest to forest plantation in Sabah, Malaysia. II. Changes of nutrient dynamics and net losses in streams due to treatments. Hydrological Processes, 8: 195–209.

*Martins, P. F. da S., Cerri, C. C., Volkoff, F., Andreux F.* and *Chauvel, A.* (1991): Consequences of clearing and tillage on the soil of a natural Amazonian ecosystem. Forest Ecology and Management, 38: 273–282.

*Nicholson, D. I.* (1958): An analysis of logging damage in tropical rainforest in North Borneo. Mal. For. XXI: 235–245.

*Nussbaum, R., Andersson, J.* and *Spencer, T.* (1995): Factors limiting the growth of indigenous tree seedlings planted on degraded rainforest soils in Sabah, Malaysia. Forest Ecology and Management, 74: 149–159.

*Nykvist, N.* (1996): The above-ground biomass growth of secondary vegetation after the great "Borneo fire" of 1983. Journal of Tropical Ecology, 12: 307–312.

*Nykvist, N., Grip, H., Sim, B. L., Malmer, A.* and *Wong, F. K.* (1994): Nutrient losses in forest plantations in Sabah, Malaysia. Ambio, 23(3): 210–215.

*Nykvist, N., Sim, B. L.* and *Malmer, A.* (1996): Effects of tractor logging and burning on biomass production and nutrient accumulation in Acacia mangium plantations in Sabah, Malaysia. Journal of Tropical Forest Science, 9(2): 161–183.

*Sim, B. L.* and *Nykvist, N.* (1991): Impact of forest harvesting and replanting. Journal of Tropical Forest Science, 3(3): 251–284.

*Van der Plas* and *Bruijnzeel, L. A.* (1993): Impact of mechanized selective logging on topsoil infiltrability in the Upper Segama area, Sabah, Malaysia. IAHS Publ. 216: 203–211.

*Van der Weert, R.* (1974): Influence of mechanical forest clearing on soil conditions and the resulting effect on root growth. Trop. Agric. (Trinidad), 51(2): 325–331.

*Wierda, A., Veen, A. W. L.* and *Hutjes, R. W. A.* (1989): Infiltration at the Tai rainforest (Ivory Coast): measurements and modeling. Hydr. Proc., 3: 371–382.

*Woods, P.* (1989): Effects of logging, drought and fire on structure and composition of tropical forests in Sabah, Malaysia. Biotropica, 21(4): 290–298.

# Index

# Springer and the environment

At Springer we firmly believe that an international science publisher has a special obligation to the environment, and our corporate policies consistently reflect this conviction.

We also expect our business partners – paper mills, printers, packaging manufacturers, etc. – to commit themselves to using materials and production processes that do not harm the environment. The paper in this book is made from low- or no-chlorine pulp and is acid free, in conformance with international standards for paper permanency.

Springer

Druck: Strauss Offsetdruck, Mörlenbach
Verarbeitung: Schäffer, Grünstadt